Lecture Notes in Computer Science 16868

Founding Editors

Gerhard Goos
Juris Hartmanis

AF183851

Editorial Board Members

Elisa Bertino, *Purdue University, West Lafayette, IN, USA*
Wen Gao, *Peking University, Beijing, China*
Bernhard Steffen, *TU Dortmund University, Dortmund, Germany*
Moti Yung, *Columbia University, New York, NY, USA*

The series Lecture Notes in Computer Science (LNCS), including its subseries Lecture Notes in Artificial Intelligence (LNAI) and Lecture Notes in Bioinformatics (LNBI), has established itself as a medium for the publication of new developments in computer science and information technology research, teaching, and education.

LNCS enjoys close cooperation with the computer science R & D community, the series counts many renowned academics among its volume editors and paper authors, and collaborates with prestigious societies. Its mission is to serve this international community by providing an invaluable service, mainly focused on the publication of conference and workshop proceedings and postproceedings. LNCS commenced publication in 1973.

Elena Troubitsyna · David de Andrés ·
Friedemann Bitsch
Editors

Computer Safety, Reliability, and Security

45th International Conference, SAFECOMP 2026
Valencia, Spain, September 23–25, 2026
Proceedings

Springer

Editors
Elena Troubitsyna
KTH Royal Institute of Technology
Stockholm, Sweden

David de Andrés
ITACA, Universitat Politècnica de València
Valencia, Spain

Friedemann Bitsch
Hitachi Rail GTS Deutschland GmbH
Ditzingen, Germany

ISSN 0302-9743 ISSN 1611-3349 (electronic)
Lecture Notes in Computer Science
ISBN 978-3-032-34866-1 ISBN 978-3-032-34867-8 (eBook)
https://doi.org/10.1007/978-3-032-34867-8

© The Editor(s) (if applicable) and The Author(s), under exclusive license
to Springer Nature Switzerland AG 2027

This work is subject to copyright. All rights are solely and exclusively licensed by the Publisher, whether the whole or part of the material is concerned, specifically the rights of translation, reprinting, reuse of illustrations, recitation, broadcasting, reproduction on microfilms or in any other physical way, and transmission or information storage and retrieval, electronic adaptation, computer software, or by similar or dissimilar methodology now known or hereafter developed.
The use of general descriptive names, registered names, trademarks, service marks, etc. in this publication does not imply, even in the absence of a specific statement, that such names are exempt from the relevant protective laws and regulations and therefore free for general use.
The publisher, the authors and the editors are safe to assume that the advice and information in this book are believed to be true and accurate at the date of publication. Neither the publisher nor the authors or the editors give a warranty, expressed or implied, with respect to the material contained herein or for any errors or omissions that may have been made. The publisher remains neutral with regard to jurisdictional claims in published maps and institutional affiliations.

This Springer imprint is published by the registered company Springer Nature Switzerland AG
The registered company address is: Gewerbestrasse 11, 6330 Cham, Switzerland

If disposing of this product, please recycle the paper.

Preface

The SAFECOMP conference series traces its origins to 1979, when EWICS TC7 – the Technical Committee on Reliability, Safety and Security of the European Workshop on Industrial Computer Systems – conceived a recurring venue for the exchange of knowledge and technology among academia, industry, research centres, and licensing bodies. Since 1985 the International Conference on Computer Safety, Reliability and Security (SAFECOMP) has been held annually, establishing itself as a cornerstone of the safety-critical community.

This volume gathers the papers selected for presentation at the 45th SAFECOMP Conference, hosted in Valencia, Spain, in September 2026. The call attracted 125 high-quality submissions from 18 European and 15 extra-European countries, evidencing an impressive level of cross-border and cross-institutional collaboration. Every contribution underwent a single-blind review by at least three independent experts, and the final program was approved by the International Program Committee at its April 2026 meeting. After this rigorous evaluation, 23 papers were accepted for publication in the present proceedings and for oral presentation at the conference.

The scientific programme was enriched with keynotes provided by distinguished speakers:

- "Toward Large-Scale Fault-Tolerant Quantum Computers: Security in the Quantum Era" – Carmen G. Almudéver, Universitat Politècnica de València (Spain)
- "Physical AI Safety: A Unified Framework for Safety, Security, and AI Assurance" – Gianpiero Negri, Amazon Robotics (Italy)
- "Combining High-Performance Hardware and Software with High Software Assurance: Is it Possible?" – Leonidas Kosmidis, Barcelona Supercomputing Center (Spain)

In keeping with tradition, SAFECOMP 2026 was organised as a single-track conference, fostering intensive networking during breaks, social events, and the full set of plenary talks. The day before the main conference featured eight high-impact workshops: CSRIS, SASSUR, DECSoS, ORCAS, SafetyNXT, DAIES, SENSEI, and WAISE—each addressing distinct topics and coordinated by dedicated organising groups. The workshop contributions appear in separate Springer LNCS SAFECOMP Workshop Proceedings.

We extend our sincere gratitude to all who made this edition possible: the authors for their contributions, the invited speakers for their inspiring talks, the members of the International Program Committee and the external reviewers for their diligent assessments, EWICS and the sponsors for their support, and the Local Organization Committee for the execution of the event in Valencia.

We are confident that readers will appreciate the breadth and depth of the research presented, spanning a wide array of application domains. Moreover, we hope this volume serves as an invitation to join the EWICS TC7 and SAFECOMP communities to continue driving advances in the safety, reliability, and security of computer systems.

September 2026

Elena Troubitsyna
David de Andrés
Friedemann Bitsch

Organization

EWICS TC7 Chair

Mario Trapp	Technical University of Munich, Germany

EWICS TC7 Vice-chairs

Francesca Saglietti	University of Erlangen-Nuremberg, Germany
Uwe Becker	Drägerwerk AG & Co KGaA, Germany

General Chair

Juan Carlos Ruiz García	Universitat Politècnica de València, Spain

Conference Program Co-chairs

Elena Troubitsyna	KTH Royal Institute of Technology, Sweden
David de Andrés	Universitat Politècnica de València, Spain

General Workshop Co-chairs

Erwin Schoitsch	AIT Austrian Institute of Technology, Austria
Joaquín Gracia-Morán	Universitat Politècnica de València, Spain

Publication Chair

Friedemann Bitsch	Hitachi Rail, Germany

Finance Chair

Juan-Carlos Baraza-Calvo	Universitat Politècnica de València, Spain

Position Papers Chair

Jérémie Guiochet	LAAS-CNRS, Université de Toulouse, France

Local Arrangements

Luis José Saiz Adalid	Universitat Politècnica de València, Spain

Web Chairs

Joaquín Gracia-Morán	Universitat Politècnica de València, Spain
Juan Carlos Ruiz García	Universitat Politècnica de València, Spain

International Program Committee

Jaume Abella	Barcelona Supercomputing Center, Spain
Magnus Albert	SICK AG, Germany
Ignacio Álvarez	Technische Hochschule Ingolstadt, Germany
Uwe Becker	Drägerwerk AG & Co KGaA, Germany
Peter G. Bishop	Adelard, UK
Friedemann Bitsch	Hitachi Rail, Germany
Andrea Bondavalli	University of Florence, Italy
Jeroen Boydens	Katholieke Universiteit Leuven, Belgium
Simon Burton	University of York, UK
António Casimiro	Universidade de Lisboa, Portugal
Andrea Ceccarelli	University of Florence, Italy
Mary Cummings	George Mason University, USA
Peter Daniel	EWICS TC7, UK
Jose Luis de la Vara	University of Castilla-La Mancha, Spain
Ewen Denney	KBR/NASA Ames Research Center, USA
Felicita Di Giandomenico	ISTI-CNR, Italy
Wolfgang Ehrenberger	Fulda University of Applied Science, Germany
Francesco Flammini	Mälardalen University, Sweden
Cody Fleming	Iowa State University, USA
Barbara Gallina	Mälardalen University, Sweden
Jérémie Guiochet	LAAS-CNRS, France
Ibrahim Habli	University of York, UK
Yan Jia	University of York, UK
Mark Lawford	McMaster University, Canada

Elena Lisova	Volvo CE, Sweden
Zoltan Micskei	Budapest University of Technology and Economics, Hungary
Leonardo Montecchi	NTNU - Norwegian University of Science and Technology, Norway
Natasha Neogi	NASA-Langley, USA
Carmen Noll	TÜV SÜD Rail GmbH, Germany
Ganesh Pai	KBR/NASA Ames Research Center, USA
Philippe Palanque	ICS-IRIT, University of Toulouse, France
Maria Panou	Hellenic Institute of Transport, Greece
Yiannis Papadopoulos	University of Hull, UK
Peter Popov	City St George's, University of London, UK
Alexander Romanovsky	Newcastle University, UK
Matteo Rossi	Politecnico di Milano, Italy
Martin Rothfelder	Siemens AG, Germany
Francesca Saglietti	University of Erlangen-Nuremberg, Germany
Behrooz Sangchoolie	RISE Research Institutes of Sweden, Sweden
Horst Schirmeier	TU Dresden, Germany
Erwin Schoitsch	AIT Austrian Institute of Technology, Austria
Christel Seguin	Office National d'Études et de Recherches Aérospatiales, France
Evgenia Smirni	College of William and Mary, USA
Wilfried Steiner	TTTech Computertechnik AG, Austria
Mark-Alexander Sujan	Loughborough University, UK
Kenji Taguchi	National Institute of Informatics, Japan
Carsten Thomas	Hochschule für Technik und Wirtschaft Berlin, Germany
Stefano Tonetta	Fondazione Bruno Kessler, Italy
Martin Törngren	KTH Royal Institute of Technology, Sweden
Mario Trapp	Technical University of Munich, Germany
Valeria Vittorini	Università Federico II di Napoli, Italy
Marcus Völp	University of Luxemburg, Luxembourg

Sub-reviewers

Zsófia Ádám	Budapest University of Technology and Economics, Hungary
Pierre Bieber	Office National d'Études et de Recherches Aérospatiales, France
Sina Borrami	Alstom, Sweden
Laure Buysse	Katholieke Universiteit Leuven, Belgium

Silvano Chiaradonna	ISTI-CNR, Italy
Franca Corradini	Dalle Molle Institute for Artificial Intelligence (IDSIA USI-SUPSI), Switzerland
Riccardo D'Elia	Dalle Molle Institute for Artificial Intelligence (IDSIA USI-SUPSI), Switzerland
Mathieu Dario	LAAS-CNRS, France
Yash Deo	University of York, UK
Allan Espindola	Universidade de Lisboa, Portugal
Tom Van Eyck	Katholieke Universiteit Leuven, Belgium
Martin Farkas	Budapest University of Technology and Economics, Hungary
Camille Fayollas	ICS-IRIT, University of Toulouse, France
Shaun Feakins	University of York, UK
Gianfilippo Fornaro	KTH Royal Institute of Technology, Sweden
José Manuel Gaspar Sánchez	KTH Royal Institute of Technology, Sweden
Mattia Gianotti	KTH Royal Institute of Technology, Sweden
Maria Gomes da Silva	Universidade NOVA de Lisboa, Portugal
Nicholas Gordon	Barkhausen Institut, Germany
Simone Guarino	Università Campus Bio-Medico di Roma, Italy
Nathan Hughes	University of York, UK
Mir Riyanul Islam	KTH Royal Institute of Technology, Sweden
Kim Littler	University of York, UK
Yiyang Lu	College of William and Mary, USA
Tloul Mohammed	Katholieke Universiteit Leuven, Belgium
Per Natzschka	TU Dresden, Germany
Marcus Nolte	KTH Royal Institute of Technology, Sweden
Carlos Paradis	KBR/NASA Ames Research Center, USA
Sotirios Foivos Pournaropoulos	University of Thessaly, Greece
Bertalan Zoltán Péter	Budapest University of Technology and Economics, Hungary
Kate Preston	University of York, UK
Krystian Radlak	Warsaw University of Technology, Poland
Stefan Ramdhan	McMaster University, Canada
Jenni Reuben	Saab, Sweden
Viktor Reusch	TU Dresden, Germany
Leonie Seelisch	TU Dresden, Germany
Irfan Sljivo	KBR/NASA Ames Research Center, USA
Bowen Sun	College of William and Mary, USA
Robin Thunig	TU Dresden, Germany
Jens Vankeirsbilck	Katholieke Universiteit Leuven, Belgium
Francesco Vitale	Università Federico II di Napoli, Italy
Tianhao Wang	TU Dresden, Germany

Alan Wassyng	McMaster University, Canada
Ikue Warren	National Institute of Informatics, Japan
Ármin Zavada	Budapest University of Technology and Economics, Hungary

SafeComp Sponsors

University Research Institute of Information and Communication Technologies (ITACA)

Departamento de Informática de Sistemas y Computadores, UPV

SafeComp Supporting Institutions

European Workshop on Industrial Computer Systems
Technical Committee 7 on Reliability, Safety and Security

Universitat Politècnica de València

KTH Royal Institute of Technology

Austrian Institute of Technology

Hitachi Rail GTS Deutschland GmbH

Lecture Notes in Computer Science (LNCS)
Springer Science + Business Media

European Research Consortium for Informatics and Mathematics

Technical Group ENCRESS in GI and ITG

Austrian Computer Society

Austrian Software Innovation Association

Electronics and Software Based Systems (ESBS) Austria

Contents

Fault Tolerance and Diagnosis

Safety Protection

Cybersecurity and Mixed-Criticality in Safety-Critical Systems

Safety of AI-Based Systems

Industrial Safety Engineering in the AI-Era

Engineering Safety Systems

Practice of Hazard Analysis and Risk Assessment

Fault Tolerance and Diagnosis

OpenFI4ASIC: An Open-Source Fault Injection Framework for ASIC Designs via FPGA-Based Rapid Prototyping

Jasper Homann[1(✉)], Eike Trumann[1], Umut Durak[2], and Guillermo Payá-Vayá[1]

[1] Chair for Chip Design for Embedded Computing, Technische Universität Braunschweig, Braunschweig, Germany
{jasper.homann,eike.trumann,g.paya-vaya}@tu-braunschweig.de
[2] Institute of Flight Systems, German Aerospace Center (DLR), Braunschweig, Germany
umut.durak@dlr.de

Abstract. With shrinking structure sizes, radiation-induced faults in microelectronic components are a concern for the design of high-reliability systems. As radiation testing is expensive, non-reproducible, and only possible after a chip has been fabricated, tools for analyzing the effect of faults on ASIC designs during the design phase are needed. FPGA-based fault-injection analysis can provide a speed-up compared to simulation, enabling higher fault-space coverage. This work presents OpenFI4ASIC, a modular open-source framework for performing fault-injection campaigns on layout netlists of ASIC designs. Faults are injected by replacing standard-cell instances with instrumented instances connected via a scan chain. This instrumentation is performed automatically by the OpenFI4ASIC framework, which also provides the hardware interfaces and software drivers to facilitate fault-emulation campaigns on an SoC FPGA. As a case study, a configurable open-source RISC-V core is evaluated for five implementations, from single-cycle to five-stage pipelined, using three standard benchmarks (ISQRT, BUBBLESORT, and CRC32). The cores are synthesized for 45nm NanGate technology, and the resulting netlists are used as input for OpenFI4ASIC and emulated on a Zynq-7000 SoC FPGA. The generated FPGA projects enable fault-injection campaigns at up to 6878.76 fault-injection runs per second, allowing full coverage of the single-bit fault within minutes. The results highlight the importance of considering the architecture implementation, showing differences of up to 30.3% and 18.8% in the number of program executions yielding incorrect results among different RISC-V ISA implementations when injecting faults into ISA registers only and into all implemented flip-flops, respectively.

Keywords: fault injection · hardware reliability · FPGA

© The Author(s), under exclusive license to Springer Nature Switzerland AG 2027
E. Troubitsyna et al. (Eds.): SAFECOMP 2026, LNCS 16868, pp. 3–17, 2027.
https://doi.org/10.1007/978-3-032-34867-8_1

1 Introduction

Faults in microelectronic systems can occur for a variety of reasons, including manufacturing defects and radiation exposure. For microelectronics deployed in radiation environments such as space, aircraft, or nuclear power plants, an evaluation of the potential effects of radiation-induced faults is mandatory. As natural background radiation affects any component operated on Earth, safety-critical systems should generally be evaluated concerning their fault tolerance. This is supported by the ISO 26262 standard for automotive electronics, which requires fault-injection (FI) testing.

In the context of this work, a fault is an event or condition that causes a component of a microelectronic system to behave differently from how it is specified. This includes, among others, manufacturing defects, radiation-induced faults, and electromagnetic interference. Faults can occur in unused parts of a system or be masked by following logic; therefore, not all faults cause errors. Errors are deviations from a system's intended operation that can lead to failures when the system loses its intended functionality. The process by which a fault becomes a failure is also known as fault propagation and might be stopped by fault masking or fault-tolerance mechanisms. Faults in digital logic are typically modeled as bit errors or stuck-at faults. An FI campaign consists of multiple fault injection runs to generate statistics about the failures arising. The fault space to be evaluated is the product of all possible fault locations, times at which a fault may occur, and fault types. While an extensive coverage of the fault space is desirable, often compromises must be made for larger systems, as full fault coverage quickly becomes infeasible.

This paper presents `OpenFI4ASIC`, a novel open-source FPGA-based fault injection framework that enables fully automated fault-injection campaigns through FPGA emulation. It can cover the complete single-bit fault space of medium-complexity designs, such as RISC-V cores, within minutes to hours when executing programs with thousands to tens of thousands of executed instructions. The main contributions of this paper are:

- A modular FPGA-accelerated fault injection framework for post-synthesis ASIC netlists is proposed, in contrast to traditional RTL-level injection.
- A fully open-source implementation of the framework is provided, including a scriptable software interface (FI-SHELL) that enables automated and flexible experimentation. Host driver and analysis software support streamlined data collection and evaluation. In contrast to other FI tools, `OpenFI4ASIC` requires only an MPSoC FPGA, thereby minimizing hardware dependencies.
- A case study on a configurable RISC-V family demonstrates the applicability of the methodology to complex systems and the influence of microarchitecture on fault tolerance, highlighting that ISA-level fault injection alone is insufficient.

This paper is organized as follows. Section 2 provides an overview of related work and introduces a classification of existing fault injection tools. Section 3 introduces the proposed framework. In Sect. 4, a case study evaluates the

Table 1. Overview of Emulation-Based Fault Injection Platforms.

Reference	Platform	Granularity	Evaluation Target	Design Level
[6] (2021)	ZCU-102	FF	SCR1 RISC-V	RTL
[20] (2024)	Kintex Ultrascale	FF	C906 RISC-V	RTL
[15] (2002)	Virtex-1000E	FF	LEON3 SPARC	RTL
[8] (2018)	Zynq-7000	FF	2 RISC-V Cores	generated Verilog
[18] (2025)	ZCU-104	Register	ROCKET RISC-V	generated Verilog
[13] (2011)	Virtex-5	FF	-	Netlist
[16] (2019)	Virtex-7	multiple	ARM Cortex M0	Netlist
[21] (2017)	Virtex-5	Gate	ITC HDL circuits	Netlist
[7] (2023)	Zynq-7000	Gate	ITC HDL circuits	layout-level
[14] (2015)	Virtex-6	Gate	8-bit CPU	layout-level
OpenFI4ASIC	Zynq-7000	Gate or FF	custom RISC-V	Netlist

framework across different RISC-V architectures. Finally, Sect. 5 concludes the paper.

2 Related Work

Fault injection (FI) campaigns can be carried out through different technical means and to varying levels of detail. An overview of representative solutions is given in Table 1. These solutions can be classified according to three categories: execution platform, design level, and granularity.

Based on the **execution platform**, three subcategories can be distinguished:

- *Simulation-based FI* uses a model of the device under test (DUT), in which fault injection is possible through the simulation environment. It can be used for any circuit. The granularity of the fault injection can be chosen arbitrarily, and fault models can be defined freely. The model can be derived from an instruction set architecture (ISA), an RTL description, or already synthesized netlists, with or without timing and layout information. Since available computing power limits simulation speed, the coverage of the fault space may be restricted. Simulation-based FI is available in commercial EDA software [1–3]. Open-source tools on the ISA level include UN-FIT [4] and FAIL* [17].
- *Emulation-based FI* uses an FPGA to emulate the system behavior of the DUT. The DUT can be represented as RTL or netlist, depending on the tool used. In addition, various fault models can be included. FPGA emulation is limited to digital circuits and cannot emulate ASIC timing. Compared to simulation-based FI, the execution time is significantly shorter, enabling the exploration of a larger portion of the fault space.
- *Hardware-based FI* can include either modified hardware that allows for fault injection, e.g., through a physically implemented scan chain [12], or external

fault sources such as radiation or laser applied to the DUT [5]. A primary concern is that the observability is limited through the available external ports. The cost is typically vastly higher than for other types of FI evaluations.

The **design level** describes the steps toward an ASIC that have been made before a design is used in the fault injection. At the *RTL*, the original design hardware description language implementation is used for the fault injection [6,15,20]. In Table 1, additionally, the category *generated Verilog* can be found [8,18] for designs generated from a Chisel description. Here, an automatic conversion between two RTL descriptions is included, with the actually simulated design being an RTL description. With an *ASIC netlist*, the evaluated DUT is closer to the final implementation because of optimization steps carried out during the synthesis process, such as register retiming, and also the exact primitives chosen in the technology mapping are included in the evaluation [21]. The **VLSI layout level** additionally includes placement information, which allows for multi-bit-upsets to be modeled [7,14].

At this point, it is worth noting that the use of FPGA-emulation-based FI platforms can introduce important inconsistencies during FI analysis. In such platforms, an RTL design-level entry is typically used, where the hardware functionality is mapped to native FPGA components. This mapping can lead to significant differences between the ASIC implementation and the FPGA emulation. The tools listed in the first part of Table 1 operate at the RTL level and may therefore introduce deviations from fault injection results for ASIC technologies [20]. In contrast, the tools in the second part of the table (including `OpenFI4ASIC`) operate from the ASIC netlist and only use a re-synthesis for FPGA emulation, which ensures that the functionality of the ASIC cells is emulated accurately on the FPGA. Moreover, FPGA-emulation-based FI systems may either perform fault injection at the FPGA configuration level or instrument the ASIC circuit with additional logic elements, often referred to as *saboteurs*, that enable fault injection. Fault injection at the FPGA configuration level modifies the FPGA representation of the circuit rather than the actual components of the ASIC system under investigation. In contrast, the saboteur-based approach directly targets components of the ASIC netlist and therefore provides a more accurate representation of faults occurring in ASIC technology.

The **granularity** of the fault injection describes which circuit components are available as targets for the fault injection. The most fine-grained systems allow injection at the **gate level**, which is suitable for analyzing the effects of any possible defects, including manufacturing defects. Fault injection tools at the **flip-flop (FF) level** enable injection into all internal memory cells of the design. This is considered best practice for evaluating single-event radiation effects, as these rarely affect individual single logic gates but typically induce bit-flip errors in memory cells [8]. Finally, it is also possible for specific programmable designs to inject faults only into the **program-accessible registers**, which greatly reduces the design instrumentation for fault injection. This can be achieved either through a simulation or emulation environment, allowing an external manipulation of the register state, or through a modified program that includes additional instructions to alter the system state. The latter is also known as *software-based FI.* In [16], the differences between different granularities are evaluated. Additionally, the types of faults supported can be distinguished. Most tools support bit-flips, i.e., spontaneous changes in the value of a memory cell. Additionally, stuck-at may be supported, where, once a fault occurs, the output of a gate remains permanently stuck at a specific value.

This paper presents the open-source **OpenFI4ASIC** tool, which enables **FPGA emulation-based FI**. The tool allows fault injection at the **gate** or **flip-flop (FF)** level in ASIC netlists. It uses the scan chain approach from [14] to enable high-speed emulation with low FPGA resource usage and further optimizes it to reduce instrumentation overhead and increase the injection speed. Most of the fault-injection platforms mentioned in this section were used exclusively in the preparation of the respective papers and have never been made publicly available. Notable open-source tools include the simulation-based tool FAIL* [17] and the ISA-Level simulation-based UN-FIT [4]. **OpenFI4ASIC** is released under the permissive MIT license at https://github.com/tubs-eis/OpenFI4ASIC.

3 OpenFI4ASIC Framework

An overview of the **OpenFI4ASIC** flow is shown in Fig. 1. The flow contains four primary parts: The fault-instrumentation tool and instrumented standard cell library, the fault-emulation hardware framework for controlling the FI operations, the fault-emulation runtime software for orchestrating the FI emulation campaign, and the host driver software for communication with the FPGA.

OpenFI4ASIC uses a scan-chain mechanism to address the individual standard-cells of the design targeted for fault injection. For example, when SEUs are to be emulated, all FFs in the design are placed on one or more scan-chains, allowing the control software to access each individual FF. With this approach, large designs can be instrumented without requiring large amounts of combinational logic and routing to access the targeted cells.

Fault Instrumentation of ASIC Standard-Cell Netlists. The input to the **OpenFI4ASIC** flow is a post-synthesis or post-layout ASIC *standard-cell netlist* of the DUT. The **OpenFI4ASIC** *fault instrumentation* python tool takes this *standard-cell netlist*, the components of the design to target, along with the list of the standard-cells to be instrumented, and generates an *instrumented netlist.* In the *instrumented netlist*, all instances of the targeted standard-cells are replaced by *instrumented cells*; these instrumented standard-cells contain scan FFs, which the tool automatically connects to form scan chains. Information about the position of the instrumented cells along the scan chains is reported in JSON format.

Figure 2 shows an example of an instrumented DFF cell from the Nangate 45 nm library optimized for efficient implementation on FPGAs. The DFF cell is a D-flip-flop cell without a reset or enable signal. In the instrumented DFF

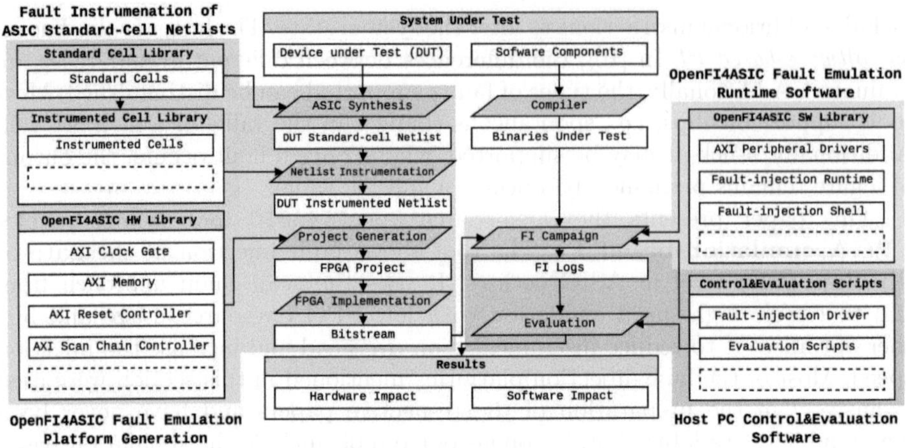

Fig. 1. OpenFI4ASIC Fault-Emulation Flow.

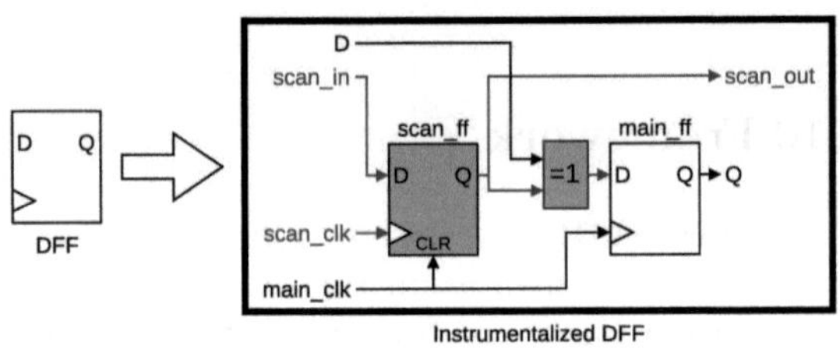

Fig. 2. Exemplary FPGA-optimized instrumented D-Flip-Flop.

cell, a scan FF controls whether the D input is stored normally or inverted, simulating a fault resulting in a bit-flip. This work extends [14] by decreasing the control overhead for bit-flip faults. The main FFs clock signal is also connected to the scan FFs asynchronous clear pin, immediately resetting the entire scan chain once the bit-flip is stored. An *instrumented standard cell* is described by two files. A JSON file containing information about the ports of the default and instrumented cell, and a model of the instrumented cell's behavior. By providing these, the functionality of OpenFI4ASIC can be extended to cover different standard cells (e.g., combinational cells) and other (proprietary) design blocks.

For FI into memory blocks, a different approach is used. Because the cells of a memory array are already addressable on the word-level, bit-flips can be emulated by reading, modifying, and writing back words from the memory while the design's clock is gated. To enable this approach, the DUT has to be prepared by moving all memory blocks in which FI is to be performed outside the design and adding the corresponding memory ports to the RTL top-level.

OpenFI4ASIC Fault Emulation Platform Generation. OpenFI4ASIC provides a *library of hardware components* for efficient fault-emulation on FPGAs, which comprises a set of *AXI peripherals* that can be connected to a processor to initialize and control the *instrumented Netlist* during the execution of an FI campaign on an FPGA device. This library was designed and validated on an AMD Zynq 7000 FPGA, but is portable to other vendors and FPGAs that support the use of AXI and UART as communication protocols. The library contains, among other parts, *instrumented dual-port memories* that can be used to instrument single-port memories of the input netlist. Therefore, the second port can be connected to the processor core through an AXI interface to support initialization, observability, and FI into memories. The memories can be configured either as synchronous or asynchronous, using blockRAMs or LUTRAMs, respectively. Due to its modular design, `OpenFI4ASIC` also supports gate-level FI, which is out of scope in this paper. By providing appropriate *instrumeted cells*, arbitrary fault models, such as stuck-at and intermittent, can be modeled. To control the FI process, three main components are used: (a) the *AXI scan chain controller*, which writes arbitrary values into the `scan_ffs` of the scan chain (see Fig. 2), (b) the *AXI reset controller*, which asserts the reset of the instrumented ASIC design, and (c) the *AXI clock-gate controller*, which controls whether a clock is activated. In a typical `OpenFI4ASIC` fault-emulation platform, two clock gates are used to control the design's `main_clk` and `scan_clk`. Enhanced clock gating for both `main_clk` and `scan_clk` is used to reduce control signal overhead when emulating the instrumented netlist, injecting faults through the scan chain controller, and resetting the instrumented netlist via the reset controller (e.g., DMEM initialization).

Example *FPGA project generation scripts* for the Zynq-7000 SoC FPGA are provided in the open-source framework. The resulting *FPGA project* can be implemented using standard FPGA tools supplied by the FPGA vendor.

OpenFI4ASIC Fault Emulation Runtime Software. The *SW library* contains software functions for design initialization and orchestration of FI campaigns, intended to run on the SoC ARM core or a softcore processor. It is organized in a three-layer architecture: (1) the bottom layer contains the *AXI peripheral drivers* for the individual peripherals described above; (2) the *fault-injection runtime* forms the middle layer, performing higher-level functions, like system reset, memory initialization, and injection of a fault into a specific FF at a specific time. It is also responsible for comparing results from a fault-free reference run to the results with FI and performing failure classification; and (3) the top layer handles communication with the host system and exposes the *fault-injection shell* via UART, enabling interaction with the fault-emulation platform either manually or automatically through the host PC software.

The *fault-injection shell* is a text-based UART interface that enables interaction with the fault-emulation platform. The interface uses human-readable commands and space-separated arguments, which are parsed at runtime; invalid commands are rejected with human-readable error messages. Binary blobs, such as memory initialization data, are transmitted in hexadecimal to avoid

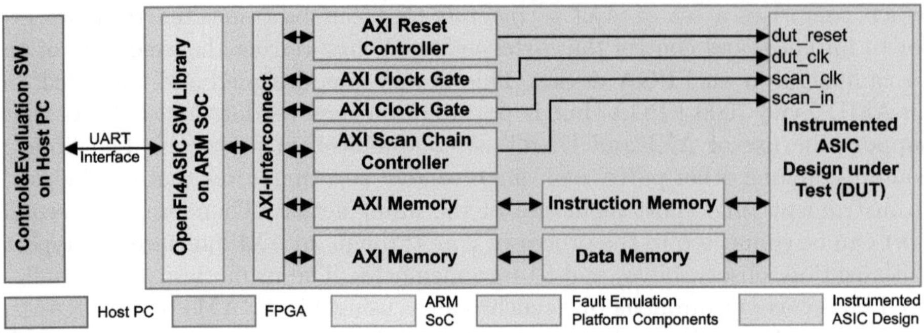

Fig. 3. Block Diagram of an exemplary OpenFI4ASIC Design Project.

non-printable characters, resulting in readable log files that improve experiment traceability and reproducibility.

The *SW library* is written in C and relies exclusively on standard library functions for memory management and for the UART I/O. Interaction with the *hardware library peripherals* is carried out entirely through reads and writes to volatile pointers mapped into the peripherals' address space. This approach results in a software platform that is portable to any processor capable of running C code with memory management, standard I/O, and AXI bus access. All layers of the software can be easily extended or adjusted to control new AXI peripherals, to perform different fault-injections and FI analyses, or to add new commands to the *fault-injection shell*.

Host PC Control&Evaluation Software. To support automated FI campaigns and data analysis, OpenFI4ASIC provides *host-side driver software* that automates communication through the *fault-injection shell* via UART. The collected data for each FI run (i.e., the injected FF, the injection cycle, and the resulting failure classification) are stored in plain text files. OpenFI4ASIC also provides example *evaluation scripts* that combine the collected data with the position of FF in the design (i.e., in the scan chain) to compute failure rates for FI into particular parts of the design.

By comparing the measured failure rates between FI runs for different hardware and software configurations, both the *hardware* and the *software impact* can be analyzed. That is, both the effect of software changes, like implementing redundancy in time, and hardware changes, like implementing triple modular redundancy, can analyzed based on the reported data.

Exemplary OpenFI4ASIC Design Project. Figure 3 shows an exemplary OpenFI4ASIC design project for FI analysis, which includes the *instrumented ASIC netlist* of a *DUT*, containing two memories (e.g., a data and an instruction memory). The exemplary design project comprises (1) a host PC, which controls the FI campaign and the FI data analysis extraction, (2) the FPGA's

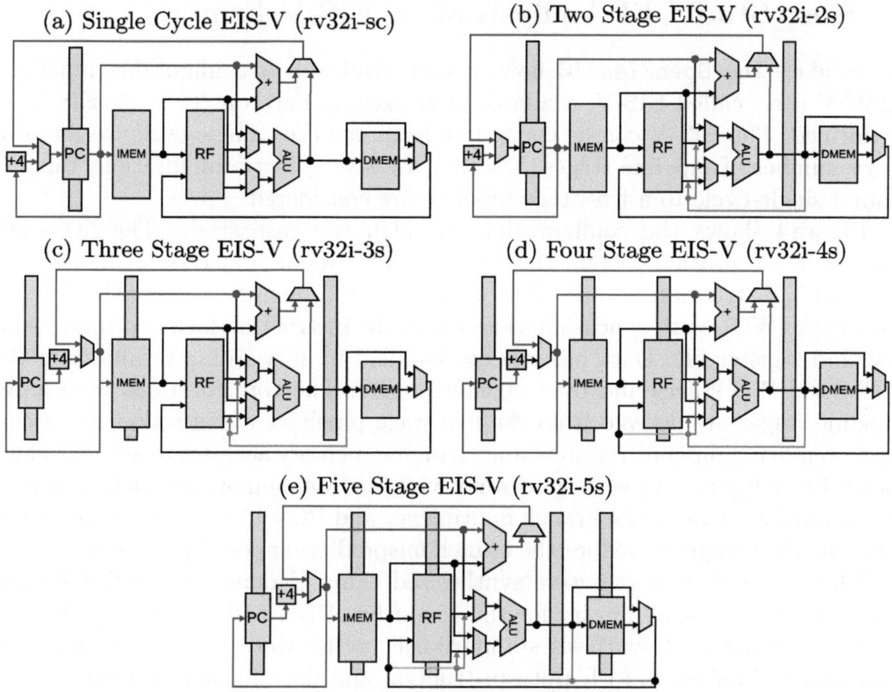

Fig. 4. EIS-V configurations from one stage (i.e., single-cycle) (a) to five stages (e). The pipeline registers (green bars) and PC control path (red), as well as forwarding (blue) and bypass (yellow) paths, are highlighted. (Color figure online)

SoC processor to interact with the DUT through a set of AXI modules from the `OpenFI4ASIC` *HW library*, and (3) the instrumented ASIC netlist of the DUT. The host PC sends control commands via UART to the SoC, e.g., to provide memory initialization data to be written into a memory before every FI run or to start an FI campaign with certain parameters. The SoC ARM software then orchestrates the AXI peripherals to control the operation of the instrumented design. While the concrete steps depend on the DUT, the general steps to perform an FI run are: (1) resetting the DUT via the *AXI Reset Controller*, (2) initializing memory via the *AXI Memory*, (3) running the DUT to the target time via the *AXI Clock Gate*, (4) injecting a fault into a FF or memory via the `AXI Scan Chain Controller` or the *AXI Memory*, (5) running the DUT until program execution is finished either for a set amount of time or until some termination condition is detected, and (6) detecting failures caused by the injected fault by, e.g., comparing the contents of a memory to a known reference.

4 Case Study: FI Analysis of an EIS-V Family

To validate the `OpenFI4ASIC` flow, a case study on a configurable pipelined RISC-V core, called EIS-V, available at https://github.com/tubs-eis/EIS-V, is performed. The EIS-V can be configured from one (i.e., single-cycle) to an arbitrary number of pipeline stages. For the case study, five configurations ranging from a single-cycle to a five-stage pipeline are considered.

Figure 4 shows the configurations used in the case study. The five-stage pipeline configuration follows the traditional RISC organization of instruction fetch (IF), instruction decode (ID), execute (EX), memory access (MEM), and writeback (WB), with synchronous access to the instruction memory, data memory, and register file. Data hazards are handled by forwarding results from the MEM and WB stages and by a register bypass. The configurations with fewer pipeline stages are derived from the five-stage pipeline configuration by successively removing pipeline registers and changing memory accesses to asynchronous mode. In configurations with synchronous instruction memory access (i.e., three-, four-, and five-stages), the branch multiplexer and PC+4 incrementer are moved after the PC register to reduce the branch misprediction penalty to a single cycle.

The five configurations were synthesized using Cadence Genus 211 for the NanGate standard-cell library for the 45 nm FreePDK technology [19]. All FFs in the resulting post-synthesis standard-cell netlist were replaced by FF cells instrumented for bit-flip FI. Both instruction and data memory are instrumentalized by means of the provided `HW library` and accessible for FI and analysis.

The case study uses three common benchmarks to evaluate the EIS-V configurations: Two benchmarks, ISQRT and CRC_32, have been adapted for bare metal systems from the MiBench benchmark suite [11], and BUBBLESORT has been adapted from [9]. Clang 20 with the command line `--target=riscv32--none-eabi -march=rv32i -O3 -flto` was used to compile the benchmarks.

FI Campaign Analysis Results. To evaluate the design, a full coverage FI campaign for single-bit faults is performed on each EIS-V configuration, i.e., the program is executed once for every combination of FF in the design and cycle in the program execution. After each execution, the computed result and final PC value are compared to the values from a fault-free reference run. If the computed result differs from the reference, it is classified as an *Output Result* failure; similarly, if the final PC differs from the reference, it is classified as a *Control Flow* failure. Therefore, the failure rates for *Output Result* and *Control Flow* are computed by dividing the number of runs with the aforementioned failures by the product of the total number of FFs and cycle counts. The *Total* failure rate considers the runs with any of the two types of failures. The FFs in the design are classified based on functionality. All FFs in the single-cycle configuration are classified as *ISA* FFs, as they are directly observable by an executed program. The rest of FFs added for the pipelined configurations are classified as *pipeline* registers. The ISA FFs are further differentiated into 32 FFs for the *PC* and 992 FFs for the 31 32-bit writable general-purpose registers (i.e., *RF*).

Table 2. EIS-V Fault-injection Failure Rate Results

Benchmark			CRC32			ISQRT			BUBBLESORT		
Conf.	FF Cat.	#FFs	Total	Output Result	Control Flow	Total	Output Result	Control Flow	Total	Output Result	Control Flow
sc	RF	992	17.3%	14.2%	8.9%	16.2%	13.1%	5.9%	32.7%	29.7%	4.4%
	PC	32	93.6%	92.5%	87.1%	92.6%	88.7%	86.2%	90.6%	82.7%	85.2%
	All	1024	19.7%	16.6%	11.4%	18.6%	15.5%	8.4%	34.5%	31.3%	7.0%
2 s	RF	992	15.1%	12.0%	8.7%	14.4%	11.4%	5.7%	31.2%	28.2%	4.4%
	PC	32	93.6%	92.6%	87.4%	92.6%	88.4%	86.2%	90.6%	82.8%	85.2%
	ISA	1024	17.6%	14.5%	11.1%	16.9%	13.8%	8.2%	33.1%	29.9%	6.9%
	Pipeline	108	29.1%	28.7%	4.1%	19.4%	19.2%	2.8%	18.4%	18.0%	2.4%
	All	1132	18.7%	15.9%	10.5%	17.1%	14.3%	7.7%	31.7%	28.8%	6.5%
3 s	RF	992	15.1%	12.0%	8.6%	14.4%	11.4%	5.7%	31.1%	28.2%	4.4%
	PC	32	85.0%	84.0%	79.4%	86.4%	82.3%	79.8%	69.7%	63.4%	65.2%
	ISA	1024	17.3%	14.2%	10.9%	16.7%	13.6%	8.0%	32.3%	29.3%	6.3%
	Pipeline	175	35.8%	35.5%	7.8%	23.9%	23.3%	5.6%	27.4%	25.8%	11.5%
	All	1199	20.0%	17.4%	10.4%	17.7%	15.0%	7.7%	31.6%	28.8%	7.0%
4 s	RF	992	14.2%	11.1%	8.8%	12.9%	9.8%	5.7%	30.9%	27.7%	5.5%
	PC	32	79.3%	78.4%	76.3%	80.9%	75.9%	78.3%	58.0%	51.9%	55.7%
	ISA	1024	16.2%	13.2%	11.0%	15.0%	11.9%	7.9%	31.7%	28.5%	7.1%
	Pipeline	375	26.4%	26.2%	7.9%	20.3%	19.7%	6.2%	22.9%	21.7%	9.9%
	All	1399	18.9%	16.7%	10.1%	16.4%	14.0%	7.5%	29.3%	26.7%	7.8%
5 s	RF	992	13.2%	10.1%	8.6%	11.7%	8.7%	5.5%	29.7%	26.5%	5.4%
	PC	32	78.2%	77.1%	76.6%	80.8%	76.1%	78.2%	58.0%	51.9%	55.7%
	ISA	1024	15.2%	12.2%	10.7%	13.9%	10.8%	7.7%	30.6%	27.3%	7.0%
	Pipeline	452	25.9%	25.8%	7.5%	20.0%	19.4%	5.8%	22.1%	21.1%	8.8%
	All	1476	18.5%	16.4%	9.7%	15.8%	13.4%	7.1%	28.0%	25.4%	7.5%

Table 2 shows the failure rates from all benchmarks, EIS-V configurations, and FF categories. A comparison with [4] reveals closely aligned results, with a PC total failure rate of approximately 93% for the ISQRT benchmark, while our results indicate 92.58%. The RF failure rate was reported at ~13.5% compared to 16.21% for our `rv32i-sc`. For an ARM core, [16] reports ~27% total failure rate for the bubble sort benchmark at FF level, with our `rv32i-3s` showing 31.61%.

With all benchmarks, the same general trend is observed: Increasing the number of pipeline stages leads to a reduction in the failure rate for FIs into the ISA FFs. This reduction can be explained by two microarchitectural features: (1) Starting in the three-stage pipeline configuration, the branch target is computed from a PC value stored in one of the pipeline registers instead of the PC register. This leads to all taken branches masking any bit-flips occurring in the PC. (2) Similarly, data hazard resolution through forwarding causes some values to be read directly from the later pipeline registers, masking bit-flips in the register file. However, it should be noted that even though the total failure rate decreases slightly when going from a single cycle to a five-stage pipeline configuration, it does so at the expense of adding more registers, which increases the risk of radiation events hitting any register.

When considering PC and register file separately, it is observed that the location of an added pipeline register influences its effect on the failure rate. The pipeline register between ALU and DMEM access, added for the two-stage configuration, and the synchronous DMEM access, added for the five-stage configuration, have little impact on the failure rates for PC FIs. Similarly, the syn-

Table 3. Maximum Clock Frequencies and FI Performance per Configuration

Config.	Clock Freq. (MHz)	ISQRT	CRC32	BUBBLESORT
rv32i-sc	`scan_clk` 160.00	4498.73 FIs/s	4273.02 FIs/s	4080.00 FIs/s
	`main_clk` 32.00	107.21 s	157.93 s	141.30 s
rv32i-2 s	`scan_clk` 200.00	6878.76 FIs/s	6435.31 FIs/s	6240.61 FIs/s
	`main_clk` 50.00	77.84 s	127.18 s	102.12 s
rv32i-3 s	`scan_clk` 200.00	5541.40 FIs/s	5182.61 FIs/s	5028.04 FIs/s
	`main_clk` 40.00	102.56 s	167.50 s	134.49 s
rv32i-4 s	`scan_clk` 200.00	5457.28 FIs/s	4990.04 FIs/s	4842.50 FIs/s
	`main_clk` 40.00	130.74 s	239.15 s	200.21 s
rv32i-5 s	`scan_clk` 200.00	5450.30 FIs/s	5061.61 FIs/s	4830.41 FIs/s
	`main_clk` 40.00	137.30 s	230.08 s	211.76 s

Table 4. Resource utilization with and without FI-instrumentation (Instr.) and overhead (OH)

Config	LUTs	Instr. LUTs (OH)	FFs	Instr. FFs (OH)	Slices	Instr. Slices (OH)
sc	2300	3261 (1.42x)	1024	2048 (2.00x)	714	988 (1.38x)
2 s	2541	3658 (1.44x)	1132	2264 (2.00x)	806	1084 (1.34x)
3 s	2568	3711 (1.45x)	1212	2421 (2.00x)	800	1137 (1.42x)
4 s	2387	3407 (1.43x)	1399	2801 (2.00x)	777	1122 (1.44x)
5 s	2528	3739 (1.48x)	1478	2958 (2.00x)	855	1194 (1.40x)

chronous IMEM added for the four-stage configuration has little impact on the failure rates for register file FIs, due to the reduced number of forwarding paths.

The results show that fault analysis requires taking the entire microarchitecture into consideration. This can be seen when comparing failure rates observed for the single-cycle configuration, which is equivalent to only injecting faults at the instruction-set level, and the five-stage pipeline configuration. When considering all FFs, the observed failure rates differ by up to 18.8% (total failure rate, BUBBLESORT), when only considering ISA FFs for the five-stage configuration, the difference is up to 30.3% (result failure rate, ISQRT). Furthermore, the total failure rate of up to 35.8% observed when injecting faults into pipeline FFs (3 s, CRC32) shows that it is insufficient to consider only FI on the RF and PC for such FI analysis. To obtain accurate results, all FFs in the design should be instrumented and targeted for FI.

Fault-Emulation Platform Resource Utilization and Performance. The case study can additionally be used to estimate the performance of the generated fault-emulation platform and to evaluate the methodology's FPGA resource utilization overhead. Table 4 compares the FPGA resource utilization in terms of LUTs, FFs, and slices of the five EIS-V configurations with and without OpenFI4ASIC instrumentation. While the FF overhead is close to 2x in all five configurations, because for each FF, one additional scan chain FF is added, the LUTs and slices incur a lower overhead in the range of 1.42x-1.48x and 1.38x-1.44x, respectively. This indicates that the additional logic required to inject the faults can, in many cases, be merged into the existing FFs, and that the synthesis tool can place many of the additional FFs and LUTs into slices already occupied in the non-instrumented core. Table 3 shows the maximum clock frequencies obtained for the five EIS-V configurations, the total runtime of each benchmark, and the resulting performance expressed as FIs per second. The time taken for each benchmark was measured on the PC running the control software, such that the communication overhead over the UART interface is included in the measured times. All benchmarks complete in under four minutes, showing that full FI for short programs is feasible using OpenFI4ASIC.

As a comparison, the FPGA-based FSFFIT tool [20] reports a duration of 12.4 ms for a single FI run for a matrix computation, equaling 81 FIs/s. The Hierarchical Fault Injection System presented by Bai et al. [6] reaches a peak performance of 25000 FIs/s on the CRC32 benchmark, compared to around 5000 FIs/s (depending on configuration, see Table 3) for OpenFI4ASIC, by instrumenting registers at the RTL level, which, as explained in Sect. 2, introduces a divergence from the results produced at the ASIC netlist level due to optimizations such as register retiming intruduced during ASIC netlist synthesis. The FT-UNSHADES2 [13] tool, which uses an FPGA reconfiguration mechanism to inject faults into the FF from an ASIC netlist, can reach 7407 FIs/s on a small test workload, but requires sophisticated hardware composed of 3 FPGA systems (controller, golden reference, and target) to do so, while at the same time only allowing FF injection.

5 Conclusion

OpenFI4ASIC is a modular, extensible, open-source tool to build FPGA-based fault-emulation platforms for arbitrary ASIC designs. The tool supports SEU fault-injection into FreePDK NanGate 45nm standard-cell netlists and can be extended to other PDKs, standard-cell libraries, and fault models.

A case study on a configurable pipelined RISC-V processor validates the methodology and shows that performance is sufficient to obtain results in the range of minutes for short-running programs. For longer-running programs, where full fault coverage is infeasible, the true failure rate can be sufficiently approximated using randomized fault-injection. The case study highlights the shortcomings of instruction-set-level fault injection, and fault injection into subsets of the microarchitecture is insufficient to accurately determine the failure rate of the system. OpenFIASIC and all additional files and scripts to reproduce the results from the case study are published at https://github.com/tubs-eis/OpenFI4ASIC.

Future work includes utilizing fault-masking terms [10] to reduce the number of fault injections required for full coverage of the fault space and employing checkpointing to speed-up campaigns on long-running programs by initializing the system to a starting state later in program execution.

Acknowledgment. This work was partly funded by the German Federal Ministry of Research, Technology, and Space (BMFTR) under project number 16ME0960K (DI-GATE-V).

Disclosure of Interests. The authors have no competing interests to declare that are relevant to the content of this article.

References

1. Functional Safety — Cadence. https://www.cadence.com/en_US/home/solutions/automotive-solution/functional-safety.html#digital
2. Questa Analyze Fault — Siemens Software. https://eda.sw.siemens.com/en-US/ic/questa-one/formal-verification/analyze-fault/
3. VC Z01X: Fault Simulation & Injection — Synopsys. https://www.synopsys.com/verification/simulation/vc-z01x.html
4. Aponte-Moreno, A., Restrepo-Calle, F., Pedraza, C.: Reliability evaluation of RISC-V and ARM microprocessors through a new fault injection tool. In: 2021 IEEE 22nd Latin American Test Symposium (LATS), pp. 1–6 (2021). https://doi.org/10.1109/LATS53581.2021.9651874
5. Arlat, J., Crouzet, Y., Karlsson, J., Folkesson, P., Fuchs, E., Leber, G.H.: Comparison of physical and software-implemented fault injection techniques. IEEE Trans. Comput. **52**(9), 1115–1133 (2003). https://doi.org/10.1109/TC.2003.1228509
6. Bai, J., Wang, X., Zhang, Z., Cai, C., Chen, G.: A hierarchical fault injection system for RISC-V processors targeting single event upsets in flip-flops. Proc. of Int. Conf. on ASIC (2021). https://doi.org/10.1109/ASICON52560.2021.9620299
7. Chen, X., et al.: FPGA-based cross-Hardware MBU Emulation Platform for Layout-Level Digital VLSI. In: Proceedings of the Asian Test Symposium (2023)
8. Cho, H.: Impact of microarchitectural differences of RISC-V processor cores on soft error effects. IEEE Access **6**, 41302–41313 (2018)
9. Dietrich, C., Bargholz, M., Loeck, Y., Budoj, M., Nedaskowskij, L., Lohmann, D.: SailFAIL: model-derived simulation-assisted ISA-level fault-injection platforms. In: Computer Safety, Reliability, and Security, pp. 207–221. Springer, Cham (2022). https://doi.org/10.1007/978-3-031-14835-4_14
10. Dietrich, C., Schmider, A., Pusz, O., Vayá, G.P., Lohmann, D.: Cross-layer fault-space pruning for hardware-assisted fault injection. In: 2018 55th ACM/ESDA/IEEE Design Automation Conference (DAC), pp. 1–6 (2018). https://doi.org/10.1109/DAC.2018.8465787
11. Guthaus, M.R., Ringenberg, J.S., Ernst, D., Austin, T.M., Mudge, T., Brown, R.B.: MiBench: A Free, Commercially Representative Embedded Benchmark Suite
12. Liu, Z., Shanmugam, D., Schaumont, P.: FaultDetective: Explainable to a Fault, from the Design Layout to the Software. IACR Trans. Cryptographic Hardware Embedded Syst. **2024**(4), 610–632 (2024). https://doi.org/10.46586/tches.v2024.i4.610-632

13. Mogollon, J.M., Guzmán-Miranda, H., Nápoles, J., Barrientos, J., Aguirre, M.A.: Ftunshades2: A novel platform for early evaluation of robustness against see. In: 12th European Conference on Radiation and Its Effects on Components and Systems (2011). https://doi.org/10.1109/RADECS.2011.6131392
14. Nowosielski, R., Gerlach, L., Bieband, S., Payá-Vayá, G., Blume, H.: FLINT: layout-oriented FPGA-based methodology for fault tolerant ASIC design. In: 2015 Design, Automation & Test in Europe Conf. & Exhibition (DATE), pp. 297–300 (2015). https://doi.org/10.7873/DATE.2015.0278
15. Rebaudengo, M., Sonza Reorda, M., Violante, M.: Analysis of SEU Effects in a Pipelined Processor. In: Proceedings of the Eighth IEEE International On-Line Testing Workshop (IOLTW 2002), pp. 112–116 (2002). https://doi.org/10.1109/OLT.2002.1030193
16. Schirmeier, H., Breddemann, M.: Quantitative cross-layer evaluation of transient-fault injection techniques for algorithm comparison. In: 2019 15th European Dependable Computing Conference (EDCC), pp. 15–22 (2019). https://doi.org/10.1109/EDCC.2019.00016
17. Schirmeier, H., Hoffmann, M., Dietrich, C., Lenz, M., Lohmann, D., Spinczyk, O.: FAIL*: an open and versatile fault-injection framework for the assessment of software-implemented hardware fault tolerance. In: 2015 11th European Dependable Computing Conf. (EDCC), pp. 245–255 (2015). https://doi.org/10.1109/EDCC.2015.28
18. Sehm, R., Ewert, C., Buchty, R., Berekovic, M., Mulhem, S.: Nail: Not Another Fault-Injection Framework for Chisel-generated RTL (2025). https://arxiv.org/abs/2508.06344
19. Stine, J.E., et al.: FreePDK: An Open-Source Variation-Aware Design Kit. In: 2007 IEEE International Conference on Microelectronic Systems Education (MSE'07), pp. 173–174 (2007). https://doi.org/10.1109/MSE.2007.44
20. Sun, Y., Wang, Z., Huang, Z., Liang, L.: An FPGA-based emulation platform for functional safety verification in automotive SoC systems. Proc. of the Asian Test Symp. (2024). https://doi.org/10.1109/ATS64447.2024.10915395
21. Ullah, A., Sanchez, E., Sterpone, L., Cardona, L.A., Ferrer, C.: An FPGA-based dynamically reconfigurable platform for emulation of permanent faults in ASICs. Microelectron. Reliab. **75**, 110–120 (2017). https://doi.org/10.1016/J.MICROREL.2017.06.032

Two-Layer Adaptive Fault Tolerance Framework for Internet of Medical Things: A Weighted Autoencoder with Reinforcement Learning-Based Threshold Adaptation

Abdelhammid Bouazza[1(✉)], Hichem Debbi[1], and Hicham Lakhlef[2]

[1] Department of Computer Science, Laboratory of Informatics and its Applications of M'sila (LIAM), University of M'sila, PO Box 166, Ichebilia, M'sila 28000, Algeria
{abdlhammid.bouazza,hichem.debbi}@univ-msila.dz

[2] CNRS-LaBRI, University of Bordeaux/Bordeaux INP, P.O. Box 99, Talence Cedex, 33400 Talence Cedex, France
hicham.lakhlef@u-bordeaux.fr

Abstract. Fault tolerance in Internet of Medical Things (IoMT) systems is challenging due to heterogeneous architectures, stringent safety requirements, and concurrent exposure to cyberattacks and performance degradation. This paper proposes a two-layer adaptive fault tolerance framework integrating lightweight intrusion detection with domain-aware anomaly quantification. The first layer employs a machine learning-based intrusion detection system to produce a binary safety score. The second layer uses a *domain-weighted autoencoder* to compute a continuous Health Index (HI) from Quality-of-Service metrics and the safety score. The principal contribution is the systematic demonstration that *feature weighting within the reconstruction loss plays a central role* in safety-critical detection, improving Critical Recall from 7.3% (uniform weighting) to 100% (safety-first weighting) on the evaluated dataset. Decision thresholds are optimized via an ϵ-greedy Q-learning agent, and a Teacher–Student knowledge distillation pipeline compresses cloud-level intelligence for edge deployment, achieving a 34.6× reduction in computational cost while preserving Critical Recall. Experiments on the WUSTL-EHMS-2020 dataset validate the framework against ablation baselines and a structurally comparable state-of-the-art method.

Keywords: Internet of Medical Things · Fault tolerance · Weighted autoencoder · Reinforcement learning · Knowledge distillation

1 Introduction

The Internet of Medical Things (IoMT) integrates heterogeneous edge–cloud architectures, biosensors, and intelligent communication protocols to enable real-time physiological monitoring and autonomous clinical decision-making. These

© The Author(s), under exclusive license to Springer Nature Switzerland AG 2027
E. Troubitsyna et al. (Eds.): SAFECOMP 2026, LNCS 16868, pp. 18–31, 2027.
https://doi.org/10.1007/978-3-032-34867-8_2

systems generate high-frequency biomedical data streams where even minor disruptions can compromise patient safety, making continuous, secure, and fault-resilient operation a critical research imperative [5].

Existing fault-tolerance mechanisms face three key limitations in IoMT environments. First, malicious cyberattacks and benign performance degradation manifest through similar Quality-of-Service (QoS) degradation patterns, yet many anomaly detectors do not explicitly distinguish them—risking either false negatives (security breaches) or false positives (disrupted clinical workflows). Second, most frameworks use static thresholds calibrated offline, which degrade under concept drift from fluctuating loads and evolving attacks. Third, many autoencoder-based detectors use uniform loss functions that treat all features equally, under-prioritizing safety-critical features (e.g. intrusion indicators) relative to performance metrics.

To address these gaps, this paper makes the following specific contributions:

1. **Domain-aware weighted reconstruction loss:** We introduce a weighted Mean Squared Error (MSE) loss for the autoencoder that explicitly encodes feature criticality. Through a systematic ablation across three weighting schemes, we provide evidence that this design choice plays a *central role* in safety performance—improving Critical Recall from 7.3% to 100% on the evaluated dataset—suggesting that feature weighting should be treated as a primary architectural decision rather than a peripheral hyperparameter (Sect. 4.4).
2. **Two-layer security–performance disambiguation:** We propose a hierarchical architecture where a lightweight Intrusion Detection System (IDS) generates a binary safety score that is injected as an explicit input feature into the autoencoder's Health Index computation, providing semantic context that enables the fault tolerance engine to distinguish cyberattacks from benign congestion (Sect. 3).
3. **Adaptive RL-based thresholding:** We formulate threshold optimization as a Markov Decision Process (MDP) and deploy an ϵ-greedy Q-learning agent that dynamically adjusts warning and alarm boundaries, improving Accuracy from 76.2% (static) to 82.8% without increasing false alarms on the evaluated dataset (Sect. 4.5).
4. **Teacher–Student distillation for edge deployment:** A knowledge distillation pipeline compresses the cloud-level autoencoder+RL system into a lightweight Random Forest classifier, achieving a 34.6× reduction in training cost while preserving Critical Recall (Sect. 4.6).

We emphasize that while individual components (autoencoders, RL, distillation) exist independently, the *contribution lies in their principled integration* under a unified safety-aware objective. This work is an exploratory architectural study of the interaction between safety-aware feature weighting and adaptive thresholding, with the observation that domain-informed loss weighting—rather than architectural complexity—appears central to safety-critical IoMT fault detection.

The remainder of this paper is organized as follows. Section 2 reviews related work. Section 3 describes the proposed framework. Section 4 presents the experimental evaluation. Section 5 concludes the paper.

2 Related Work

Fault-tolerant control (FTC) has evolved from classical model-based techniques—including ON/OFF regulation [3], PID control [18,20], sliding mode control [4], and robust adaptive control—toward intelligent, learning-driven frameworks. Early methods relied on hardware redundancy with voting and backup sensors, but required complete model knowledge and offered limited robustness in nonlinear environments. Neural networks and fuzzy inference systems then introduced intelligent approximation for estimating uncertain dynamics and adaptive fault compensation [1,11,17]; Kharrat [12] added online disturbance estimation, and constraint-aware strategies using barrier Lyapunov functions [10,14,15] enabled FTC to enforce safety constraints, marking the shift from hardware to analytical redundancy.

Recent advances integrate machine learning, digital twins, and reinforcement learning into unified architectures: RL and fixed-time adaptive control achieve rapid fault recovery [13,19], digital twin-driven frameworks improve diagnostic precision via virtual–real sensor fusion [21,22], and model predictive control with adaptive Kalman filtering addresses actuator/sensor uncertainties [2]. Learning-based IoT fault detection using deep learning and multi-criteria decision analysis was proposed in [5], and reviews [7,8,16,23] identify AI, RL, and digital twins as primary enablers of hybrid fault-tolerant systems.

Despite this progress, several open challenges remain for IoMT: (1) limited real-time threshold adaptation to separate minor degradation from critical faults; (2) lack of proactive prevention at early warning stages; and (3) limited adaptive routing or node isolation to prevent cascading failures. The present work addresses these through a two-layer framework combining domain-weighted anomaly detection, RL-based adaptive thresholding, and knowledge distillation for edge deployment.

3 Methodology

This section describes the proposed two-layer adaptive fault tolerance framework for IoMT environments. As illustrated in Fig. 1, the architecture comprises an upper Intrusion Detection System (IDS) layer that generates a binary safety score and a lower Fault Tolerance (FT) engine that quantifies node health through domain-aware anomaly detection. The framework addresses four challenges: (1) distinguishing cyberattacks from performance faults, (2) computing a continuous Health Index, (3) adapting thresholds via reinforcement learning, and (4) enabling edge deployment through knowledge distillation.

Scope Clarification. The framework is a real-time *detection* system: it classifies each node's current state (Healthy, Warning, or Critical) from the most recent observation $\mathbf{x}_t$. "Proactive" refers to the Warning state triggering preventive mitigation *before* full failure, not to forecasting future faults from time-series extrapolation.

Terminology. We distinguish *security* (defence against malicious cyberattacks, the role of Layer 1) from *safety* (preservation of correct clinical operation, the system-level objective), and treat security as a *means* toward safety. A detection is *safety-critical* when the node enters the **Critical** state ($\mathrm{HI}_t \leq \theta_{\mathrm{alarm}}$)—i.e. when degradation is severe enough to compromise reliable physiological monitoring, whether caused by a confirmed intrusion ($S_t = 1$) or by severe benign degradation. Critical Recall measures how reliably the framework flags these states.

3.1 Problem Formulation and Notation

At each discrete time step t, we construct a four-dimensional feature vector:

$$\mathbf{x}_t = [L_t, J_t, P_t, S_t]^T \in \mathbb{R}^4 \tag{1}$$

where L_t, J_t, and P_t denote synthesized QoS metrics for Latency, Jitter, and Packet Loss, respectively, and $S_t \in \{0, 1\}$ is the binary safety score produced by Layer 1.

3.2 Layer 1: Lightweight IDS for Safety Score Generation

The first layer is a lightweight classifier (Random Forest, selected empirically in Sect. 4.3) trained for binary classification of network traffic as `benign` ($S_t = 0$) or `attack` ($S_t = 1$):

$$S_t = \mathrm{IDS}_{\mathrm{L1}}(\mathrm{flow\_data}_t) \in \{0, 1\} \tag{2}$$

Feature Pipeline Clarification. The WUSTL-EHMS-2020 dataset contains 44 raw features per flow instance. Layer 1 (IDS) operates on all 44 raw features to maximize intrusion detection accuracy. Layer 2 (FT engine) operates on a derived 4D vector $\mathbf{x}_t$: three QoS channels synthesized from raw features (described below) plus the binary S_t from Layer 1. This two-stage design decouples high-dimensional intrusion classification from low-dimensional health assessment, enabling efficient edge deployment of the Layer 2 model.

The intuition behind this separation is a *division of labour*: Layer 1 answers the binary security question ("is this flow malicious?") in the full 44-feature space, where a Random Forest excels at sharp boundaries, while Layer 2 answers the graded safety question ("how healthy is this node?") in a compact 4D space suited to a lightweight autoencoder. Folding the Layer 1 verdict back as the single feature S_t injects security *context* without re-processing the raw features, which is what keeps Layer 2—and the distilled edge model—small.

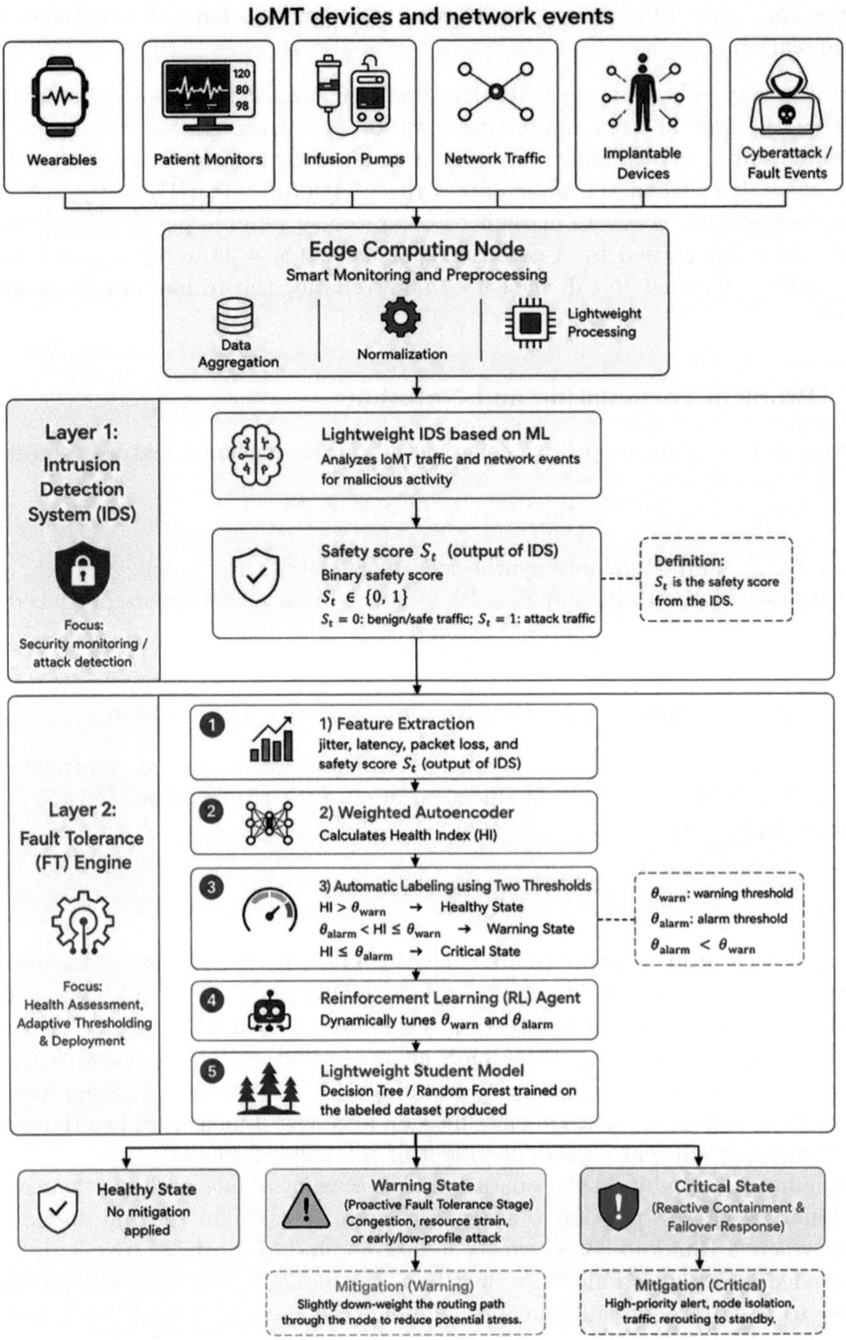

Fig. 1. Proposed two-layer adaptive fault tolerance framework.

3.3 Layer 2: Fault Tolerance Engine

The FT engine receives $\mathbf{x}_t$ and comprises three components: data preprocessing, anomaly detection, and adaptive decision-making.

QoS Feature Engineering. Three QoS channels are synthesized from raw network flow attributes: Latency (L_t), the median of inter-packet arrival times (SIntPkt, DIntPkt, SrcGap, DstGap); Jitter (J_t), the maximum variance in packet timing (SrcJitter, DstJitter); and Packet Loss (P_t), the maximum loss rate (pLoss, pSrcLoss, pDstLoss). All four features are z-score normalized using statistics computed exclusively on "Healthy" training data.

Domain-Weighted Autoencoder. A symmetric feed-forward autoencoder (encoder: $4 \to 16 \to 8 \to 2$; decoder: $2 \to 8 \to 16 \to 4$) is trained exclusively on normal-operation data. The central design choice is the **domain-aware weighted loss**:

$$\mathcal{L}_{\text{weighted}} = \frac{1}{N} \sum_{n=1}^{N} \sum_{i \in \{L,J,P,S\}} w_i \cdot (x_{i,n} - \hat{x}_{i,n})^2 \tag{3}$$

where w_i is the importance weight for feature i. A high weight w_S on the safety score forces the autoencoder to allocate disproportionate representational capacity to the intrusion channel, amplifying reconstruction errors for attack-induced anomalies relative to benign QoS fluctuations.

Health Index (HI). The per-sample reconstruction error e_t is mapped to a continuous, normalized Health Index $\text{HI}_t \in [0, 1]$ via robust min-q95 scaling:

$$\text{HI}_t = \text{clip}\left(1 - \frac{e_t - e_{\min}}{q_{0.95} - e_{\min}},\ 0,\ 1\right) \tag{4}$$

where $e_{\min}$ and $q_{0.95}$ are the minimum and 95th-percentile errors on a normal validation set. Values near 1 indicate nominal operation; values near 0 indicate critical failure.

3.4 Adaptive Thresholding via Reinforcement Learning

Static thresholds degrade under concept drift. We cast threshold optimization as an MDP $\mathcal{M} = \langle \mathcal{S}, \mathcal{A}, \mathcal{R}, \gamma \rangle$ with the following components:

State Space ($\mathcal{S}$). Each state $s_t \in \mathbb{R}^8$ encodes: (1) rolling mean μ_{HI} and (2) variance σ^2_{HI} over $w = 100$ samples; (3) instantaneous HI gradient ∇HI_t; (4) empirical false-alarm rate FAR_t over 500 samples; (5) current θ_{warn} and (6) θ_{alarm}; (7) time since last critical event τ_{critical}; (8) warning-to-critical ratio $\rho_{w/c}$.

Action Space ($\mathcal{A}$). Joint threshold increments:

$$(\Delta_{\text{warn}}, \Delta_{\text{alarm}}) \in \{-0.05,\ -0.01,\ 0,\ +0.01,\ +0.05\}^2 \tag{5}$$

yielding $|\mathcal{A}| = 25$ discrete actions. Thresholds are updated as:

$$\theta_{\text{warn}}^{t+1} = \text{clip}\big(\theta_{\text{warn}}^{t} + \Delta_{\text{warn}},\ 0.4,\ 0.9\big), \tag{6}$$

$$\theta_{\text{alarm}}^{t+1} = \text{clip}\big(\theta_{\text{alarm}}^{t} + \Delta_{\text{alarm}},\ 0.3,\ \theta_{\text{warn}}^{t+1} - 0.1\big), \tag{7}$$

enforcing $\theta_{\text{alarm}} < \theta_{\text{warn}}$ with a minimum safety margin of 0.1.

Reward Function ($\mathcal{R}$). The reward penalizes three operational risks:

$$r_t = -\left(\lambda_1 \text{FAR}_t + \lambda_2 \text{FNR}_t + \lambda_3 \text{Delay}_t\right) \tag{8}$$

with $\lambda_1 = 100$ (false positive penalty), $\lambda_2 = 1000$ (false negative penalty, 10× higher to prioritize safety), and $\lambda_3 = 10$ (detection latency). The asymmetric weighting ($\lambda_2 \gg \lambda_1$) explicitly encodes the clinical requirement that missed critical faults are far more costly than false alarms.

Q-Learning. The action–value function is updated as:

$$Q(s_t, a_t) \leftarrow (1-\alpha)\, Q(s_t, a_t) + \alpha\Big[r_t + \gamma \max_{a'} Q(s_{t+1}, a')\Big] \tag{9}$$

with $\alpha = 0.1$ and $\gamma = 0.95$. Continuous state features are discretized into 10 equiprobable bins; this is intentionally coarse to keep the Q-table small and interpretable for edge operation, and is tolerable because several state dimensions co-vary (e.g. μ_{HI} and σ^2_{HI}), lowering the effective dimensionality. Training runs for up to 10,000 episodes (length $T = 1{,}000$) with ϵ-greedy exploration ($\epsilon_t = \max(0.01,\ \beta^t)$, $\beta{=}0.995$), converging when the maximum Bellman residual falls below 10^{-4} for 100 consecutive episodes, typically within 3,500 iterations. At deployment, thresholds are updated every 100 samples with $\epsilon{=}0.01$.

3.5 Teacher–Student Knowledge Distillation

The Teacher (autoencoder + RL thresholds) achieves high detection accuracy at $\mathcal{O}(n^2 d)$ complexity.

Label Generation Rationale. Because WUSTL-EHMS-2020 provides only binary labels (normal/attack), the intermediate "Warning" state is absent from the raw annotations. To build a three-class training corpus for the Student, the fully trained Teacher (domain-weighted autoencoder + converged RL agent) acts as a label generator: it computes the Health Index HI_t for each historical observation, and the RL-optimized boundaries $\theta_{\text{warn}}, \theta_{\text{alarm}}$ partition the HI space into Healthy ($\text{HI}_t > \theta_{\text{warn}}$), Warning ($\theta_{\text{alarm}} < \text{HI}_t \le \theta_{\text{warn}}$), and Critical ($\text{HI}_t \le \theta_{\text{alarm}}$). These labels inherit the safety bias of the reward (Eq. 8, $\lambda_2 = 10\lambda_1$), so "Warning" marks an early-degradation state preceding critical failure—a safety-informed reference for the lightweight Student.

A Student model (Random Forest classifier) is trained on the Teacher's predicted state labels (Healthy/Warning/Critical), reducing complexity to $\mathcal{O}(n \log n)$ while preserving the Teacher's decision boundaries. This enables real-time inference on resource-constrained edge devices.

4 Experimental Evaluation

4.1 Dataset and Experimental Setup

All experiments use the **WUSTL-EHMS-2020** dataset [9], a public benchmark for IoMT cybersecurity and reliability assessment. The dataset contains 16,318 labeled network flow instances: 87.5% normal and 12.5% attack (Man-in-the-Middle and data injection). Each instance comprises 44 features spanning QoS metrics (latency, jitter, packet loss), biometric signals (heart rate, SpO_2), and traffic flow statistics (packet size, inter-arrival time).

WUSTL-EHMS-2020 is, to our knowledge, the principal public benchmark collected on a real-time healthcare-monitoring testbed that jointly records *network flow metrics* and *patient biometrics* under realistic attacks [9]—precisely the pairing our two-layer design exploits (QoS channels feed the Health Index; attack labels supervise Layer 1). Its binary labelling motivates the Teacher-based label generation of Sect. 3.5. The proposed mechanism is not specific to medical traffic: it applies to any IoT setting where a sparse, high-priority indicator must not be masked by higher-variance performance metrics.

A temporal 70/15/15 train/validation/test split emulates live deployment conditions. The autoencoder is trained exclusively on Healthy instances from the training set; Layer 1 uses all 44 raw features while Layer 2 uses the derived 4D vector (Sect. 3).

4.2 Evaluation Metrics

We employ four metrics: *Critical Recall (CR)*, the proportion of critical faults correctly detected (our primary safety metric); *Macro F1-Score* across all classes; *False Alarm Rate (FAR)*, the proportion of healthy states incorrectly flagged; and *Training Time* on a standard workstation (Intel i7, 32 GB RAM, no GPU), as a proxy for deployment feasibility.

4.3 Layer 1: Intrusion Detection Evaluation

Table 1 compares three lightweight classifiers on the full 44-feature space. Random Forest achieves the best accuracy (0.994) and F1-score (0.994) with the lowest inference latency (0.08 ms/sample), supporting its suitability for real-time edge-level IDS.

Table 1. Layer 1 Intrusion Detection Performance (44 raw features).

Model	Accuracy	F1-Score	Inference (ms/sample)
k-Nearest Neighbors	0.972	0.971	1.45
Support Vector Machine	0.981	0.981	0.92
Random Forest	**0.994**	**0.994**	**0.08**

4.4 Impact of Loss Function Weighting

To isolate the effect of domain-aware feature weighting, we evaluate three representative configurations spanning the weighting spectrum (Table 2): (i) a *uniform* configuration used as a controlled ablation reference (not as a competing method), (ii) a *QoS-dominant* scheme reflecting network-centric priorities, and (iii) a *safety-first* scheme that maximally weights the safety-score channel. These schemes are not the result of exhaustive hyperparameter search; rather, they represent three qualitatively distinct engineering philosophies whose contrasting performance reveals the *mechanism* by which weighting governs detection. The weights are assigned by a simple, reproducible heuristic rather than tuned for the test set: features are ranked by domain criticality (safety score > timing-based QoS > loss-based QoS) and the per-group budget is shifted accordingly, with all weights summing to one. A systematic sweep (e.g., Bayesian optimization over the weight simplex) is left for future work, though the large performance gap (7.3% vs. 100% CR) between the uniform reference and the safety-first configuration suggests that the qualitative finding is robust.

Table 2. Autoencoder weighting schemes and fault detection performance.

Scheme	w_L	w_J	w_P	w_S	CR	FAR
I – Uniform	0.25	0.25	0.25	0.25	0.073	0.142
II – QoS-Dominant	0.35	0.35	0.25	0.05	0.979	0.089
III – Safety-First	0.20	0.30	0.10	0.40	**1.000**	0.238

CR = Critical Recall, FAR = False Alarm Rate.
Weights: $[w_{\text{Latency}}, w_{\text{Jitter}}, w_{\text{PacketLoss}}, w_{\text{Safety}}]$

Analysis. The uniform reference (Scheme I) achieves only 7.3% Critical Recall despite 70.6% Accuracy. A plausible explanation is that QoS metrics exhibit higher temporal variance than the binary S_t under normal operation, so the autoencoder allocates representational capacity to QoS reconstruction, effectively masking intrusion-related anomalies within benign network dynamics.

Scheme III (Safety-First, $w_S = 0.40$) achieves 100% Critical Recall by forcing the autoencoder to prioritize safety-score reconstruction, at the cost of an elevated FAR of 23.8%. We do *not* advocate this configuration for deployment: a false-alarm rate near one-in-four would, as the alarm-fatigue literature warns, risk desensitizing clinical staff and ultimately cause genuine alarms to be missed. Scheme III is reported as the *upper bound* of the weighting mechanism—evidence that the safety channel alone suffices to recover every critical event.

Scheme II (QoS-Dominant, $w_S = 0.05$) achieves 97.9% CR with only 8.9% FAR, revealing that malicious intrusions induce correlated QoS degradation, so security can be inferred indirectly from performance anomalies even with minimal direct safety weight. We therefore adopt Scheme II as the *recommended operating point* and as the default in all subsequent experiments (Tables 3–4): it sacrifices ∼2 points of Critical Recall while cutting FAR by roughly 2.7× relative to Scheme III, directly mitigating alarm fatigue. Residual false alarms are further reduced by the RL thresholding of Sect. 3.4, whose reward (Eq. 8) explicitly penalizes FAR ($\lambda_1 = 100$). In effect, the framework exposes an interpretable safety/false-alarm dial—weight choice for the coarse setting, RL thresholds for online fine-tuning—and our recommendation occupies the high-recall, moderate-FAR region rather than the 100%-recall extreme.

Across schemes, the 7.3%–100% CR range indicates that *feature weighting plays a central role in safety-critical detection within this architecture*—arguably more so than model depth or latent dimensionality—and should be treated as a primary design decision encoding domain safety requirements.

4.5 Ablation: Safety Weighting and Adaptive Thresholding

Table 3 presents a component-wise ablation isolating the contributions of (a) the safety-aware weighting and (b) the RL-based adaptive thresholding. The Scheme II (QoS-Dominant) weighting is used as the default for the full system.

Table 3. Ablation study on key architectural components.

Configuration	CR	Accuracy	Time (s)
AE + Static Threshold (no S_t)	0.953	0.771	22.37
AE + Static Threshold (with S_t)	1.000	0.762	0.01
AE + RL Threshold (with S_t)	**0.979**	**0.828**	**35.62**

Recall. "S_t" denotes the safety score feature from Layer 1

Effect of Safety Feature. Removing S_t from the input (row 1 vs. row 2) degrades Critical Recall from 1.000 to 0.953, suggesting that the binary safety score provides additional discriminative information beyond what QoS metrics alone can capture.

Effect of RL Thresholding. Comparing static vs. RL thresholds (row 2 vs. row 3): the static baseline reaches high CR but lower Accuracy (0.762), suggesting fixed thresholds are overly sensitive and over-alert in practice. The RL agent raises Accuracy to 0.828 by learning context-dependent boundaries that adapt to QoS variability. The extra training time (35.62 s vs. 0.01 s) is a one-time offline cost; at deployment the learned Q-table updates thresholds in < 1 ms per step.

Figure 2 illustrates the Health Index distribution across operational states. The visible separation between Healthy, Warning, and Critical clusters suggests that the HI provides a meaningful continuous degradation metric, supporting proactive mitigation in warning states before system failure.

4.6 Knowledge Distillation and Deployment Efficiency

Table 4 compares the Teacher (AE+RL) with two Student models trained on Teacher-generated labels.

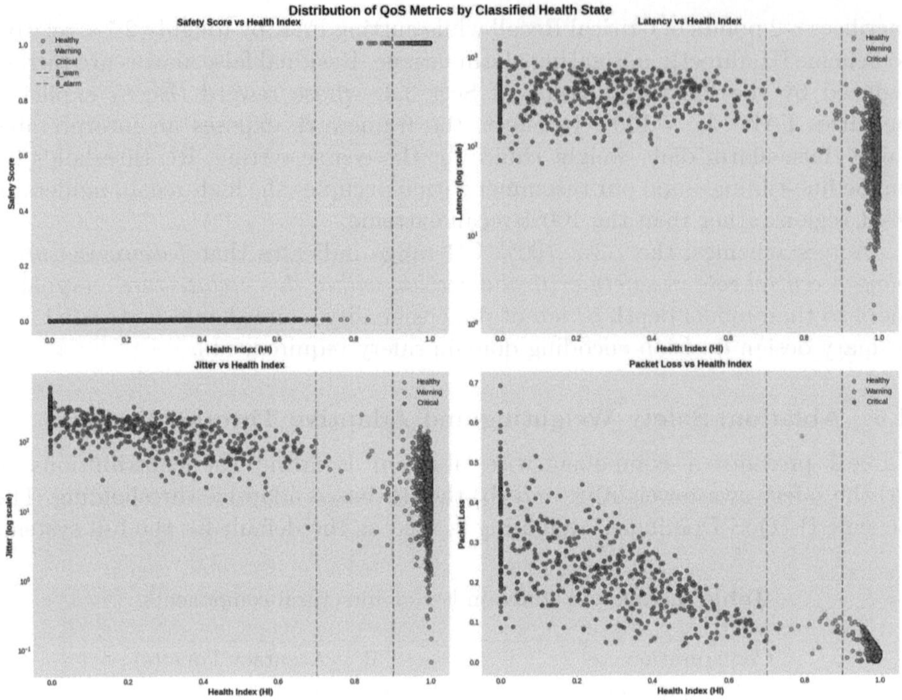

Fig. 2. Distribution of QoS metrics across Health Index states.

Table 4. Teacher vs. Student model performance (knowledge distillation).

Model	CR	Macro F1	Training Time (s)
Teacher (AE+RL)	0.979	0.643	35.62
Student (Random Forest)	**0.979**	**0.672**	1.03
Student (SVM-RBF)	0.958	0.669	0.24

The Random Forest Student preserves full Critical Recall (0.979) while achieving a 34.6× reduction in training time. Notably, the Student's Macro F1 (0.672) *exceeds* the Teacher's (0.643), suggesting that the Random Forest's decision boundaries generalize the Teacher's per-sample decisions more cleanly than the autoencoder's continuous error distribution. This result provides initial evidence that knowledge distillation is a viable compression strategy for safety-critical IoMT models, though further validation on additional datasets would strengthen this conclusion.

Interpreting the Reported Times. The Teacher and Student costs in Table 4 are reported separately, not summed, because they arise at different lifecycle stages. The Teacher (autoencoder + RL) is trained *once*, offline and centrally, solely to generate the Healthy/Warning/Critical labels and is never shipped to the edge; the Student is the deployed artefact, retrained per node. The 34.6× figure thus compares the recurring per-node cost (1.03 s) against the Teacher (35.62 s). Counted end-to-end, the one-time pipeline costs $35.62 + 1.03 \approx 36.65$ s, after which each additional node or retraining cycle costs only 1.03 s—so the benefit grows with the number of deployments.

Deployment Considerations. Training time here is a computational *proxy*; it is not a full edge benchmark with memory, energy, and latency profiling on real IoMT hardware (e.g., Raspberry Pi, ARM Cortex-M). Still, the Student requires only 4 input features and ~100 trees (< 1 MB), enabling sub-millisecond inference. Comprehensive edge profiling is planned for future work.

4.7 Comparison with State-of-the-Art

To contextualize the proposed framework, we compare it with three representative methods that cover complementary aspects of anomaly-driven fault tolerance in IoT and IoMT environments. These studies were selected for architectural relevance to the main components of our system—multi-layer anomaly detection, adaptive decision-making, and lightweight deployment—rather than as an exhaustive benchmark of all detection paradigms. We emphasize that the most direct quantitative comparison is with our own prior deep-learning baseline [5], which we re-evaluate on the identical WUSTL-EHMS-2020 split; the other two methods are compared at the architectural level because re-running them under our exact protocol is left to future work. Compared with [5], our framework differs in three main respects: it introduces a domain-aware weighted reconstruction loss that explicitly encodes feature criticality, improving Critical Recall from 7.3% under uniform weighting to 100% under safety-first prioritization; it replaces fixed decision boundaries with RL-based adaptive thresholding, yielding a 6.6%-point improvement in Accuracy (76.2% → 82.8%); and it incorporates a Teacher–Student distillation pipeline that supports edge deployment at 34.6× lower computational cost.

The other two baselines address complementary dimensions but differ at the detection layer. IoT-AD [24] couples anomaly detection with containment to prevent fault propagation, but relies on uniform feature treatment and static detection logic. The Isolation Forest-based self-healing approach in [6] performs unsupervised scoring and automated reconfiguration, but its equal treatment of all dimensions can let sparse, high-priority safety signals be masked by higher-variance QoS behaviour. In contrast, our weighted autoencoder encodes safety priorities directly in the loss while adaptive thresholding improves robustness to changing conditions.

5 Conclusion

This paper presented a two-layer adaptive fault tolerance framework for IoMT environments that integrates security-aware anomaly detection, domain-informed feature weighting, and reinforcement learning-based adaptability. The primary contribution lies in the system-level integration of intrusion detection, safety-aware health assessment, and adaptive thresholding within a unified architecture designed for IoMT environments.

The hierarchical combination of a lightweight IDS with a weighted autoencoder helps disambiguate cyberattacks from benign degradation, while the RL agent improves overall Accuracy by 6.6 percentage points over static thresholds. A Teacher–Student distillation pipeline achieves comparable Critical Recall at 34.6× lower computational cost.

Future directions include automated weight and reward optimization, cross-dataset validation to establish generality beyond WUSTL-EHMS-2020, an explicit channel for accidental (non-malicious) faults, and integration of federated learning for cross-institutional scalability. The proposed architecture provides a promising direction for further research on safety-aware monitoring in IoMT systems.

References

1. Amin, A.A., Hasan, K.M.: A review of fault tolerant control systems: Advancements and applications. Measurement **143**, 58–68 (2019). https://doi.org/10.1016/j.measurement.2019.04.083
2. Asfihani, T., Maulana Syafi'i, A., Hasan, A.: Fault-tolerant model predictive control for unmanned surface vehicles. Int. J. Syst. Sci. **56**(4), 680–695 (2025). https://doi.org/10.1080/21642583.2025.2469598
3. Belchior, C.A.C., Araújo, R.A.M., Landeck, J.A.C.: Dissolved oxygen control of the activated sludge wastewater treatment process using stable adaptive fuzzy control. Comput. Chem. Eng. **37**, 152–162 (2012). https://doi.org/10.1016/j.compchemeng.2011.09.011
4. Blanke, M., Kinnaert, M., Lunze, J., Staroswiecki, M.: Diagnosis and Fault-Tolerant Control. Springer, 3rd edn. (2016). https://doi.org/10.1007/978-3-662-47943-8
5. Bouazza, A., Debbi, H., Lakhlef, H.: Ensuring iot system fault tolerance using deep learning and multi-criteria decision analysis. In: Advanced Information Networking and Applications, Lecture Notes on Data Engineering and Communications Technologies, vol. 247. Springer, Cham (2025). https://doi.org/10.1007/978-3-031-87769-8_2
6. Devi, K., Thenmozhi, R., Kumar, D.S.: Self-healing IoT sensor networks with isolation forest algorithm for autonomous fault detection and recovery, pp. 451–456 (2024). https://doi.org/10.1109/autocom60220.2024.10486184
7. Edwards, C., Lombaerts, T., Smaili, H. (eds.): Fault Tolerant Flight Control: A Benchmark Challenge, Lecture Notes in Control and Information Sciences, vol. 399. Springer, Cham (2010). https://doi.org/10.1007/978-3-642-11690-2
8. Fatorachian, H., Kazemi, H.: Ai-enhanced fault-tolerant control and security in transportation and logistics systems. Int. J. Control Eng. Syst. (2024). https://doi.org/10.20517/ces.2024.35
9. Hady, A.A., Ghubaish, A., Salman, T., Unal, D., Jain, R.: Intrusion detection system for healthcare systems using medical and network data: A comparison study. IEEE Access **8**, 106576–106584 (2020). https://doi.org/10.1109/ACCESS.2020.3000421

10. Huang, S., Wang, Y., Luo, H.: A dual-branch generative adversarial network with self-supervised enhancement for robust auditory attention decoding. Eng. Appl. Artif. Intell. (2025). https://doi.org/10.1016/j.engappai.2025.111122
11. Joo-Hwa, T., Xiyue, Z.: Dissolved oxygen control in activated sludge process using a neural-fuzzy approach. Water Res. **34**(10), 2689–2696 (2000). https://doi.org/10.1016/S0043-1354(99)00394-6
12. Kharrat, M.: Neural network-based adaptive fault-tolerant control for nonlinear systems with unknown dynamics. Commun. Nonlin. Sci. Numer. Simul. **141** (2025). https://doi.org/10.1016/j.cnsns.2024.108478
13. Kharrat, M., Alhazmi, H.: Neural networks-based adaptive fault-tolerant control for a class of nonstrict-feedback nonlinear systems with actuator faults and input delay. AIMS Math. **9**(6), 13689–13711 (2024)
14. Li, D., Han, H.G., Qiao, J.F.: Fuzzy-approximation adaptive fault tolerant control for nonlinear constraint systems with actuator and sensor faults. IEEE Trans. Fuzzy Syst. **32**(5), 2614–2624 (2024)
15. Li, D., Liu, Y., Huang, D.: Development of semi-supervised multiple-output soft sensors with co-training and tri-training mpls and mrvm. Chemom. Intell. Lab. Syst. **199**, 103970 (2020). https://doi.org/10.1016/j.chemolab.2020.103970
16. Muzakkir Quamar, M., Nasir, A.: Review on fault diagnosis and fault-tolerant control scheme for robotic manipulators: Recent advances in ai, machine learning, and digital twin, arXiv preprint (2024).
17. Narendra, K.S., Parthasarathy, K.: Identification and control of dynamical systems using neural networks. IEEE Trans. Neural Netw. **1**(1), 4–27 (1990). https://doi.org/10.1109/72.80202
18. Shtessel, Y., Edwards, C., Fridman, L., Levant, A.: Sliding Mode Control and Observation. Control Eng. Birkhäuser (2014). https://doi.org/10.1007/978-0-8176-4893-0
19. Tahoun, A.: Neural network-based adaptive fault-tolerant control for nonlinear systems with uncertainties. ISA Trans. **152**, 217–228 (2024). https://doi.org/10.1016/j.isatra.2024.07.003
20. Vilanova, R., Visioli, A.: PID Control in the Third Millennium: Lessons Learned and New Approaches. Advances in Industrial Control. Springer, Cham (2012). https://doi.org/10.1007/978-1-4471-2425-2
21. Yang, C., et al.: Digital twin-driven fault diagnosis method for composite faults by combining virtual and real data. J. Ind. Inf. Integr. **33**, 100469 (2023)
22. Yang, C., et al.: Cross-validation enhanced digital twin driven fault diagnosis methodology for minor faults of subsea production control system. Mech. Syst. Signal Process. **204**, 110813 (2023)
23. Yang, Y., Zhu, J., Shen, Q.: Adaptive neural fault-tolerant consensus tracking for input-delayed multi-agent systems. Neurocomputing **621**, 129299 (2025). https://doi.org/10.1016/j.neucom.2024.129299
24. Zahan, H., Azad, W.A., Ali, I., Mastorakis, S.: IoT-AD: A framework to detect anomalies among interconnected IoT devices. IEEE Internet Things J. (2023). https://doi.org/10.48550/arXiv.2306.06764

Diagnosis of Runtime Property Violations

Marco Bozzano[1(✉)], Alessandro Cimatti[1], Alberto Sambrotta[1,2], and Stefano Tonetta[1]

[1] Fondazione Bruno Kessler, Trento, Italy
{bozzano,cimatti,asambrotta,tonettas}@fbk.eu
[2] Università degli Studi di Udine, Udine, Italy
sambrotta.alberto@spes.uniud.it

Abstract. Runtime verification checks, during execution, whether a system satisfies a given property. When a violation is detected, identifying the underlying causes is essential for fault recovery. However, existing approaches to model-based diagnosis do not fully address this problem. Model-Based Diagnosis (MBD) explains deviations between observed and nominal behaviors, but does not target property violations specifically. Model-Based Safety Analysis (MBSA), on the other hand, explains possible causes of a top-level event, but does not exploit runtime observations. In this work, we address the Property Violation Diagnosis (PVD) problem, which lies at the intersection of runtime verification and model-based diagnosis: given a system description, a monitored property, and a set of observations that indicate its violation, we aim to compute the set of faults that are relevant to explain the violation. We provide a formal characterization of PVD in propositional logic and propose symbolic algorithms for generating the relevant diagnoses. Our experimental evaluation demonstrates the applicability of the proposed approach.

Keywords: Property Violation · Model-Based Diagnosis · Model-Based Safety Analysis · Fault Diagnosis

1 Introduction

Safety-critical systems require the continuous monitoring of specific properties to ensure correct operation [7]. When a violation is detected, a recovery mechanism must be triggered [6,12,16]. However, an effective recovery requires identifying precisely which faults caused the property violation. A classic example of such requirements can be found in the aerospace domain [1,22,24], where Fault Detection, Identification, and Recovery (FDIR) software is heavily utilized to continuously monitor the system state and trigger appropriate recovery mechanisms upon fault detection. Furthermore, depending on the specific application and the criticality of the fault, the time available to initiate recovery may be subject to strict real-time constraints.

Standard Model-Based Diagnosis (MBD) [9] identifies fault sets that render the system model consistent with observations. However, MBD explains general

© The Author(s), under exclusive license to Springer Nature Switzerland AG 2027
E. Troubitsyna et al. (Eds.): SAFECOMP 2026, LNCS 16868, pp. 32–46, 2027.
https://doi.org/10.1007/978-3-032-34867-8_3

deviations from nominal behavior rather than specific property violations. This is insufficient for our context, as we must isolate the faults specifically responsible for the violation to trigger the right recovery. The core issue with applying standard MBD here is the inclusion of "irrelevant" faults. From observations - i.e., information about the state of the system - we may infer the occurrence of a fault that is completely unrelated to the violation of the monitored property.

We term this specific challenge Property Violation Diagnosis (PVD). PVD lies at the intersection of runtime verification [2] and model-based diagnosis: given a system description, a monitored property, and a set of observations indicating a violation, we aim to compute the set of faults that are strictly relevant to explaining that violation.

In this paper, we first provide a formal characterization of PVD in propositional logic. To ensure this definition captures the intuitive notion of "blame" in partial observability, we validate our formalization against a set of motivating examples. These scenarios are designed to illustrate cases where standard diagnosis fails to distinguish between faults that caused the violation and those that were merely coincidentally present. In the related work, we also discuss why other similar approaches such as model-based safety analysis, causal analysis and abductive reasoning are not suitable to solve the PVD problem.

We then translate this formalization into a quantified Boolean formula and propose a solution based on propositional quantifier elimination. Our algorithm leverages symbolic methods to compute the set of relevant diagnoses.

Note that, although our presentation focuses on propositional logic and uses BDD-based quantifier elimination [18], the definitions and algorithms extend naturally to richer theories in first-order logic, and can be instantiated using SMT solvers with only modest changes to the underlying reasoning procedures.

The main contributions of this paper are the following:

- We formally define the PVD problem and we validate this definition through varied scenarios involving partial observability and irrelevant faults.
- We provide a symbolic algorithm based on propositional quantifier elimination to compute relevant diagnoses.
- We provide an experimental evaluation demonstrating the applicability of the proposed approach, suitable for an online application.

The rest of the paper is organized as follows. Section 2 reviews the background on MBD. Section 3 presents the motivating examples used to validate our definition. Section 4 discusses related work. Section 5 details the formalization of the PVD problem. Section 6 presents the symbolic approach to solve PVD and Sect. 7 presents the experimental evaluation. Finally, Sect. 8 concludes the paper.

2 Background

We assume standard background in propositional logic (formulae, models, satisfiability, entailment), which forms the basis of our propositional formalization. Only the notions required for diagnosis are summarized below [17].

Definition 1 (set of cutsets). *Let β be a formula in propositional logic, W a set of models and X a set of variables. The set of cutsets for β in the set of models W, over the variables in X is defined as*

$$CS(X, W, \beta) = \{cs \subseteq X \mid \exists w \in W.\ w \models \beta,\ \forall x \in cs.\ w \models x,\ \forall x \notin cs.\ w \models \neg x\} \tag{1}$$

Definition 2 (minimal/maximal cutsets). *Let S be a set of cutsets. The set of minimal/maximal cutsets in S is defined as*

$$MinCS(S) = \{cs \in S \mid \nexists cs' \in S.\ cs' \subset cs\} \tag{2}$$
$$MaxCS(S) = \{cs \in S \mid \nexists cs' \in S.\ cs \subset cs'\} \tag{3}$$

In the following sections it will be necessary to represent cutsets as propositional formulae. Given a cutset cs, its formula representation is defined as $\varphi(cs) = \bigwedge_{f \in cs} f \wedge \bigwedge_{f \notin cs} \neg f$.

Model-Based Diagnosis (MBD). MBD is defined for first-order logic, but it can be applied to a variety of logics; in this paper we focus on the propositional logic. Let $COMPS = \{c_1, ..., c_n\}$ be the *set of components*, represented by constants. For a component c, its *nominal behaviour* is defined by the propositional formula $B(c)$, and variable f_c determines whether c is faulted or nominal. The *system description* is defined by the formula $SD := \bigwedge_{c \in COMPS} \neg f_c \rightarrow B(c)$. An observation OBS is an assignment to the variables of SD representing the input/outputs of the components. The system is not behaving nominally iff $SD \wedge OBS \wedge \bigwedge_{c \in COMPS} \neg f_c$ is unsatisfiable. A *diagnosis* for $\langle SD, COMPS, OBS \rangle$ is a minimal set $\Delta \subseteq COMPS$ such that $SD \wedge OBS \wedge \bigwedge_{c \in \Delta} f_c \wedge \bigwedge_{c \in COMPS \setminus \Delta} \neg f_c$ is satisfiable.

We adapt MBD to our notation as follows: $COMPS$ is the set of state Boolean variables V, SD is a propositional formula defined over V and fault variables F, and OBS is a truth assignment over a set of observable variables O. The system is not behaving nominally iff $SD \wedge OBS \wedge \bigwedge_{f \in F} \neg f \models \bot$, and the set of diagnoses for $\langle SD, COMPS, OBS \rangle$ is defined as $MinCS(CS(F, SD \wedge OBS, \top))$.

3 Motivating Examples

Consider a system with two rooms, A and B. In room A, there is a permanent emergency light and three lights controlled by a single switch. Room B has one light that turns on if the switch is on and motion is detected. The breakage of the switch turns off all connected lights permanently, while a broken light only affects itself. Additionally, Room B's light has a sensor to detect if it's malfunctioning. And last, we say that room A is lit if at least one of the lights in it is on. The described system can be formalized as the following propositional formula:

$$\begin{aligned}
SD := &\ \text{power_on} \leftrightarrow (\text{switch_on} \wedge \neg\text{switch_faulted}) \wedge \\
&\text{emergency_light_on} \leftrightarrow \neg\text{emergency_light_faulted} \wedge \\
&(\wedge_{i=1}^{3}\text{light}_{Ai}\text{_on} \leftrightarrow (\text{power_on} \wedge \neg\text{light}_{Ai}\text{_faulted})) \wedge \\
&\text{light}_{B1}\text{_on} \leftrightarrow (\text{power_on} \wedge \neg\text{light}_{B1}\text{_faulted} \wedge \text{motion_detected}) \wedge \\
&\text{roomA_dark} \leftrightarrow ((\wedge_{i=1}^{3}\neg\text{light}_{Ai}\text{_on}) \wedge \neg\text{emergency_light_on}) \wedge \\
&\text{observe_light}_{B1}\text{_broken} \leftrightarrow \text{light}_{B1}\text{_faulted}
\end{aligned}$$

Suppose that we want to monitor three different properties and that for each one we are given different information about the system:

- "$PROP_1 :=$ switch_on $\rightarrow$ (light$_{A1}$_on $\wedge$ light$_{A2}$_on)", we want to monitor the nominal behaviour of the first and second light in room A, meaning that if the switch is on then both lights must be on:
  - "$OBS_1 :=$ ¬light$_{A1}$_on ∧ light$_{A2}$_on ∧ light$_{A3}$_on". Since we can observe that only the first light is off, we can say with certainty that it is broken;
  - "$OBS_2 :=$ ¬light$_{A1}$_on ∧ ¬light$_{A2}$_on ∧ light$_{A3}$_on". The observation shows that only the third one is on, while the first and second light are off, meaning that they must be broken, but since the property is violated with either one of them broken they are both relevant faults;
  - "$OBS_3 :=$ ¬light$_{A1}$_on ∧ light$_{A2}$_on ∧ ¬light$_{A3}$_on". We know that only the second light is on, while the first and third lights are off, meaning that they are broken, but only the breakage of the first light is relevant, since even if the third light was on (thus not broken), the violation would have still occurred;
  - "$OBS_4 :=$ switch_on ∧ ¬light$_{A1}$_on ∧ ¬light$_{A2}$_on ∧ ¬light$_{A3}$_on". We can observe that the switch is on but all of the lights in room A are off. The breakage of the switch can be a cause, since we can't observe any light on, and the breakage of the third light is not relevant, as in previous case, while the breakage of either the first or the second light can be considered as relevant faults.
- "$PROP_2 :=$ emergency_light_on", we want to monitor the correct behaviour of the emergency light, stating that it must stay on no matter what:
  - "$OBS_1 :=$ switch_on ∧ roomA_dark." Here the switch is on but there is no light in the room. If the system was nominal then there would be light in the room, but since it is not illuminated we can conclude all of the lights are off. But since the property states that just the emergency light must be on then its breakage must be relevant to the violation.[1]
- "$PROP_3 :=$ switch_on $\rightarrow$ (light$_{A1}$_on ∧ (motion_detected $\rightarrow$ light$_{B1}$_on))", we want to monitor the nominal behaviour of the first light in room A and the light in room B, stating that if the switch is on then the first light of room A must be on, and if there is motion in room B, then also the light in room B must be on:

[1] Note that if we observed ¬roomA_dark then we wouldn't have been able to deduce the violation, since we don't know if the emergency light is on.

- "$OBS := \text{switch\_on} \wedge \neg \text{light}_{A1}\text{\_on} \wedge \neg \text{light}_{B1}\text{\_on} \wedge \neg \text{motion\_detected} \wedge \text{observe\_light}_{B1}\text{\_broken}$". The switch is on, both of the lights are off, the light in room B is broken, and finally there is no motion in room B. Even if there are scenarios in which the breakage of the light in room B can violate the property, in this one it didn't contribute, since the sensor didn't detect any movement. Thus, the faults relevant to the violation are either the breakage of the first light or the switch.

4 Related Work

In the following, we compare our work with Halpern and Pearl definition of Actual Causality, MBD, and MBSA (which are the most relevant work w.r.t. our approach), and then we discuss other works of the state-of-the-art and compare them with our work.

Our notion of relevance is broadly related with causality. Halpern and Pearl, see [14], have defined (several variants of) the notion of (actual) causality based on structural equations. Causes are taken from a generic set of endogenous variables, whereas we defined explanations as sets of faults. We inherit from them the concept of counterfactual reasoning. However, in all cases our notion of causality is different: in the original definition, causes may be restricted to single conjuncts, whereas we consider sets; in the modified definition [14] conjunctive and disjunctive scenarios yield different results w.r.t. our definition. Our work is rooted in the setting of Model-Based Diagnosis [9] and extends it to compute relevant explanations for the violation of a given property. Starting from the observations, we compute minimal sets of causes that do not contain faults that are irrelevant (i.e., they do not contribute to the violation). To the best of our knowledge, this is the first work that addresses the notion of relevance for diagnosing property violations, for partially observable systems. Model-Based Safety Analysis (MBSA) [4,5,15] looks for explanations in terms of faults, however the analysis is done at design time rather than at runtime, without considering observability.

In [13] the authors define causality for Discrete Event Systems, where causes are subsequences of events consistent with the system and the observations. Their notion is similar to ours, however minimization is carried out on a larger set of events, other than faults, which, as the authors admit, may result in explanations including irrelevant events, e.g., in case of system initialization. In [11] the authors look for temporal explanations of property violation on a given execution trace (no system description is given). The framework is very expressive, it covers both MTL and LTL, and causality is rooted in the notion of *prime implicant*, which is a generalization of *cut set*, however they do not consider the notion of irrelevance. [3] aims to pinpoint the possible explanations for an LTL property on a given execution trace (no system description is given). Causality is defined using counterfactual reasoning and a notion of contingency inspired from [14]. However, causes may only contain signals occuring in the property, whereas we consider faults, that are generally unobservable and do not appear in properties. Moreover, they consider individual causes only, whereas we consider sets. [25] aims to compute instant causes of the violation of a property expressed in Signal Temporal Logic (STL) in the context of hybrid systems (no system description is given). The authors quantify the relevance of each instant, however their notion of causality is local to each time instant, and is tailored to signal evaluation in hybrid systems.

Finally, there is a close relationships between Model-Based Diagnosis and abduction [10,20]. Our theory of diagnosis could be equivalently formulated in the realm of abduction. As in the case of diagnosis, the theory of abduction does not address the notion of irrelevance that is central in our work.

5 Problem Formalization

In this section, we formalize the PVD problem.

Definition 3 (Monitor Setting). *The monitor setting $MS = \langle V, O, F, SD, PROP, OBS\rangle$ is a tuple where: V is the set of state variables, $O \subseteq V$ is the set of observable variables, F is the set of fault variables, $SD(V, F)$ a formula representing the system description, $PROP(V)$ a formula representing the property to be monitored, and $OBS(O)$ a conjunction of assignments over each variable $o \in O$ representing an observation on the system.*

We will often refer to assignment to fault variables as *fault configuration*, and $\neg PROP$ as Top-Level Event (TLE). Throughout the rest of the paper, the term *occurrence of faults* is used to denote a set of fault variables assigned to $\top$. We now present three desired properties of a monitor setting that are assumed to be true for the rest of the paper and then, at the end of the section, show how the results are affected if these assumptions don't hold.

Assumption 1 (Nominal behaviour). *Let SD, $PROP$, and F be as in Definition 3. We assume that the following holds:*

$$SD \models \left(\bigwedge_{f \in F} \neg f \right) \rightarrow PROP \tag{4}$$

Informally, if the system is behaving nominally, then the property is satisfied, implying that if $(SD \wedge OBS) \models TLE$ holds, then the system is not behaving nominally, meaning that it can be diagnosed.[2]

Assumption 2 (TLE event monotonicity w.r.t. faults). *Let SD, TLE, and F be as in Definition 3. We assume that the following holds:*

$$SD \models \forall \Delta \in CS(F, SD, TLE).\ \forall \Delta^+.\ (\Delta \subseteq \Delta^+ \rightarrow \Delta^+ \in CS(F, SD, TLE)) \tag{5}$$

[2] If $(SD \wedge OBS) \models PROP$ holds, then the system is nominal. If $(SD \wedge OBS) \not\models PROP$ and $(SD \wedge OBS) \not\models \neg PROP$ both hold, then we don't know if the property is satisfied or not.

Informally, if the *TLE* is satisfied with the fault configuration Δ, then there are no possible additions to Δ such that they restore the property.

Definition 4 (Irrelevant subsets of faults). *Let MS be a monitor setting such that $(SD \wedge OBS) \models TLE$ is satisfied, and let $\Delta \in CS(F, SD \wedge OBS, \top)$. A subset $\Delta' \subseteq \Delta$ is irrelevant iff by taking $\Delta^- = \Delta \setminus \Delta'$ there exists a conjunction OBS' of a subset of literals in OBS such that:*

$$(SD \wedge OBS' \wedge \varphi(\Delta^-)) \not\models \bot \text{ and } (SD \wedge OBS' \wedge \varphi(\Delta^-)) \models TLE \quad (6)$$

The intuition is to use the counterfactual approach to filter out faults in a cutset consistent with the observation that did not contribute to the *TLE*. By assuming Assumption 2, only the occurrence of faults can be considered as a diagnosis. Given $\Delta \in CS(F, SD \wedge OBS, \top)$ we can "flip" the faults of a subset Δ' in Δ and check whether the *TLE* still occurs or not in the context given by the observation. It could be possible that modified cutset $\Delta^- = \Delta \setminus \Delta'$, makes the formula $SD \wedge OBS \wedge \Delta^-$ unsatisfiable. In order to avoid this problem, we search for a subset OBS' of OBS such that $SD \wedge OBS' \wedge \Delta^-$ is satisfiable. The idea is to find a "similar" situation to the one given by the observation in order to consider a different fault configuration.

Definition 5 (Relevant Diagnosis). *Let MS be a monitor setting such that $(SD \wedge OBS) \models TLE$ is satisfied. Δ_r is a relevant diagnosis for the TLE if there exists $\Delta \in CS(F, SD \wedge OBS, \top)$ such that $\Delta_r \subseteq \Delta$ and it does not contain subsets of irrelevant faults.*

Note that *"it does not contain subsets of irrelevant faults"* is a minimality condition, meaning that the aim is to remove the largest subsets of irrelevant faults.

Finally, we can define the PVD problem as follows

Definition 6 (PVD Problem). *Let MS be a monitor setting such that $(SD \wedge OBS) \models TLE$ is satisfied. The PVD problem aims to find all relevant diagnosis for the TLE.*

Clearly, MBD doesn't solve the PVD problem, since it is independent from the property. However, it is useful in our approach since a diagnosis consistent with the observation can be used as a starting point.

Remark on the Properties Assumed on the System. If Assumption 1 doesn't hold, then *PROP* can be violated when everything is nominal, i.e., the empty cutset is a potential cause of the TLE, and by minimality, it would be the only diagnosis returned. If Assumption 2 doesn't hold, then situations in which the non-occurrence of a fault is a potential cause of the TLE are allowed, but Definition 5 focuses only on occurrence of faults, i.e., it ignores the non-occurrence of faults.

6 Algorithms

In the following we propose two BDD-based algorithms: one that enumerates the cut sets of MBD and prunes irrelevant faults and one that is fully symbolic based just on BDD operations. Both algorithms are rooted in knowledge compilation concepts [8], meaning that we pre-compute a parameterized BDD to query at runtime substituting the parameters with the actual values.

Parameters. Let *SD*, *TLE*, *V*, *O*, *F* be defined as in Definition 3, *Obs* be an assignment to the variables in O representing OBS (i.e., $OBS = \bigwedge_{o \in O} o \leftrightarrow Obs(o)$), $\Delta \in CS(F, SD \wedge OBS, \top)$, A_O and PH_O be two sets of Boolean variables, where $a_o \in A_O, ph_o \in PH_O$ are renaming of the variable o for each $o \in O$, and a set of Boolean variables PH_F, where $ph_f \in PH_F$ is a renaming of the variable f for each $f \in F$.

We refer to variables A_O as *activation variables* and we use them to represent the subset OBS' of Definition 4. The variables PH_O and PH_F act as symbolic parameters that allow us to compile the BDDs offline without knowing the actual runtime observations (*Obs*) or the specific fault configuration (Δ) ahead of time; at runtime, these placeholders are simply replaced with concrete assignments. Specifically, PH_O is required for both proposed approaches, whereas PH_F is used only by the enumerative algorithm.

6.1 Enumerative Algorithm

Enumerative Relevant Diagnoses Encoding. The following formula encodes Definition 5:

$$\begin{aligned}\text{relevant}(F, PH_O, PH_F) = &\exists V, A_O.\ (\text{lhsE}(V, F, A_O, PH_O, PH_F) \wedge \\ &(\forall V'.\ \text{lhsE}(V', F, A_O, PH_O, PH_F) \rightarrow TLE))\end{aligned} \tag{7}$$

where:

- $\text{lhsE}(V, F, A_O, PH_O, PH_F) = SD(V, F) \wedge \text{act\_obs}(A_O, O, PH_O) \wedge \text{sub\_of}(F, PH_F)$, represents the formula $(SD \wedge OBS' \wedge \varphi(\Delta^-))$;
- $\text{act\_obs}(A_O, O, PH_O) = \bigwedge_{o \in O} a_o \rightarrow (o \leftrightarrow ph_o)$, stating that for each $o \in O$, activation variable a_o controls the variables O that take the same value as defined by ph_o (i.e., they belong to OBS');
- $\text{sub\_of}(F, PH_F) = \bigwedge_{f \in F} f \rightarrow ph_f$, stating that the assignment to the variables in F must be a subset of the assignment to the variables in PH_F.

The minimality condition is encoded into Formula (7) by adding the constraint $\nexists V', F', A'_O.(\text{lhsE}(V', F', A'_O, PH_O, PH_F) \wedge (\forall V''.\ \text{lhsE}(V'', F', A'_O, PH_O, PH_F) \rightarrow TLE) \wedge \text{sub\_of}(F', F) \wedge \bigvee_{f \in F} \neg(f \leftrightarrow f'))$.

Let ϕ_E denote the BDD of Formula (7). Notice that ϕ_E is independent from *Obs* and Δ, meaning that it can be computed statically for any given set O and

Algorithm 1. Relevant diagnoses

1: **function** COMPUTE_RELEVANT_DIAGNOSES(V, O, F, SD, Obs, ϕ_E)
2: $\mathcal{D} = \emptyset$
3: $CS = \exists V.\ SD \wedge \bigwedge_{o \in O} o \leftrightarrow Obs(o)$
4: **for** $\Delta \in CS$ **do**
5: $\phi'_E = \phi_E[ph_{o_1}/Obs(o_1), ..., ph_{o_m}/Obs(o_m), ..., ph_{f_1}/\Delta_{f1}, ..., ph_{f_n}/\Delta_{f_n}]$
6: $\mathcal{D} = \mathcal{D} \cup \Delta_r$ **for** Δ_r **in** enumerate_models(ϕ'_E)
7: **end for**
8: return $\mathcal{D}$
9: **end function**

propositional formula $PROP$. In the following algorithm we assume that ϕ_E has been pre-computed.

Algorithm 1 computes every cutset $\Delta \in CS(F, SD \wedge OBS, \top)$ (line 3). Then, for each Δ, it computes ϕ' by substituting ph_o with $Obs(o)$ for each $o \in O$, and ph_f with Δ_f for each $f \in F$ (line 5). Then it calls the enumerate_models function which takes a BDD as a parameter and it returns all of its satisfying assignments. Finally, it returns the set of relevant diagnoses (line 8). We can say that $\Delta_r \subseteq F$ is a relevant diagnosis iff $\varphi(\Delta_r) \models \phi'_E$, meaning that Δ_r is a model of ϕ'_E.

Algorithm 1 computes the set $CS(F, SD \wedge OBS, \top)$, which can be exponential in size. To optimize this enumeration, we note the following property.[1]

Lemma 1. *If Δ_r is a relevant diagnosis in Δ, then Δ_r is a relevant diagnosis for each Δ^+ such that $\Delta \subseteq \Delta^+$.*

Lemma 2. *The set obtained by the union of the relevant diagnoses of each $cs \in CS(F, SD \wedge OBS, \top)$ is equal to the set obtained by the union of the relevant diagnoses of each $mcs \in MaxCS(CS(F, SD \wedge OBS, \top))$.*

Lemma 2 allows to check only cutsets in $MaxCS(CS(F, SD \wedge OBS, \top))$.

6.2 Fully Symbolic Approach

Symbolic Relevant Diagnoses Encoding The following formula encodes Definition 5:

$$\exists V, F'.\ (SD(V, F') \wedge \bigwedge_{o \in O} o \leftrightarrow ph_o \ \wedge\ \text{sub\_of}(F, F') \ \wedge$$
$$(\exists V', A'_O.\ \text{lhsS}(V', F, A'_O, PH_O) \ \wedge\ (\forall V''.\ \text{lhsS}(V'', F, A'_O, PH_O) \rightarrow TLE))) \qquad (8)$$

where:

- $\text{lhsS}(V, F, A_O, PH_O) = SD(V, F) \ \wedge\ \text{act\_obs}(A_O, O, PH_O)$, represents the formula $(SD \wedge OBS' \wedge \varphi(\Delta^-))$;

[1] All proofs are available in the extended version online at https://es-static.fbk.eu/people/sambrotta/safecomp26/extended_version.pdf.

Similarly to Formula (7), the minimality condition is encoded into Formula (8).

Let ϕ_S denote the BDD of Formula (8). As before, ϕ_S can be computed statically for any given set O and propositional formula $PROP$, since it is independent from Obs and Δ. Let $\phi'_S = \phi_S[ph_{o_1}/Obs(o_1), ..., ph_{o_m}/Obs(o_m)]$ be the BDD obtained from ϕ_S by substituting PH_O with Obs. This approach does not require to compute beforehand the set $CS(F, SD \wedge OBS, \top)$, because the first line of Formula (8) is dedicated to the computation of the model-based diagnoses. Then, $\Delta_r \subseteq F$ is a relevant diagnosis iff $\varphi(\Delta_r) \models \phi'_S$, meaning that Δ_r is a model of ϕ'_S. In this case, the diagnoses for a given observation can be computed by performing the above substitution on ϕ_S and by calling the enumerate_models function on ϕ'_S.

6.3 Correctness[1]

Theorem 1. *The set of minimal diagnoses of Formula (8) is equal to the set of relevant diagnoses defined by Definition 5.*

Theorem 2. *The set of minimal diagnoses of Algorithm 1 is equal to the set of relevant diagnoses defined by Definition 5.*

7 Experimental Evaluation

In this section we give a brief description of the systems used for benchmarking, data about the performance of our implementation and some insight about the obtained results.

7.1 Benchmarks

We tested our implementation on two benchmarks: the house electrical circuit example from Chap. 5.5.3 of [19] and the Triple Modular Generator (TMG) model [4].

The house electrical circuit consists of a series of wires, switches, and circuit breakers that connect the power source to lights and power outlets. The faults can affect only circuit breakers and switches. The properties to be monitored define the correct behaviour of the system - e.g., wire 2 is powered if wire 1 is powered and the switch that connects wire 1 and 2 is up.

The TMG consists of three generators that power three buses through six circuit breakers. For a bus to function, it must receive electricity from exactly one input; receiving power from two or more sources simultaneously will cause it to break. Generators and circuit breakers are controlled via commands (i.e., on/off and open/close respectively) but these components can fail: generators can get stuck at off, and breakers can get stuck in either the open or closed position. We stripped away the fault recovery modules to focus purely on detection and identification. The properties to be monitored follow the rule *"if a subset of generators is on, then a corresponding subset of buses must be working"*. Finally, an oracle provides the commands required to satisfy these conditions, assuming the system is operating without any faults.[4]

The algorithms were implemented in Python 3.12.3, using the dd library, which is a wrapper for CUDD [23]. The tests were conducted on an AMD EPYC 7413 CPU with 16 GB of RAM and a 30 min timeout for both the construction of the BDD of the relevant diagnoses formula and the diagnosis process, and we built the BDDs of SD beforehand without a time limit. Since both benchmarks can be scaled up, we tested our algorithms on the TMG with 3, 4, and 5 generators and buses, and 3 to 12 houses electrical circuits, all connected to the same power source with sizes of O ranging from 10% to 100% the size of V, with a 10% step. The instances of the problems (assignments to V and F) were randomly generated, ensuring that $(SD \wedge OBS) \models TLE$ holds, and both algorithms were tested on the same instances.

7.2 Performance

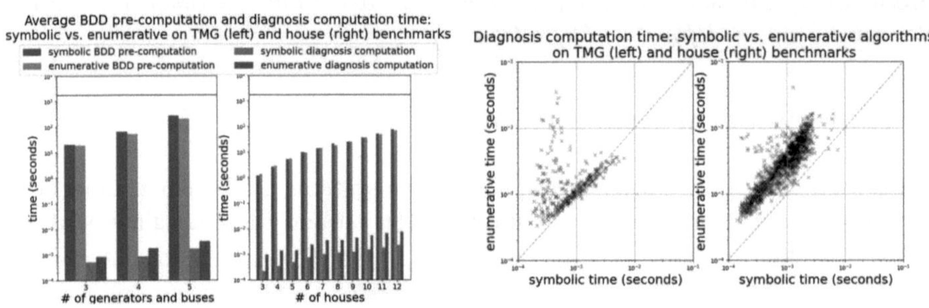

Fig. 1. Comparison between average BDD pre-computation and diagnosis computation time (left) and comparison of diagnosis computation time between both algorithms (right) on both benchmarks.

Figure 1 shows the time execution comparison between symbolic and enumerative algorithms, on the TMG and house benchmarks, for both of BDD pre-computation and diagnosis computation tasks. All problems were solved within the 30 min time limit. Specifically, on the TMG benchmark, execution times ranged from 1.11 s to 1254.19 s for BDD pre-computation and from 0.0001 s to 0.03 s for the diagnosis task. On the house benchmark, these tasks required 0.26 s to 278.15 s and 0.0001 s to 0.04 s, respectively. The execution time for the BDD pre-computation task is roughly equivalent for both algorithms across all benchmarks. We hypothesize that although the enumerative encoding is simpler than

[4] The properties take the form *premise* → *implication*, meaning that if the premise is false then the property is satisfied and the diagnosis is not needed. For this reason the oracle is implemented in such a way that it computes only sets of commands that, in absence of faults, imply that the premise is true.

the symbolic one, its lower complexity is offset by the larger number of variables - i.e. PH_F variables unique to the enumerative approach. In the diagnosis the symbolic algorithm takes the lead, because it doesn't enumerate explicitly all of the maximal model-based cutsets.

Table 1. Percentage of model-based diagnoses affected and number of faults removed from them.

SD	# vars	# mbds	% unaffected	% affected	avg	median	min	max
house3	76	122	0.59	0.41	1.6	2	1	3
house4	101	123	0.76	0.24	2.03	2	1	3
house5	126	132	0.52	0.48	2.58	2	1	6
house6	151	120	0.54	0.46	1.74	2	1	4
house7	176	120	0.39	0.61	1.95	2	1	5
house8	201	126	0.35	0.65	3.21	3	1	7
house9	226	122	0.39	0.61	2.86	3	1	7
house10	251	120	0.38	0.62	2.33	2	1	6
house11	276	120	0.35	0.65	3.69	3	1	9
house12	301	120	0.36	0.64	2.40	2	1	5
TMG3	195	107	0.24	0.76	1.62	1	1	4
TMG4	260	141	0.2	0.8	2.77	2	1	7
TMG5	325	175	0.22	0.78	3.67	3	1	8

Table 1 shows three main sets of information on both benchmarks: the global number of model-based diagnoses on all instances for a given system, the percentage of them that were affected/unaffected, and the average, median, minimum and maximum number of faults removed from the affected model-based diagnoses. As shown, on the house benchmark on average the number of faults removed doesn't follow a specific trend, but the median stays around 2 and 3 faults removed, while on the other hand on the TMG benchmark both average and median follow an ascending trend.

These results show that our approach is able to remove irrelevant faults (when present) from the model-based diagnoses, making it easier to trigger the correct recovery mechanism in order to restore the property. Also, as shown in Fig. 1 it is able to return diagnoses in a matter of milliseconds, which is a key requirement in the runtime verification setting, by querying a pre-computed BDD.

7.3 Insights

In this section we show that the situations identified by the motivating examples can occur in a real system as the TMG. In the following paragraphs we refer to components by their initial - i.e., G for generator, B for bus, GB for generator breaker, and BB for bus breaker.

Suppose that $PROP := (G1\_is\_on \vee G2\_is\_on) \rightarrow B1\_is\_working$. The commands given to the system are $G1$ on, $G2$ and $G3$ off, $GB1$ and $GB2$ closed, $GB3$ open, $BB1$ and $BB3$ closed, and $BB2$ open. Let $O = \{G1, B1, GB2\}$, and $OBS = \{G1\_is\_on, \neg B1\_is\_working, \neg GB2\_is\_closed\}$. $PROP$ is clearly violated, since $G1$ is on and $B1$ is not working, and the MBD process returns $\{GB1\_stuck@open, GB2\_stuck@open\}$. Fault $GB2\_stuck@open$ would be sufficient in other scenarios to cause the property violation (e.g., one where $G2$ is on instead of $G1$) but not in this one, since $G2$ is off, meaning that the breakage of $GB2$ is not desired in the relevant diagnosis since it didn't contribute in this scenario. In fact, the only relevant diagnosis found in this scenario is $\{GB1\_stuck@open\}$ as expected.

Now suppose that $O = \{G1, B1, GB2, BB3\}$ and $OBS = \{G1\_is\_on, \neg B1\_is\_working, \neg GB2\_is\_closed, \neg BB3\_is\_closed\}$. The observation from the previous case has been augmented with information about $BB3$, that is open even though its command is close. This failure simply does not interfere with the property, so it should be discarded from the relevant diagnosis, and in fact this is what happens, since the only relevant diagnosis remains $\{GB1\_stuck@open\}$.

Suppose that the system must always be powered, meaning that at least one of the generators must be on. Instead of having one sensor per generator, let's say that there is only one sensor that detects if there is power in the system. For this purpose, the constraint $\text{system\_powered} \leftrightarrow (G1\_is\_on \vee G2\_is\_on \vee G3\_is\_on)$ is added to the system description. Let $PROP := G1\_is\_on$, the commands to the generators are $G1$ and $G2$ on, and $G3$ off, $O = \{\text{system\_powered}\}$ and $OBS = \{\neg\text{system\_powered}\}$. $G1$ and $G2$ are both off, but since the property states that only $G1$ must be on, the breakage of $G2$ is not relevant, meaning that the only relevant diagnosis should be $\{G1\_stuck@off\}$, and this is what our definition returns.

8 Conclusion

In this paper, we addressed the problem of diagnosing runtime property violations. We highlighted that standard MBD and other related notions are not suitable for this task because they focus on the consistency between the model and observations, rather than explaining the specific violation of a monitored property. This often leads to diagnoses that include "irrelevant" faults, i.e., faults that are present but did not contribute to the violation.

To overcome this limitation, we introduced the PVD problem. We provided a formal characterization of the problem in propositional logic, distinguishing it from MBD and other causal frameworks. Central to our approach is the notion of "relevance": we apply counterfactual reasoning to filter out fault configurations that are not strictly necessary to explain the violation. We then proposed a symbolic algorithm based on propositional quantifier elimination to compute these relevant diagnoses efficiently. Our experimental evaluation confirmed the applicability of the approach, demonstrating its ability to isolate relevant causes in systems with partial observability.

There are several promising directions for future research. While this work focused on the combinational setting, extending the framework to dynamic systems where the properties are expressed in Linear Temporal Logic (LTL) like in runtime verification [2] is a natural next step. We plan to investigate more complex fault models, including transient faults and common cause failures, to better capture real-world scenarios. Another direction is to extend diagnosability checking [21] to determine if we have enough observation to identify the relevant diagnosis of a property violation. Incorporating probabilistic information could allow for ranking diagnoses based on likelihood, helping operators prioritize recovery actions. Another promising future direction consists in refining the diagnoses by returning also information about the context in which the diagnosis occurs. Finally, we aim to explore advanced optimization techniques to further improve the scalability of the symbolic algorithms for large-scale systems.

References

1. Aiello, M., Berryman, J., Grohs, J., Schierman, J.: Run-time assurance for advanced flight-critical control systems. In: AIAA Guidance, Navigation, and Control Conference, p. 8041 (2010)
2. Bauer, A., Leucker, M., Schallhart, C.: Runtime verification for LTL and TLTL. ACM Trans. Softw. Eng. Methodol. **20**(4), 14:1–14:64 (2011)
3. Beer, I., Ben-David, S., Chockler, H., Orni, A., Trefler, R.: Explaining Counterexamples Using Causality. In: Proceedings of CAV. LNCS, vol. 5643, pp. 94–108 (2009)
4. Bozzano, M., Cimatti, A., Gario, M., Jones, D., Mattarei, C.: Model-based safety assessment of a triple modular generator with xSAP. Formal Aspects Comput. **33**(2), 251–295 (2021)
5. Bozzano, M., Villafiorita, A., Åkerlund, O., et al.: ESACS: an integrated methodology for design and safety analysis of complex systems. In: Proceedings of European Safety and Reliability Conference (ESREL 2003), pp. 237–245. Balkema Publisher (2003)
6. Bozzano, M., Cimatti, A., Gario, M., Tonetta, S.: Formal design of asynchronous fault detection and identification components using temporal epistemic logic. Logical Methods Comput. Sci. **11** (2015)
7. Cimatti, A., Tian, C., Tonetta, S.: Assumption-Based Runtime Verification with Partial Observability and Resets. In: Finkbeiner, B., Mariani, L. (eds.) RV 2019. LNCS, vol. 11757, pp. 165–184. Springer, Cham (2019). https://doi.org/10.1007/978-3-030-32079-9_10
8. Darwiche, A., Marquis, P.: A knowledge compilation map. J. Artif. Intell. Res. **17**, 229–264 (2002)
9. De Kleer, J., Kurien, J.: Fundamentals of model-based diagnosis. In: IFAC Proceedings Volumes, vol. 36, pp. 25–36 (2003)
10. Dupré, D.T.: Abductive and Consistency-Based Diagnosis Revisited: a Modeling Perspective. CoRR cs.AI/0003016 (2000). https://arxiv.org/abs/cs/0003016
11. Ferrère, T., Maler, O., Nickovic, D.: Trace diagnostics using temporal implicants. In: Proceedings of ATVA. LNCS, vol. 9364, pp. 241–258 (2015)
12. Gao, Z., Cecati, C., Ding, S.X.: A survey of fault diagnosis and fault-tolerant techniques–part i: Fault diagnosis with model-based and signal-based approaches. IEEE Trans. Industr. Electron. **62**(6), 3757–3767 (2015)

13. Gössler, G. and Mari, T. and Pencolé, Y. and Travé-Massuyès, L.: Towards causal explanations of property violations in discrete event systems. In: Proceedings of DX, pp. 1–8 (2019)
14. Halpern, J.: A modification of the Halpern-Pearl definition of causality. In: Proceedings of IJCAI, pp. 3022 – 3033 (2015)
15. Joshi, A., Miller, S., Whalen, M., Heimdahl, M.: A proposal for model-based safety analysis. In: Proceedings of AIAA / IEEE Digital Avionics Systems Conference (DASC) (2005)
16. Misra, A., Sztipanovits, J.: A model-based failure detection, isolation and recovery system (1996)
17. Mozetič, I.: Model-based diagnosis: An overview. In: Mřík, V., Štěpánková, O., Trappl, R. (eds.) Advanced Topics in Artificial Intelligence, pp. 419–430. Springer, Cham (1992)
18. Pan, G., Vardi, M.Y.: Symbolic techniques in satisfiability solving. J. Autom. Reason. **35**(1), 25–50 (2005)
19. Poole, D.L., Mackworth, A.K.: Artificial Intelligence: Foundations of Computational Agents. Cambridge University Press (2010)
20. Preist, C., Eshghi, K., Bertolino, B.: Consistency-based and abductive diagnoses as generalised stable models. Ann. Math. Artif. Intell. **11**, 51–74 (1994)
21. Sampath, M., Sengupta, R., Lafortune, S., Sinnamohideen, K., Teneketzis, D.: Diagnosability of discrete-event systems. IEEE Trans. Autom. Control **40**(9), 1555–1575 (1995)
22. Schierman, J., et al.: Run-time verification and validation for safety-critical flight control systems. In: AIAA Guidance, Navigation and Control Conference and Exhibit, p. 6338 (2008)
23. Somenzi, F.: Cudd: Cu decision diagram package release 2.4. 2. University of Colorado at Boulder (2009)
24. Space Avionics Open interface Architecture: SAVOIR FDIR Handbook, vol. SAVOIR-HB-003. European Space Agency (2019). https://savoir.estec.esa.int/SAVOIRDocuments.htm
25. Zhang, Z., et al.: Online Causation Monitoring of Signal Temporal Logic. In: Proceedings of CAV. LNCS, vol. 13964, pp. 62–84 (2023)

Safety Protection

Defeaters from Accidents and Incident Reports: Discovery, Extraction, Identification

Tihomir Rohlinger[1,2(✉)], Daniel Ratiu[1], Carmen Carlan[3], and Stefan Wagner[2,4]

[1] CARIAD SE, Munich, Germany
[2] University of Stuttgart, Stuttgart, Germany
t.rohl.edu@gmail.com
[3] TÜV SÜD Rail GmbH, Munich, Germany
[4] Technical University of Munich, Heilbronn, Germany

Abstract. Safety cases rely on expert experience often formalized in natural language reports about accidents, incidents, and defects. Discovering relevant information in this growing body of knowledge is complex. We present a traceable, closed-corpus hybrid retrieval-augmented generation (RAG) pipeline for automatic defeater generation from autonomous driving accident reports. It combines sparse term frequency - inverse document frequency (TF-IDF) and dense semantic retrieval to extract relevant counter evidence from incidents issued by the NTSB, NHTSA, and Cruise. The pipeline parses the excerpts to a large language model (LLM) for generating defeaters. We evaluated our approach using the Open Autonomous Safety Case Framework's (OASCF) high-level safety argument. Three domain experts assessed independently the generated defeaters based on their logical relevance. Two experts independently evaluated the same subset of about 167 generated defeaters on a scale of 0–2. The inter-rater agreement was measured using Cohen's kappa (k = –0.01), indicating poor agreement. A third domain expert provided a qualitative in-depth analysis of the defeaters and an overall discussion, confirming the conceptual soundness of the approach. The ratio of strongly identified defeaters to weak and non-defeaters was about 1:3. The method is proposed for safety engineers examining the argument coverage in the preliminary safety case stage, reducing manual corpus review efforts while maintaining full traceability to the extracted source evidence.

Keywords: Defeater Discovery · Hybrid RAG · NLP

1 Introduction

Safety-critical systems necessitate rigorous assurance to operate safely within their intended environments, a process formalized and potentially monitored

© The Author(s), under exclusive license to Springer Nature Switzerland AG 2027
E. Troubitsyna et al. (Eds.): SAFECOMP 2026, LNCS 16868, pp. 49–63, 2027.
https://doi.org/10.1007/978-3-032-34867-8_4

through the construction of safety cases. These safety cases compile documented evidence and arguments to demonstrate that a system is acceptably safe for operation [2]. Yet, the integrity of safety cases can be compromised by implicit or too strong assumptions, incomplete decomposition of goals into sub-goals, or too weak evidence. To increase confidence in the soundness and completeness of safety arguments, a common practice endorsed by Rushby [29] is to systematically search for defeaters associated with the main argument claims and cover them explicitly. The identification of tangible defeaters requires broad background knowledge and expertise accumulated over years of experience with building and operating similar systems.

A substantial body of knowledge related to assurance aspects of operating safety-critical systems is documented through incidents and accident reports that contain detailed information about mishaps, their context of occurrence, recommendations and lessons learned. These reports, for example, detail real-world occurrences, near-misses, and accidents containing invaluable insights; underlying causes and contributing factors of safety-critical events [21–23]. We assume that using safety reports to identify potential counterarguments—which are statements, vulnerabilities, or evidence that challenge the validity of a safety claim—can significantly strengthen the robustness of safety cases, thereby supporting data-driven improvements.

Methods from natural language processing and understanding (NLP and NLU) offer promising automation for the extraction of such counterarguments from large repositories of incident data [26]. Specifically, advanced NLP techniques, such as semantic search and information retrieval models, can be employed to systematically identify and categorize relevant historical incidents that contradict or weaken existing safety arguments. This enhances the proactive identification of potential failure modes and refines risk assessments [30].

Contributions In this paper, we present a novel approach that uses Hybrid Retrieval-Augmented Generation and Large Language Models to leverage the knowledge contained in accident and incident reports to extract evidence associated with a certain safety argument. The automatically generated defeaters are traceable to relevant parts of the reports corpus, which represent concrete information leading to their identification. We demonstrate the approach using the Open Autonomy Safety Case Framework (OASCF) [9] and accident and incident reports concerning autonomous driving vehicles. Three domain experts reviewed the generated defeaters to evaluate the approach. We analyze the LLM-generated outputs and the level of agreement between expert judgments. The results suggest that the proposed method effectively extracts evidence from report corpora to generate defeaters, taking subjective uncertainties into account. Additionally, emphasizing citation frequency improves auditability of source relevance and provides stronger analytical support for evaluating safety arguments.

2 Related Work

Argument Mining from Natural Language Documents. The broader field of *argument mining* [19,32] provides relevant foundations for our task. Argument mining aims to automatically identify argumentative discourse units—claims, premises, and rebuttals—in natural language text. Safety incident reports written by domain experts in highly condensed, domain-specific language differ substantially from debate transcripts, making the direct transfer of argument mining models non-trivial. Our work bridges this gap by framing defeater extraction as a targeted counterargument-discovery task anchored in the assurance-case structure [5] rather than as open-ended argument mining.

Defeater Discovery in Safety Reports. The analysis of safety incident reports has increasingly leveraged computational methods, particularly natural language processing, to find meaningful insights from incidents to support safety assessments (Hughes et al., 2019 [14]), holding the potential of discovering unconsidered defeaters. The application of NLP to safety-critical document corpora has a growing literature. In aviation, the NASA Aviation Safety Reporting System (ASRS) corpus has been widely used for causal factor extraction, anomaly detection, and risk classification [6,17].

Retrieval/Extraction Methods in Safety Engineering. Prompt-based approaches using large language models (LLMs) improved scalability but introduced a high noise-to-signal ratio: while LLMs can enumerate plausible defeaters fluently, they frequently generate defeaters that are either semantically redundant, logically unsound, or ungrounded in the provided system description [11,33]. Alternatives were proposed to improve the quality of LLM outputs with Retrieval-Augmented Generation (RAG) [18]. RAG depends on retrieval quality, source reliability, and the model's reasoning. To mitigate dependent issues, knowledge-graph-augmented RAG (KG-RAG) was introduced, ingesting domain knowledge as structured entity–relation graphs rather than flat text chunks [3]. While this supports multi-hop reasoning and verifiable context for LLMs [1], the subject–relation–object triplet structure proved to be too rigid for flexible generalization of assurance-case claims in our work. HybridRAG [34] addresses this by combining graph-based reasoning with BM25 lexical and dense-vector retrieval, introducing additional flexibility in the information extraction and retrieval process while improving diagnostic precision in aircraft fault analysis. RAG frameworks have transformed information retrieval for complex domain-specific texts such as safety reports [4]. By grounding responses in external knowledge, they reduce factual inaccuracies and outdated information in large language model outputs [1].

Defeater Identification. A central method for achieving and maintaining confidence in assurance cases is the identification of *defeaters*—conditions or evidence that undermine, rebut, or undercut individual argument steps [29]. Eliminative Argumentation (EA) [12] addresses this challenge directly by framing assurance-case construction as a process of systematically eliminating candidate defeaters through defeasible reasoning: a claim is considered sufficiently assured only when all credible defeaters have been explicitly countered or resolved. Maintaining system assurance [5] in a continuously evolving operational design domain (ODD) for example for autonomous systems is challenging. Field exposure reveals unforeseen failure modes that necessitate frequent identification and mitigation processes [13,24,27]. Recently, several papers have been published about using LLMs to generate defeaters at scale [11,33]. In the current work, we advance in the direction of defeater generation for dedicated safety cases by leveraging the information contained in posed incident and accident reports.

3 Method

We address the challenge of discovering and extracting evidence-grounded defeaters from large-scale incident report corpora for safety assurance cases using Hybrid RAG-based retrieval [34] with sparse lexical BM25 and dense semantic matching. Unlike prompt-only approaches or prior LLM-based defeater generation without retrieval grounding [16,33], our method ensures evidence traceability and prevents reference blending across documents. In the following, we will describe the different parts of the method.

3.1 Corpus Preparation

The knowledge corpus contains all relevant documents for the target domain. It is closed, so nothing can be added or deleted during the generation process. Its contents can vary from accident reports and comments to investigation documentation.

Source Documents. Safety reports serve as the primary knowledge corpus for our safety case coverage analysis, forming the foundation of all retrieval indices. Reports from the National Highway Traffic Safety Administration (NHTSA), National Transportation Safety Board (NTSB), and Quinn Emanuel Urquhart & Sullivan's Cruise report provide in-depth legal analysis, investigative findings, and strategic insights into accidents, product defects, transportation risks, and complex litigation matters—offering clear evaluations of risk, regulatory exposure, and compliance considerations.

Text Chunking. Each segment in the text follows a consistent logic: page-wise separation combined with paragraph-level markers. Paragraphs preserve rhetorical structure and local context better than sentence splitting, which can fragment multi-sentence explanations common in technical reports [34]. This allows to preserve important passages that contain a coherent context. A minimal chunk size determines the lower boundary for these segments. Each chunk is stored as raw text and includes a document ID, page, and paragraph index.

3.2 Hybrid Index Construction

Dual indices provide complementary retrieval signals, sparse with Term Frequency Inverse Document Frequency (TF-IDF), and Best-Matching (BM25) for Document Length Normalization (DLN) [31] ranking for lexical precision and dense embeddings [10] for semantic recall. The sparse component allows for precise lexical matching, for example, to technical terms. The dense component provides semantic retrieval of paraphrased or implicit text evidence.

Sparse TF-IDF Component. TF-IDF provides keyword-based matching through standard preprocessing: lowercasing, English stopword removal, and whitespace tokenization with minimum word length ≥ 2. A `TfidfVectorizer` is fitted across all N paragraph chunks to construct the vocabulary V and corresponding IDF weights, yielding a sparse TF-IDF matrix $X_{\text{TF-IDF}} \in \mathbb{R}^{N \times V}$. Rows represent chunks, columns unique terms, and the entries are the TF-IDF scores capturing lexical overlap between claims and corpus passages. While TF-IDF scores scale linearly with raw term frequency, longer documents gain a disproportionate advantage through higher term repetition. BM25 addresses this via DLN and TF saturation.

Dense Embedding Component. Dense vector embeddings are generated using a sentence-transformer encoder that maps each paragraph into a high-dimensional vector representation. At retrieval time, the safety assurance claims (the queries) are embedded into the same vector space. Each chunk is encoded into a dense vector $X_{\text{emb}} \in \mathbb{R}^{N \times d}$. These resulting dense vectors are used to compute cosine similarity scores. This allows to retrieve and rank relevant evidence that would be missed by simple keyword matching. The retrieval output structure is a dictionary that aggregates all the necessary retrieval artifacts, including raw corpus texts, document-level metadata, the TF-IDF vectorizer, the sparse matrix, and the dense embedding array.

3.3 Hybrid Retrieval

One component parses the assurance case claims from the source file. In parallel, another extracts the lexical and semantic information and ranks the chunks based on how significant the content of each claim is related.

Claim Representation. Assurance case claims are loaded from a CSV format containing goal texts and structural metadata. Each claim q produces dual representations: $\mathbf{q}_{\text{TF-IDF}}$ through the shared vectorizer and $\mathbf{q}_{\text{emb}}$ by the embedding model.

Scoring Function. For each chunk i, hybrid relevance combines sparse and dense signals:

$$s_{\text{hyb}}(q,i) = \alpha \cdot s_{\text{dense}}(q,i) + (1-\alpha) \cdot s_{\text{sparse}}(q,i) \quad . \tag{1}$$

where $s_{\text{dense}} = \cos(\mathbf{q}_{\text{emb}}, \mathbf{x}_{i,\text{emb}})$, $s_{\text{sparse}} = \cos(\mathbf{q}_{\text{TF-IDF}}, \mathbf{x}_{i,\text{TF-IDF}})$, and $\alpha = 0.65$ balances semantic ($\approx 65\%$) and lexical ($\approx 35\%$) contributions [34].

Top-*k* Retrieval and Filtering. The candidate chunks are ranked by the hybrid score s_{hyb}, and the top-$k = 5$ contexts are retrieved per claim. Retrieved contexts are post-filtered using a minimum score threshold of s_{min}. Each retained context includes the text segment, its hybrid score, and associated metadata (source, page, paragraph index).

3.4 Defeater Generation via LLM

Next, we will describe the fusing of the claim and the retrieved context.

Per-Context, Non-Blending Prompting. Unlike multi-context batch prompting, we generate *one defeater per retrieved context* with independent LLM calls. Each prompt contains exactly one evidence excerpt, enforcing strict traceability and preventing cross-document blending.

Prompt Structure

Role: Expert in safety assurance/GSN.
Input: Claim; one evidence excerpt (including document name, page, paragraph); excerpt truncated to `evidence character maximum`.
Task: Generate exactly one defeater; type $\in$ {Rebutting, Undermining, Undercutting}; 2–3 sentences; logically defeat the claim using only the excerpt.
Output: Strict JSON list a single object:

```
[{ "type": ..., "description": ... }]
```

Post-processing and Structuring. LLM responses are parsed via regex-based extraction of JSON arrays. Valid defeaters are retained and enriched with (i) a retrieval score as a confidence proxy, (ii) deterministic evidence citation, and (iii) a field for expert evaluation. Deviated responses are discarded to ensure output quality. The proposed extraction method is evaluated in a domain-specific use case in the following section.

4 Experiment

Evaluating contextual understanding in Retrieval-Augmented Generation (RAG) systems for defeater generation requires assessing both the groundedness of retrieved evidence and the quality of the generated outputs.

4.1 Evaluation Setup

To assess the effectiveness of the proposed Hybrid RAG approach, we investigate the following research questions:

- **RQ1: Retrieval Relevance.** Can the Hybrid RAG approach retrieve relevant evidence from safety incident and accident reports with respect to the analyzed GSN claims?
- **RQ2: Defeater Validity.** Are the defeaters generated from the retrieved source evidence logically valid to the given GSN claims, as judged by domain experts?

RQ1 by human expert annotations, assessed for traceability and relevance; RQ2 evaluated using expert annotations focusing on the validity of defeaters.

4.2 Dataset and Assurance Case Fragment

Knowledge Corpus. The retrieval corpus consists of 9 ADS-related accident and incident reports sourced from STANDING GENERAL ORDER ON CRASH REPORTING spanning years 2021–2025. After preprocessing and chunking, the corpus comprises 119,499 tokens. The reports cover a range of operational scenarios and failure modes relevant to ADS perception, planning, and control subsystems. Their base layout has these document structures

- NHTSA: Simple government format (A, B, C sections)
- NTSB: Technical report (1.1.1 decimal hierarchy, figures/tables)
- QuinnEmanuel: Legal investigation (I.A.1.a.i multi-level hierarchy, footnotes)

Assurance Case Fragment. The experimental target case OASCF [9] provides a templated open-source safety argumentation compliant with UL 4600 and development arguing against the ISO 26262/21448 standards. The fragment comprises 46 goal nodes, and 25 strategy nodes expressed in Goal Structuring Notation (GSN).

4.3 System Configuration

Retrieval. Safety reports are segmented into pages and paragraphs/passages. A hybrid index combines sparse TF-IDF representations with dense embeddings generated by `all-MiniLM-L6-v2` ($d = 384$) [25]. For each claim, the top-k passages are retrieved using reciprocal rank fusion [8] of the lexical and semantic signals.

Generation. Retrieved passages are provided to the open language model `meta-llama/Llama-3.1-8B-Instruct`, which generates defeaters using the structured prompt introduced in Subsect. 3.4. Generated defeaters are filtered for structural compatibility and stored as structured JSON object.

4.4 Evaluation

The Hybrid Retrieval-Augmented Generation (RAG) pipeline is evaluated in two stages by three domain experts (with 10 years of experience; in automotive, system safety engineering, and as safety assessor). First, we assess the performance of the retrieval component (RQ1) by analyzing the relevance of defeaters constructed from the retrieved evidence passages. Second, we evaluate the quality of the generated defeaters themselves through independent expert assessment (RQ2). Both stages employ a three-level composite rating scale capturing logical relevance, logical argumentative strength, and evidence grounding.

Retrieval Performance (RQ1). To evaluate the performance of the retrieval component, experts assessed the defeaters generated from retrieved passages. The rating, therefore, reflects whether the retrieved evidence enables the construction of meaningful counterarguments against the claim.

Overall Defeater Rating (0–2). Each defeater receives a composite relevance score combining *logical relevance*, *logical strength*, and *evidence grounding.*

- **2 (High)** – Directly targets the claim, presents a logically strong counterargument, and is clearly supported by the cited source. If true, the defeater would significantly weaken the claim.
- **1 (Medium)** – Logically relevant and partially supported by evidence, but lacking strong argumentative impact or clear evidential grounding. The defeater raises a plausible concern but remains incomplete.
- **0 (Low)** – Lacks logical relevance, argumentative strength, or evidential support. The defeater is tangential, speculative, or not grounded in the cited passage.

This composite scoring scheme allows the review to represent the relevance and quality in one judgment call. From these annotations, we can then compute three ranking metrics. **Mean relevance** measures the average score assigned by a reviewer across all evaluated items. **Precision@5** [20] considers the five highest-rated defeaters per claim and computes the proportion with relevance scores of at least medium (≥ 1), averaged across claims. **nDCG@5** (*Normalized Discounted Cumulative Gain at rank 5*) [15] evaluates ranking quality by rewarding highly relevant defeaters that appear earlier in the ranked list, with a normalized score of 1.0 indicating a perfectly ordered ranking.

Retrieval Results. We evaluated multiple configurations across 46 OASCF claims (target: 5 defeaters per claim). Three reviewers participated: Reviewers 1 and 2 provided full defeater ratings (R1: n=167, R2: n=168) R1 left one out; Reviewer 3 additionally validated the approach conceptually and provided an in-depth discussion analysis.

The paragraph-level + BM25 normalized configuration received (335 total) ratings, the statistics of the per-reviewer annotations are shown in (Table 1), revealing the variance:

Table 1. Per-reviewer retrieval performance

Reviewer	Mean relevance	Precision@5	nDCG@5	N
Reviewer 1	0.54	0.39	0.66	167
Reviewer 2	1.39	0.89	0.91	168
mean total	0.97	0.64	0.79	335

The resulting inter-annotator agreement between R1 and R2 (Cohen's kappa [7] $\kappa = -0.01$) is poor, reflecting the subjective and nuanced nature of claim–evidence reasoning in safety argumentation requiring an objective judgment [28].

Answer to RQ1. The Precision@5 value indicates that 64% of the top-5 generated defeaters per claim were rated at least moderately relevant. The nDCG@5 score of 0.79 further shows that highly relevant defeaters (score = 2) tend to appear earlier in the ranked list. Together, these results suggest that the hybrid retriever surfaces passages that support the generation of meaningful counterarguments, addressing RQ1.

Defeater Generation Quality (RQ2). The domain experts assessed the generated defeaters using the 0–2 logical relevance scale (Sect. 4.4), focusing on evidential grounding logical relevance and logical argumentative strength.

Answer to RQ2. The total of 335 defeaters was independently assessed by the two reviewers (Reviewer 1: $n = 167$, Reviewer 2: $n = 168$). Reviewer 1 classified 15% of the evaluated defeaters as strong (rating 2; 25/167), whereas Reviewer 2 classified 50% as strong (84/168). When aggregating the assessments across both reviewers, 33% of all generated defeaters were rated as strong (109/335), with an overall mean relevance score of 0.97. These results suggest that approximately one-third of the generated defeaters provide evidentially grounded and logically valid counterarguments for the OASCF claims, thereby supporting RQ2.

Evidence Source Analysis. To better understand the sources of evidence used for defeater generation, the distribution of citations across investigation reports was analyzed. The results are summarized in Table 2. The majority of evidence supporting the generated defeaters stems from the Cruise report and the NTSB's investigation of the Uber Highway Accident Report (har1903).

Figure 1 visualizes report page-level citation patterns in relation to expert-rated defeater quality. Pages with high citation frequency and high mean scores can be interpreted as particularly informative evidence sources, whereas frequently cited pages with low or variable scores may indicate evidence that is easy to retrieve but less consistently useful for constructing strong defeaters.

Table 2. Evidence source distribution

Source	Count	%
2024.01.24_Quinn_Emanuel_Report_Cruise.pdf	80	48%
har1903.pdf	57	34%
INIM-PE24016-13541.pdf	11	6%
INCR-EA22002-14496.pdf	9	5%
INCLA-PE23018-11022.pdf	5	3%
INIM-PE24016-13047.pdf	3	2%
INOA-PE23018-11587.pdf	3	2%

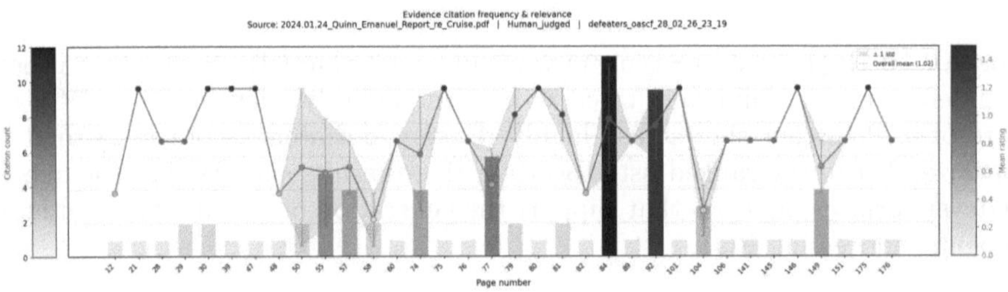

(a) Quinn Emanuel Cruise report (48% of citations)

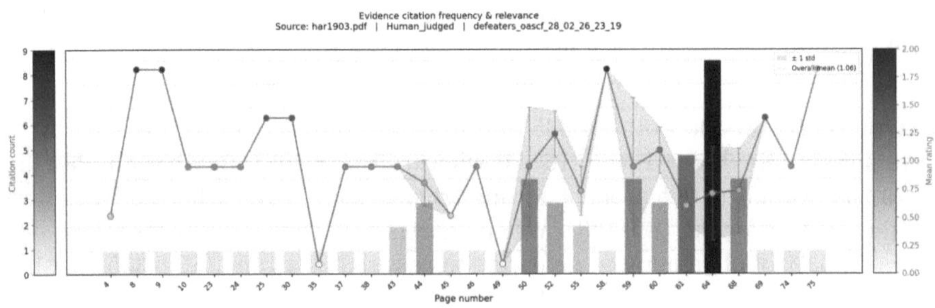

(b) NTSB Uber report (34% of citations)

Fig. 1. Page-level citation frequency plotted against the source mean defeater score (± standard deviation) for a and b, the two most frequently cited sources.

5 Discussion and Lessons Learned

The automated defeater generation experiment highlights both the strengths and limitations of Hybrid Retrieval-Augmented Generation for safety assurance case analysis. To be useful for safety engineers, generated defeaters must satisfy three criteria: they must be *evidence grounded* (e.g., traceable to factual evidence in the reports), *logically relevant* (e.g., subverting a specific GSN claim), and *logically argumentative strong* (e.g., contributing relevant insights about the claim's defeat).

5.1 Examples of Generated Defeaters

The following examples illustrate both successful and problematic defeater generation cases.

Goal 4.2.2 "We identified all mishaps that might occur during operation."
Defeater: "The existence of a previously unidentified defect that could cause a collision, followed by an unintended pull-over response, undermines the claim that all mishaps have been identified."
Discussion: This defeater is only meaningful if additional contextual knowledge about the Cruise incident is available, namely that a pedestrian became trapped underneath the vehicle during a pull-over maneuver.

Goal 4.2.3 "We identified all hazards that can lead to mishaps."
Defeater: "The report suggests that cameras can interpret lighting conditions differently than the human eye, potentially leading to unidentified hazards."
Discussion: The cited report section discusses camera perception under varying lighting conditions. Similar evidence could be used to challenge hazard analysis completeness in perception-related safety analyses.

Goal 4.2.6 "We modeled the risk that autonomy algorithms contribute to hazards."
Defeater: "The evidence from the NTSB report undermines the claim that autonomy algorithms contributed to hazards, as it highlights the limitations of the ADS system in detecting pedestrians and predicting their trajectories, which is a critical aspect of hazard contribution."
Discussion: The first part of the defeater is incorrectly phrased. However, the underlying evidence regarding pedestrian detection and trajectory prediction failures is relevant. A more accurate formulation would be that the accident suggests these algorithmic risks may not have been fully considered.

Goal 4.2.8: "We implemented the specified mitigations with a suitable Level of Rigour (LoR)."
Defeater: Defeater: "The claim that the implemented mitigations have a suitable Level of Rigour (LoR) is undercut by the evidence that the Uber ATG's initial decision to disable the Volvo FCW and AEB without replacing their full capabilities removed a layer of safety redundancy and increased risks, indicating a potential LoR deficiency in the initial mitigation implementation."
Discussion: The generated defeater correctly highlights that disabling redundant safety systems reduces mitigation robustness. However, the model fails to recognise that the disabled redundancies were external to the Uber software stack.

Goal 4.2.9: "We verified that we implemented the specified mitigations right"
Defeater: "The claim that specified mitigations were implemented is undercut by the fact that Cruise's permit was suspended due to a lack of safety procedures and safety hazard analysis involving vulnerable road users, as per the DMV's Reinstatement Questions, indicating that existing mitigations were insufficient or ineffective."

Discussion: The suspension of Cruise's operating permit represents a strong indicator that mitigation measures were insufficient. This type of evidence acts as a lagging safety performance indicator and could constitute a powerful defeater, although the generated formulation requires refinement.

5.2 Use of Defeaters in Safety Case Development

Our analysis of LLM-assisted defeater generation revealed several usage patterns that counter confirmation bias and support continuous improvement. The proposed approach enables an iterative process in which defeaters are discovered and extracted for individual GSN goals, exemplified within the OASCF argument template. Once a potential defeater is identified, it can be interpreted to assess its impact on the argument, leading to one of several possible outcomes:

- The safety engineer may verify whether the defeater is already covered in other branches of the safety argument.
- The affected goal may be refined or further decomposed to explicitly address the identified concern and improve the argument structure.
- Additional mitigation may be introduced to eliminate or reduce the impact of the defeater.
- Evidence may be imported through refined retrieval modalities or further expanded document analysis.
- The residual risk associated with the defeater may be explicitly accepted.

The proposed hybrid retrieval and LLM-based defeater generation method is a partially automated decision support tool within a human-in-the-loop safety assurance workflow. The retrieval component alone adds value by linking safety case entities to relevant report segments, but experts must still validate contextual relevance, interpret operational constraints, and detect subtle hazards not explicitly reflected in the retrieved evidence.

Compared with manually analyzing full accident reports, the workflow lets reviewers focus on pre-filtered evidence and candidate defeaters instead of processing the entire document corpus.

A substantial number of generated defeaters were discarded during review; this represents a deliberate preference toward recall over precision. In safety-critical contexts, overlooking a potentially relevant defeater is more serious and consequential than requiring additional expert filtering.

5.3 Limitations and Threats to Validity

One limitation concerns the dependency on the quality and completeness of accident and incident reports used as retrieval sources. The usefulness of generated defeaters is strongly influenced by the availability of detailed causal analysis, operational context, and mitigation information within the underlying reports. High-quality investigations enabled stronger and more contextually grounded defeaters.

An additional limitation concerns defeater coverage relative to human safety analysis. The current study evaluated the relevance and strength of generated defeaters, but did not establish whether the automated approach identifies the same defeaters as human experts, only a subset of them, or potentially complementary safety concerns.

Although multiple prompt formulations were explored during development, prompt engineering effects were not systematically evaluated in this study. Since prompts constrain how retrieved evidence is interpreted and synthesized by the language model, different prompt structures may significantly affect defeater specificity, factual grounding, and argumentative strength.

To improve factual grounding and traceability, defeaters were generated only from retrieved report excerpts explicitly included in the prompt, and the generation temperature was kept low to limit randomness. The current implementation does not automatically verify that citations match their supporting evidence; deterministic checking could be added using text matching or retrieval-index tracing.

6 Conclusion

In this paper, we presented a discovery process for identifying defeaters through evidence-based extraction from a closed, domain-specific corpus of ADS accident and incident reports. Our Hybrid RAG approach improves traceability and retrieval quality by surfacing relevant report passages as evidence for LLM-based defeater generation. The results demonstrate the potential of combining retrieval and generative AI to support the systematic identification of counter-arguments in safety assurance. The final assessment of whether a defeater is logically sound and contributes to argument coverage remains the responsibility of safety experts.

Further work could involve manually curating benchmark defeater sets derived from safety incident reports and independent expert analysis to evaluate the impact on argument completeness.

References

1. Ai, K., Karr, J.A., Jiang, M., Chawla, N.V., Wang, C.: KEO: Knowledge Extraction on OMIn via Knowledge Graphs and RAG for Safety-Critical Aviation Maintenance. arXiv:2510.05524 (2025)
2. ANSI/UL. 4600:2023 standard for safety for the evaluation of autonomous products. Technical report, Underwriters Laboratories [ul], Northbrook, IL, USA, 2023. Standard, 3rd ed.
3. Bahr, L., Wehner, C., Wewerka, J., Bittencourt, J., Schmid, U., Daub, R.: Knowledge graph enhanced retrieval-augmented generation for failure mode and effects analysis. J. Ind. Inf. Integr. **45**, 2025 (100807)
4. Bernardi, M.L., Cimitile, M., Panella, G., Pecori, R., Simoncelli, G.: Automatic generation of job safety reports with explainable RAG-based LLMs. Inf. Syst. Front. (2025)

5. Bloomfield, R.E., Rushby, J.: Assurance 2.0: A manifesto. arXiv preprint arXiv: 2004.10474 (2021)
6. Chandra, C., et al.: Aviation-bert: A preliminary aviation-specific natural language model. In: AIAA AVIATION 2023 Forum (2023)
7. Cohen, J.: A coefficient of agreement for nominal scales. Educ. Psychol. Measur. **20**(1), 37–46 (1960)
8. Cormack, G.V., Clarke, C.L.A., Buettcher, S.: Reciprocal rank fusion outperforms condorcet and individual rank learning methods. In: SIGIR '09: Proceedings of the 32nd international ACM SIGIR conference on Research and development in information retrieval, pp. 758–759, New York, NY, USA (2009)
9. Cârlan, C., Wagner, M.: The open autonomy safety case framework. Safety-Critical Syst. eJournal **3**(1) (2024)
10. Devlin, J., Chang, M.-W., Lee, K., Toutanova, K.: BERT: Pre-training of deep bidirectional transformers for language understanding. In: Burstein, J., Doran, C., Solorio, T., (eds.) Proceedings of the 2019 Conference of the North American Chapter of the Association for Computational Linguistics: Human Language Technologies, Volume 1 (Long and Short Papers), pp. 4171–4186. Association for Computational Linguistics (2019)
11. Gohar, U., Hunter, M.C., Lutz, R.R., Cohen, M.B.: CoDefeater: Using LLMs to find defeaters in assurance cases. In: Proceedings of the 39th IEEE/ACM International Conference on Automated Software Engineering, ASE '24, pp. 2262–2267. Association for Computing Machinery (2024)
12. Goodenough, J.B., Charles, W., B., Klein, A.Z.: Eliminative argumentation: A basis for arguing confidence in system properties, Carnegie Mellon University (2015). CMU/SEI-2015-TR-005 technical report
13. Hawkins, R., Ryan Conmy, P.: Identifying Run-Time Monitoring Requirements for Autonomous Systems Through the Analysis of Safety Arguments. In: Computer Safety, Reliability, and Security: 42nd International Conference, SAFECOMP 2023, Toulouse, France, September 20–22, 2023, Proceedings, pp. 11–24. Springer, Cham (2023)
14. Hughes, P., Robinson, R., Figueres-Esteban, M., van Gulijk, C.: Extracting safety information from multi-lingual accident reports using an ontology-based approach. Saf. Sci. **118**, 288–297 (2019)
15. Järvelin, K., Kekäläinen, J.: Cumulated gain-based evaluation of ir techniques. ACM Trans. Inf. Syst. **20**(4), 422–446 (2002)
16. Khakzad Shahandashti, K., et al.: Assessing the impact of GPT-4 turbo in generating defeaters for assurance cases. In: Proceedings of the 2024 IEEE/ACM First International Conference on AI Foundation Models and Software Engineering, pp. 52–56. ACM (2024)
17. Kierszbaum, S., Klein, T., Lapasset, L.: Asrs-cmfs vs. roberta: Comparing two pretrained language models to predict anomalies in aviation occurrence reports with a low volume of in-domain data available, Aerospace (2022)
18. Lewis, P., et al.: Retrieval-augmented generation for knowledge-intensive nlp tasks. In: Proceedings of the 34th International Conference on Neural Information Processing Systems, NIPS '20, Red Hook, NY, USA. Curran Associates Inc (2020)
19. Lippi, M., Torroni, P.: Argumentation mining: State of the art and emerging trends. ACM Trans. Internet Technol. **16**(2) (2016)
20. Manning, C.D., Raghavan, P., Schütze, H.: Introduction to Information Retrieval, Cambridge University Press (2008)

21. National Highway Traffic Safety Administration. Investigation: EA22002 – tesla autopilot investigation (engineering analysis opening). Technical Report INOA-EA22002-3184, U.S. Department of Transportation, National Highway Traffic Safety Administration, Office of Defects Investigation (2022). Accessed Jan 2026
22. National Highway Traffic Safety Administration. Investigation: PE23018 – cruise llc autonomous driving system (opening resume). Technical Report INOA-PE23018-11587, U.S. Department of Transportation, National Highway Traffic Safety Administration, Office of Defects Investigation (2023). Accessed Jan 2026
23. National Highway Traffic Safety Administration. Investigation: PE24016 – unexpected ADS behavior (opening resume). Technical Report INOA-PE24016-12382, U.S. Department of Transportation, National Highway Traffic Safety Administration, Office of Defects Investigation (2024). Accessed Jan 2026
24. Ratiu, D., Rohlinger, T., Stolte, T., Wagner, S.: Towards an Argument Pattern for the Use of Safety Performance Indicators. In: Ceccarelli, A., Trapp, M., Bondavalli, A., Schoitsch, E., Gallina, B., Bitsch, F. (eds.) Computer Safety, Reliability, and Security. SAFECOMP 2024 Workshops. SAFECOMP 2024. Lecture Notes in Computer Science, vol 14989. Springer, Cham (2024). https://doi.org/10.1007/978-3-031-68738-9_12
25. Reimers, N., Gurevych, I.: Sentence-bert: Sentence embeddings using siamese bert-networks. In: Proceedings of the 2019 Conference on Empirical Methods in Natural Language Processing (2019)
26. Ricketts, J., Pelham, J., Barry, D., Guo, W.: An nlp framework for extracting causes, consequences, and hazards from occurrence reports to validate a hazop study. In: 2022 IEEE/AIAA 41st Digital Avionics Systems Conference (DASC), pp. 1–8 (2022)
27. Rohlinger, T.: Automated interpretation of fleet incidents to enable system level runtime assurance. In: 35th IEEE International Symposium on Software Reliability Engineering, ISSRE 2024 - Workshops, Tsukuba, Japan, October 28-31, 2024, pp. 91–94 (2024)
28. Rohlinger, T., Ratiu, D., Wagner,S.: Automating quality assessment of llm-generated defeaters. In: 2025 9th International Conference on System Reliability and Safety (ICSRS), pp. 101–110 (2025)
29. Rushby, J.: The interpretation and evaluation of assurance cases. In: Proc. Comput. Sci. Elsevier, San Francisco, CA, USA (2015)
30. Russo, D., Orlando, G.M., Gatta, V.L., Moscato, V.: Automating AI Failure Tracking: Semantic Association of Reports in AI Incident Database. arXiv:2507.23669 (2025)
31. Sparck Jones, K.: A statistical interpretation of term specificity and its application in retrieval. J. Documentation **28**(1), 11–21 (1972)
32. Stede, M., Schneider, J.: Argumentation Mining, 1st edn. Synthesis Lectures on Human Language Technologies. Springer, Cham (2019)
33. Viger, T., et al.: AI-supported eliminative argumentation: Practical experience generating defeaters to increase confidence in assurance cases. In: 2024 IEEE 35th International Symposium on Software Reliability Engineering (ISSRE), pp. 284–294 (2024). ISSN: 2332-6549
34. Xie, X., Tang, X., Gu, S., Cui, L.: An intelligent guided troubleshooting method for aircraft based on HybirdRAG. Sci. Rep. **15**(1), 17752 (2025)

A Safety Cage for Reliable GNSS and AI-Based PNT: An Experience Report

Guillermo Gomez[1], Alberto Griggio[1(✉)], Luca Morelli[1], Theodore Russell[2], Stefano Tonetta[1], and Paweł Trybała[1]

[1] Fondazione Bruno Kessler, Trento, Italy
{ggomezarnedo,griggio,lmorelli,tonettas,ptrybala}@fbk.eu
[2] GMV, Nottingham, UK
trussell@gmv.com

Abstract. Reliable Positioning, Navigation, and Timing (PNT) is a fundamental requirement for the safe operation of autonomous vehicles. While traditional GNSS-based PNT is typically very accurate in open sky conditions, its performance can degrade significantly in environments where satellite visibility is limited or signals are obstructed. To ensure continuity, it is beneficial to complement GNSS with other methods, such as AI-enhanced Visual SLAM, which remain robust under challenging environmental conditions. However, AI-based components can produce untrustworthy results and necessitate a safety cage to monitor and filter their output. This experience report presents the VAIPOSA architecture, which introduces a safety cage to perform runtime verification and fusion of hybrid GNSS and AI-based PNT sources. The safety cage monitors a variety of parameters, including GNSS signal characteristics, AI-based precision metrics, image quality, and vehicle state information, to determine the trustworthiness of each source. We describe the integration of these monitors within a supervisor that fuses data into a reliable pose estimate and discuss its validation in the CARLA simulator using automated test case generation and fault injection scenarios.

Keywords: Safety Cage · AI based PNT · Autonomous Driving · Runtime Monitoring · Formal Methods

1 Introduction

Reliable Positioning, Navigation, and Timing (PNT) is a fundamental requirement for the safe operation of autonomous vehicles. Traditionally, Global Navigation Satellite Systems (GNSS) have served as the primary technology for this task. GNSS is a mature solution capable of computing the absolute position of

The activity was fully funded by the European Space Agency NAVISP Element 1, whose objective is to innovate PNT technologies (EL1-087 bis).

© The Author(s), under exclusive license to Springer Nature Switzerland AG 2027
E. Troubitsyna et al. (Eds.): SAFECOMP 2026, LNCS 16868, pp. 64–79, 2027.
https://doi.org/10.1007/978-3-032-34867-8_5

a system with high precision and accuracy based on satellite signals. However, in safety-critical domains, the positioning solution must be resilient to environmental challenges where satellite signals may be obstructed or distorted by multi-path effects, particularly in tunnels or dense urban "canyons" [29]. In these scenarios, it is critical to augment GNSS receivers with complementary sensors.

Visual Simultaneous Localization and Mapping (SLAM) [12,14] utilizes images from stereo cameras to compute vehicle poses and is a potential method to address these limitations. While these algorithms range from standard computer vision approaches to state-of-the-art AI-based methods, their data-driven nature poses well known safety risks. Employing such black-box components in safety-critical applications necessitates a safety layer that can monitor and filter their outputs using traditional, rule-based approaches [1]. Building on them, more complex systems propose combining different sensing modalities, such as GNSS, cameras, and inertial sensors, to improve the robustness and accuracy of PNT solutions in challenging environments. These approaches can offer either early-stage fusion, where raw sensor data is combined before processing [11], or late-stage fusion, where independent PNT estimates from each modality are computed and combined at a higher system level [3].

Many works in the literature have proposed various safety architectures where efficient but potentially less reliable solutions are encapsulated by a safety layer that monitors and corrects their output. These safety design patterns have been proposed with different names such as safety monitors, safety cages, and safety shields, and are in essence aligned with the concept of fault detection, isolation and recovery. The advent of AI made the topic even more significant and safety cages have been applied to control functions using reinforcement learning as in [9] or to perception functions using machine learning as in [24].

In this paper, we report on the design and application of a *safety cage* for a verifiable, hybrid GNSS and AI-based PNT engine, developed in the VAIPOSA project and funded by the European Space Agency. Our solution performs runtime verification and fusion of the non-AI and AI PNT sources. The safety cage monitors a diverse set of parameters, including GNSS signal characteristics (e.g., satellite count and confidence statistics), AI-based precision metrics, image quality, and physical vehicle state information (e.g., velocity and "point-on-road" checks). These monitors determine the trustworthiness of each engine, which is then translated into a dynamic weighting logic used to drive an Extended Kalman Filter (EKF) for final pose estimation.

We detail our methodology, which begins with an analysis of the potential faults for each engine and the definition of available information interfaces, from which monitoring conditions are derived. Furthermore, we discuss the validation of this architecture within the CARLA simulator [5]. Using the VIVAS framework [6], we performed automated test-case generation and fault injection to verify the resilience of the system in various unsafe scenarios, such as tunnels.

2 Related Work

The challenge of ensuring safety in AI-enabled autonomous systems has led to the adoption of architectural patterns that decouple performance from safety.

In multi-modal PNT systems, the monitoring strategy and supported fault models depend strongly on the system architecture. Early-stage fusion focuses on the integrity of raw sensor data in tightly coupled systems. Alternatively, we adopted late-stage fusion, where independent PNT estimates from different modalities are computed and combined at a higher system level, the monitoring can additionally focus on the consistency and reliability of the individual PNT estimates. In the context of GNSS, this allows us to use a simplified representation of faults (see Sect. 5.2) such as multipath [16], spoofing and jamming [18], avoiding computationally expensive physical simulations, prohibitive given the large-scale scenarios and temporal horizons considered in this work. This approach preserves the impact of GNSS degradation on the overall PNT solution while enabling the study to concentrate on monitoring the AI-based component.

A well known concept in runtime assurance is the *Simplex architecture* [22], which utilizes a high-performance but potentially untrustworthy controller alongside a high-integrity, formally verified fallback. Safety monitors have been studied by Machin et al. [13], who proposed the Safety Monitoring Framework (*SMOF*) that focuses on synthesizing monitors from safety requirements, starting from a hazard analysis and using formal verification techniques to synthesize the rules. A similar notion, the *Safety Shield*, has been proposed by Könighofer et al. [9]. The related TEMPEST tool [17] allows for the synthesis of shields from LTL specifications to ensure that an AI-based agent (often a reinforcement learning policy) never enters an unsafe state. These shields are meant to be integrated into the system to override unsafe commands pre-emptively, ensuring "correct-by-construction" behavior even when the underlying AI component is a black box. The practical application of these monitors in the automotive domain has been explored in projects such as *SMILE* (and its framework *SMIRK*). These works address the challenge of creating a complete safety case for Machine Learning in safety-critical applications, such as pedestrian emergency braking [24]. They demonstrate how runtime monitors can act as an independent safety mechanism to mitigate the uncertainties of deep neural networks.

Our safety cage implementation is a direct evolution of the above works, with the addition of a logic that balances the AI and non-AI-based PNT sources, based on qualitative and quantitative monitoring properties.

While traditional PNT safety relies on statistical bounds like the Horizontal Protection Level (HPL), modern hybrid systems require more adaptive monitoring. Recent research explores using deep reinforcement learning to optimize sensor fusion and enhance GNSS resilience against interference [27], or substituting part of Kalman filters with recurrent neural networks assuming unknown system dynamic models and noise statistics [19]. Our work contributes to this niche by using a multi-parametric Safety Cage to manage the hand-off between traditional GNSS and AI-based SLAM in urban environments.

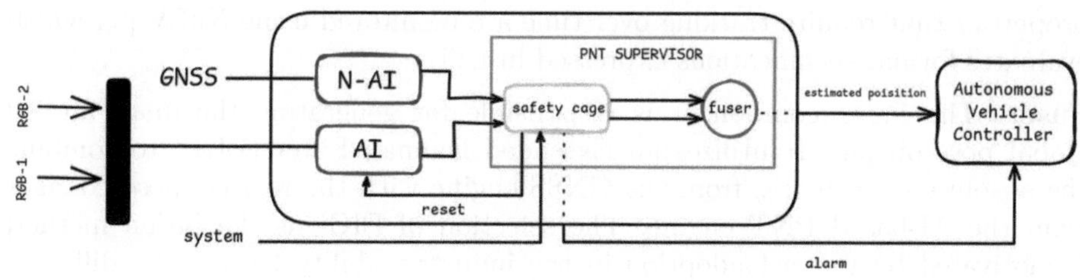

Fig. 1. The VAIPOSA safety architecture.

3 Safety Architecture

The VAIPOSA safety architecture is depicted in Fig. 1. It comprises two PNT engines, one based on GNSS and one based on visual SLAM aided by an AI module for image matching, which provide independent and redundant estimation of the pose of the vehicle, and a specialized **PNT Supervisor**, which manages the integrity and fusion of the positioning data and is composed of a Safety Cage and a Fuser component.

GNSS Engine. MSP3, a precise positioning solver, [8], provides a three-dimensional absolute position in the ECEF reference frame at approximately 1 Hz, which is then projected to the local reference frame. It processes multi-constellation satellite signals and provides essential integrity metrics, such as the Horizontal Protection Level, which statistically bounds the potential position error. A heading pseudo-measurement is additionally derived from the displacement between consecutive GNSS positions when the displacement between them exceeds the minimal threshold. The heading covariance is tied to the estimated yaw rate to account for turn-induced heading estimation uncertainty.

AI-Based Engine. A positioning component that uses COLMAP-SLAM [14, 15], a real-time Visual SLAM framework built on the COLMAP engine [20]. Roughly speaking, the algorithm works in 3 main steps: (i) feature extraction from the input image pair (coming from two RGB cameras), using a 1st NN (SuperPoint [4] or ALIKED [28]); (ii) feature matching on the stereo pair and between consecutive pairs in time, using a 2nd NN (LightGlue [10]), and successive triangulation of the features; (iii) 3D scene reconstruction and computation of the relative pose (6-vector of position and angles in 3D space) change. The inclusion of the AI components ([4, 28]) significantly enhances the robustness of pose estimation in challenging environments. (In the following, we refer to this as "AI-based" or "SLAM-based" engine.)

Safety Cage. The Safety Cage serves as the monitoring layer of the architecture. It continuously analyzes the outputs of both PNT engines alongside system-level metadata, such as camera visibility and image quality. By performing a variety of atomic and temporal checks, the Safety Cage determines the trustworthiness of each engine. These checks are translated into a weight for each source. Temporal properties that require tracking over time are monitored using NuRV [2], which evaluates formal specifications expressed in LTL.

Fuser. The Fuser component is responsible for generating the final, unified global pose output. It utilizes an Extended Kalman Filter (EKF) to combine the absolute coordinates from the GNSS engine with the relative pose changes from the AI-based PNT engine. The selection of EKF as the fusion method is motivated by its vast adoption in the industry, ability to combine different measurement modalities (i.e., absolute and relative positioning), as well as its computational efficiency for real-time applications [21,23]. The filter operates on a 15-dimensional state vector defining position, orientation, linear velocity, angular velocity, and linear acceleration. Notably, velocity and acceleration are *latent* states: no dedicated inertial sensor is fused in the current design. They are propagated by the motion model and implicitly regulated through the Kalman cross-covariance structure. Due to the application of the framework to automotive scenarios, the motion model assumes a constant-acceleration, constant-turn-rate process in the world frame [26]. The discrete process noise covariance is designed to follow a continuous-time kinematic white-noise model for a car-like motion. Finally, all measurement updates use the Joseph form of the EKF covariance update.

4 Safety Cage Design and Monitoring Properties

The design of the Safety Cage follows a systematic methodology that begins with an analysis of the PNT potential faults to then derive the monitoring properties.

4.1 Fault Analysis of PNT Engines

For the GNSS engine, performance is primarily degraded by physical signal interference, such as:

- **Radio-frequency interference**, where unintentional emitters (industrial equipment, nearby communication systems) saturate the receiver front-end and reduce correlator lock quality.
- **Antenna faults**, including partial occlusion, damaged elements, or degraded ground plane performance.
- **Satellite-related degradations**, such as unhealthy ephemeris data or temporary constellation geometry weakening.

Then, we identified the following potential faults of the AI-based PNT engine:

- **Feature starvation**, where a lack of texture in the environment (e.g., a blank wall, dense fog, smoke, or a dark tunnel) prevents the extraction of reliable keypoints.
- **Degraded luminosity contidions** caused by tunnels with low lighting or zones with less density of traffic lights.
- **Hardware faults**, such as frame drops, stuck images or sudden shut down of cameras.

4.2 Engine Interface and Monitoring Data

To detect these faults, we analyze the available information at the interface of each component. The Safety Cage acts as a consumer of both engine outputs and raw system metadata to form a comprehensive view of the system's health.

- **GNSS Interface:** Beyond the absolute position and velocity, this interface provides critical integrity data, including
  - Number of satellites in view of the reciever
  - Horizontal Protection Level (HPL). The HPL represents the radius, expressed in meters, of a circle centered at the estimated horizontal position within which the true position is guaranteed to lie with a predefined integrity risk level (in our case it is $2 \cdot 10^{-9}$).
  - Horizontal Dilution of Precision (HDOP). The HDOP is an indicator of the satellite geometry and this affects the quality of horizontal position estimation. Lower values correspond to better satellite distribution and thus higher positioning accuracy.
- **SLAM Interface:**
  - Number of features extracted per image.
  - Feature extraction time.
  - Covariance matrix of the relative pose change.
- **System Information:** the Safety Cage monitors also external system data, including the status of each sensor, the images from the stereo cameras, the information about motion such as velocity and acceleration, and may have (although it is not necessary) map information about the roads (e.g., number of lanes and their width).

4.3 Monitoring Properties and Weighting Logic

The Safety Cage evaluates the reliability of each PNT engine by assigning a dynamic weight $w \in [0, 1]$, which is then used by the Fuser to compute the final pose. This assessment follows a two-tier hierarchical logic:

1. **Qualitative Cutoffs (Hard-Gating):** A set of boolean properties derived from the identified faults. If any of these "cutoff" properties are violated, the engine is considered untrustworthy, an alarm is raised, and the engine output is blocked ($w = 0$).
2. **Quantitative Scoring (Soft-Weighting):** If no qualitative alarms are active, the Safety Cage calculates a fine-grained weight using sigmoid functions. This weight reflects the current "health" of the engine based on rolling window statistics.

To formalize the decision logic and data flow, the Safety Cage outputs a pair of real weights $(w_{\mathrm{GNSS}}, w_{\mathrm{SLAM}}) \in [0, 1]^2$, alongside a set of categorical alarm flags. The output of each engine modified by the corresponding weight is passed to the Fuser, while the alarms are sent to the system controller (see also Fig. 1). The single weight variable, one for each engine, encapsulates both evaluation tiers: the qualitative assessment acts as a binary hard-gate (forcing $w_i = 0$ upon a critical safety rule violation), while the quantitative assessment computes a continuous score within $[0, 1]$ during nominal or slightly degraded conditions.

When a qualitative cutoff occurs ($w_i = 0, i \in \{\mathrm{GNSS}, \mathrm{SLAM}\}$), the engine output is discarded. In the event of a simultaneous dual failure where both engines fail their qualitative checks ($w_{\mathrm{GNSS}} = w_{\mathrm{SLAM}} = 0$), the PNT Supervisor raises an immediate critical alarm to the vehicle controller, acknowledging that no trustworthy PNT source remains.

4.4 Qualitative Cutoff Properties

Qualitative properties are temporally extended properties expressed in LTL over the variables provided by the engines' interface. When the property is violated an alarm is sent to the vehicle controller and some recovery action is taken (either the output of the corresponding PNT engine is not sent to the Fuser or some reconfiguration of the engine is triggered). More specifically, we monitored the following conditions:

- **GNSS Properties**
  - $\mathbf{G}_{[0,3]}\ hpl < th$, where th is a threshold based on the average road width or, if the information is available, on the width of the current road the vehicle is driving. Its violation will trigger the informative alarm $high\_hpl$ and block the output of the GNSS-based PNT to the fuser.
- **SLAM Properties**
  - $\mathbf{G}_{[0,3]}\ \neg low\_luminosity$. Its violation will trigger the $swap\_feature\_extractor$ alarm. that will switch to a feature extractor better suited for low-illumination imagery.
  - $\mathbf{G}_{[0,2]}\ \neg frozen\_image_{\{left,right\}}$. This requires the image of the left/right camera not to remain frozen for 2 s. Its violation will trigger the informative alarm $frozen\_image$ that will block the output of the AI PNT to the fuser.
  - $\neg camera_{\{left,right\}}\_alive\ \mathbf{U}_{[0,10]}\ camera_{\{left,right\}}\_alive$. This requires that the left/right camera cannot remain unresponsive for up to 10 s. Its violation will trigger the $camera_{\{left,right\}}\_dead$ alarm and block the output of the AI-based PNT to the fuser.

4.5 Quantitative Weighting Properties

For engines that satisfy the qualitative cutoff properties, the Safety Cage computes a dynamic weight $w \in [0, 1]$. This approach allows the fuser to gracefully handle performance degradation without abruptly disconnecting a source.

Parametric Sigmoid Function. Each monitoring property utilizes a parametric sigmoid function to map a real-time health metric x to a weight:

$$f(x; c, \alpha) = \frac{1}{1 + e^{-k(x-c)}}$$

where c is the center of the transition and α defines the sensitivity interval $[x_1, x_2]$ (with $x_{1,2} = c \mp \alpha/2$). The slope k is derived to ensure the function reaches near-boundary values at the interval endpoints.

GNSS Weighting Logic. The GNSS weight is primarily driven by the stability of the HPL. The system computes a normalized metric $x = -hpl(t)/W_{hpl}$ where W_{hpl} is the mean of *hpl* in the last 20 steps. By applying the sigmoid function with $c = -1.5$ and $\alpha = 1.3$, the monitor effectively penalizes the GNSS source when the current uncertainty spike exceeds the recent average by a safety-tuned margin.

SLAM Weighting Logic. The weight of the AI-based engine is computed with three complementary metrics derived from the SLAM interface:

1. **Feature Drop Ratio:** This metric detects texture-less environments by monitoring the ratio $\frac{nf(t)}{W_{\text{nf}}}$ of the latest feature count to the mean of features in the last 10 steps. By applying the sigmoid function with $c = 0.5$ and $\alpha = 0.5$, the weight drops sharply if the number of detected features falls significantly below the recent trend.
2. **Extraction Time Ratio:** This metric monitors the relative change in feature extraction time (again normalizing with respect the mean of the last 10 steps). In this case, the sigmoid function has been set with parameters $c = -3$ and $\alpha = 1.5$.
3. **Covariance Spike Monitor:** This metric uses the norm of the diagonal submatrix related to the position of the SLAM covariance. In this case, the input of the sigmoid function is the deviation of current norm from the average of the last 10 steps, using $c = -3$ and $\alpha = 1.5$.

The final weight w for the AI-based PNT is the minimum value produced by the three metrics. This ensures that a single indicator of failure (e.g., a feature drop) is sufficient to reduce the trust in the engine.

Parameter Tuning. Several of the formulas above depend on specific parameters, such as c and α for the sigmoid function, the interval durations for bounded temporal operators, and the window lengths for rolling averages. We selected these values empirically by tuning them via simulation on a subset of our test scenarios (see Sect. 5) to effectively balance sensitivity and robustness in the ability of the supervisor to react to anomalous situations. A more systematic exploration of these choices, including a formal sensitivity analysis or machine-learning-based optimization, is left for future research.

5 Experimental Evaluation

We have implemented our proposed safety cage architecture in a prototype tool written in Python, and integrated it in the CARLA [5] urban driving simulation environment for experimentally evaluating its performance.

We evaluate the effectiveness of our safety cage architecture by integrating our prototype implementation in an autonomous vehicle inside CARLA, and by executing an extensive set of simulation scenarios, considering multiple environment conditions and different fault situations.

5.1 Simulation Setup

We use the standard Python API provided by CARLA to integrate our safety cage in the simulator. More specifically, we develop a custom client for the CARLA server, which instantiates an autonomous vehicle equipped with a virtual GNSS receiver (to provide data to the GNSS-based PNT engine) and two RGB cameras mounted on the front of the vehicle (which provide data to the AI-based PNT engine). The vehicle is given as input a trajectory to follow, and uses the autopilot functionalities provided by CARLA for autonomous driving.[1] For repeatability, we configure CARLA to operate in synchronous mode. At each simulation tick, the client queries the relevant sensors (GNSS receiver and RGB cameras) for data, feeds it to the corresponding PNT engines (if new data is available), computes the weights for their solutions and fuses them to obtain the PNT output. We use a duration of 0.2 s for each tick, which corresponds to the operating frequency (5 Hz) of the AI-based PNT engine. The GNSS-based engine, instead, operates with a frequency of 1 Hz. All components of our architecture were able to operate within their specified frequencies. In particular, COLMAP-SLAM took on average 88.9 ms to process each image pair in our experiments; regarding the safety cage component, the computational overhead of the NuRV monitors was almost negligible, thanks to the simplicity of the properties and the lack of assumptions.

As described earlier, we integrated a high-fidelity virtual multi-constellation GNSS receiver into the simulator. This receiver generates simulated GNSS satellite signals, incorporating models of real-world effects such as ionospheric delay, and provides them to the commercial off-the-shelf (COTS) GNSS-based PNT engine, MSP3, [8], for position estimation. A further advantage of this approach is that the GNSS signal simulator supports signal occlusion, allowing the local CARLA environment to influence GNSS signal quality. Consequently, environmental features such as trees, tunnels, buildings, and urban canyons dynamically affect GNSS reception, thereby improving realism. We selected this approach over the built-in GNSS sensor of CARLA because the latter relies on a simplistic model that merely adds noise to the vehicle ground-truth position.

To determine the set of visible satellites, at each simulation tick we compute a signal quality factor for every satellite that is above the horizon. This factor ranges from 0.0, representing complete blockage, to 1.0, representing clear-sky visibility. The value is computed by casting a ray from the ego vehicle to each satellite in the sky. Any geometry in the simulation environment that intersects the ray reduces the signal quality linearly according to a predefined attenuation factor determined by the object type. These signal quality values are then used by the GNSS simulator to model the resulting GNSS measurements more accurately.

[1] Specifically, we use an instance of the `BasicAgent` class from the `carla.agents.navigation` module for controlling the vehicle, providing it the input trajectory as a global plan to follow.

5.2 Scenario Generation and Fault Injection

For evaluating our safety cage architecture, we generate a set of simulation scenarios using an approach based on coverage-guided automated test case generation. In particular, we adopt the generic VIVAS methodology [6] that generates simulation scenarios using a formal model of the system under test, its environment (called the abstract system), and scenario exploration through symbolic reasoning techniques. Scenarios are generated by systematically exploring values for predicates expressing constraints on the main parameters affecting the computations of the two PNT solutions along the following dimensions:

Map of the World. We consider different built-in maps of CARLA, providing different environmental features (e.g. dense urban aeras, rural/forest areas, highways, narrow roads, tunnels);

Coordinates of the map on Earth, affecting the availiability of satellites. We discretize the possible coordinates corresponding to locations of some relevant cities at different latitudes;

Weather and visibility conditions, affecting the quality of the images captured by the RGB cameras and provided as input to the AI-based PNT engine. We select among the available weather conditions provided by CARLA (e.g. "clear day at noon", "foggy night", "hard rain at sunset");

Trajectory to Follow. To represent navigation tasks in a generic way, we introduce the concept of abstract Point Of Interest (POI). Conceptually, a POI consists of a location on a map with a set of semantic tags attached to it. We then define a navigation task as a sequence of POIs (global plan) that the autonomous vehicle must reach in the given order. The intermediate waypoints (local plan) between two POIs are determined by the COTS autonomous driving controller. We define different semantic tags corresponding to different features in the enviornment (e.g. "rural area", "forest area", "high buildings area", "tunnel"), and enumerate trajectories traveling through different such areas;

Fault Conditions. To evaluate the safety cage performance in non-nominal situations, our simulator supports the injection of faults that can be activated at run time. Faults are defined with a type and an interval (in simulation time) during which they are active. A fault of type "GNSS receiver" affects the GNSS-based PNT engine. For each such fault, we specify (a) which GNSS constellations should be available and (b) an upper bound N for the number of satellites that should be visible the duration of the fault. During the simulation, for all ticks when a GNSS receiver fault is active, we limit the satellites available for the computation of the GNSS measurement by randomly selecting N satellites and blocking all the remaining ones. Faults of type "RGB camera", instead, affect the AI-based PNT engine. For each such fault, we additionally specify the camera affected (left, right, or both) and the kind of defect of either (a) image completely black, or (b) image stuck at a specific frame. We constrain each generated scenario to include one fault of the above kinds.

5.3 Results

To evaluate the quality of the PNT solution computed by our engine, we use two common metrics for the evaluation of trajectory errors, namely the Absolute Trajectory Error (ATE) and the Relative Pose Error (RPE). The former (ATE) evaluates the global consistency of the estimated trajectory when compared with the ground truth, whereas the latter (RPE) measures the local trajectory accuracy, i.e., incremental changes of subsequent vehicle poses, in the vehicle reference frame. Both are widely used metrics in evaluating positioning systems when ground truth data is available (see e.g. [25]). We use the implementations available in the field, the `evo` Python package, for the evaluation of odometry and SLAM [7].

For each simulation run, we extract both the ground truth trajectory and the trajectory computed by the VAIPOSA PNT from the simulation logs, and compute the ATE and RPE using the `evo` library. We then aggregate the data by computing the median and root mean squared error (RMSE) of the two trajectory errors. We evaluate the performance of the default safety cage configuration by comparing the values for such metrics against those produced by the baseline configurations using only a single engine (either GNSS-based or AI-based). Additionally, we also evaluate a configuration using fixed values for the weights given to the output of the two PNT engines in the fusion algorithms, effectively disabling the safety cage. (In the following, we refer to such configuration as "fixed weights".)

Table 1. Summary of per-scenario statistics comparing the default configuration against the baselines (GNSS-only and SLAM-only) and the fixed-weights configurations. All the values are expressed in meters (with the best ones highlighted).

Metric	Default	GNSS only	SLAM only	Fixed weights
ATE - median	Max: 74.95 Med.: 1.51 Mean: 2.81 MAD: 1.54	Max: 291.17 Med.: 1.68 Mean: 3.64 MAD: 2.26	Max: 912.08 Med.: 7.18 Mean: 49.31 MAD: 46.07	Max: 22.91 Med.: 1.57 Mean: 3.55 MAD: 2.31
ATE - RMSE	Max: 83.77 Med.: 4.86 Mean: 8.51 MAD: 6.02	Max: 3139.90 Med.: 20.14 Mean: 62.73 MAD: 53.07	Max: 1094.30 Med.: 16.24 Mean: 68.32 MAD: 62.82	Max: 660.28 Med.: 6.10 Mean: 23.39 MAD: 19.71
RPE - median	Max: 2.01 Med.: 0.20 Mean: 0.33 MAD: 0.15	Max: 19.94 Med.: 0.30 Mean: 0.44 MAD: 0.19	Max: 2.08 Med.: 0.11 Mean: 0.52 MAD: 0.42	Max: 2.02 Med.: 0.25 Mean: 0.48 MAD: 0.30
RPE - RMSE	Max: 30.18 Med.: 0.89 Mean: 1.74 MAD: 1.11	Max: 492.14 Med.: 5.40 Mean: 13.94 MAD: 11.37	Max: 16.89 Med.: 0.42 Mean: 1.66 MAD: 1.45	Max: 205.63 Med.: 1.36 Mean: 6.34 MAD: 5.44

A summary of the results is reported in Table 1, showing different statistics (max, median, mean, and Median Absolute Deviation) of the metrics above over our data set, which consists of 360 scenarios. From the results, we can see that the default configuration is the best performing one for the majority of the indicators. This is true especially when considering the ATE errors, indicating its ability to better adapt to changing conditions and promptly react to unusual behavior of one the PNT solutions, compared to the individual engines, in terms of global positioning accuracy. Specifically, greatly lower ATE RMSE values indicate the engine robustness to outliers at different difficulty levels. The lowest MAD of this metric demonstrates also the consistency of the quality of results delivered by the default configuration. Despite the lack of the "global" aspect, in the autonomous vehicle navigation, RPE metrics also play a critical role in assuring the appropriate quality of positioning in the local scope, which directly interferes with the real-time vehicle control and planning. The results show that the VAIPOSA configuration outperforms GNSS-only and fixed weights configurations across almost all statistics, and is very close to the SLAM-only configuration, which directly provides relative pose estimations, while increasing the overall resilience to outliers in the provided PNT results.

In order to gain additional insights, we study the variations in performance in the presence of faults. By construction, each simulation scenario in our data set contains one fault activation (either of the GNSS receiver or of one or both on-board cameras, as described in Sect. 5.2 above). To evaluate the impact of such faults, for each simulation run we extract two additional trajectories: (i) a "pre fault" trajectory, consisting of the prefix of the simulation execution up to the time point at which the fault is activated; and (ii) a "during fault" trajectory, consisting of the subsequence of time points during which the fault is active.

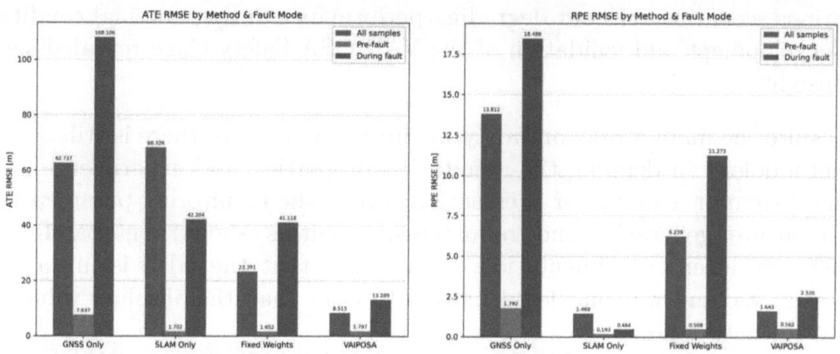

Fig. 2. Impact of faults on ATE and RPE RMSE for the different configurations.

An overview of the results is shown in Fig. 2, displaying histograms of the RMSE for the ATE and RPE values obtained with the four different configurations over our data set. As expected, from the results we can conclude that the safety cage is especially important during non-nominal situations, substantially reducing the observed trajectory errors (particularly for the ATE case). In particular, the performance of the PNT engine is not degraded during the regular use (ATE, RPE in all/pre-fault conditions), but is significantly improved when introducing faults to any of the engines. In the current configurations, experiments showed greater and more consistent improvement in global positioning capabilities (ATE during fault), while in terms of relative positioning, the median error does improve over the baselines, but still retains a visible tail of higher errors. We conjecture this could be potentially addressed by further expanding the safety cage capabilities of detecting and acting on SLAM faults.

Finally, we wish to comment on the different trend in results obtained by the SLAM configuration compared to the others, where the measures for "all samples" are worse than those for "during fault" situations. This occurs because SLAM systems naturally accumulate error and drift over time. Since the "during fault" trajectories are shorter sub-trajectories of the full "all samples" runs, they only capture a fraction of the total drift. The "all samples" metrics evaluate the entire duration, allowing error to compound continuously and ultimately resulting in the worst overall performance. Additionally, some cases producing a high RMSE value for the SLAM-only configuration are due not to scenarios containing faults, but simply to difficult-to-reconstruct 3D scenes.

6 Lessons Learned and Conclusion

In this report, we presented the design and evaluation of a safety cage for hybrid PNT systems. Our results demonstrate that the VAIPOSA architecture successfully maintains positioning integrity by dynamically weighting GNSS and AI-based PNT sources, significantly reducing the impact of engine-specific faults in various scenarios, without degrading performance during nominal conditions. The development and validation of the VAIPOSA Safety Cage revealed several key insights:

- Despite the many works on safety monitors and shields, there is still a lack of methodology in defining the monitoring properties, and experimenting with the system or a simulator is essential to tune the monitoring parameters.
- To ensure generality and robustness, monitors should be stateful. For instance, after experimentation, we realized that the HPL is initially too pessimistic and we considered its trend rather than the absolute value until a stabilization point.
- There is an inherent trade-off between minimizing absolute position error and maintaining trajectory smoothness. While our current filter minimizes error, it can lead to discontinuous trajectories, whereas the AI-based PNT solution offers smoothness but lacks absolute global accuracy.
- Fusing global and local PNT solutions as well as exploiting an extensive interface with the PNT engines were key to enable robust performance in a wide range of conditions, even using simple filter-based fusion algorithms.

- Using a simulator with 3D environment models, jointly with an automatic testing procedures of VIVAS, was essential for debugging and created a realistic and high-performance benchmarking environment that can be used also for evaluating other PNT solutions for autonomous vehicles. However, we explicitly acknowledge that our current evaluation is simulation-only and based on a limited set of faults, while a more realistic validation setup on physical target platforms is required to fully capture the complexity of physical hardware degradation in real-world deployments.
- The GNSS-based PNT engine used in our prototype already performs internal filtering. We conjecture that having access to "raw" GNSS measurements and performing the fusion only after the safety cage would provide better control and potentially superior results.

Future research will focus on evolving the safety cage into a fully stateful monitor capable of providing formal safety certificates and protection levels for the fused output. Additionally, to address a key limitation of our current configuration, which lacks an Inertial Measurement Unit (IMU), we plan to explore the integration of inertial sensors to further improve resilience during extended periods of GNSS and visual faults.

References

1. Deliverable of PASSIONS (Software Product Assurance For Autonomous On-Board Software) - Executive Summary Report (2023). https://nebula.esa.int/sites/default/files/neb_tec_studies/3154/public/PASSIONS_Executive_Summary_Report_v1.0.pdf, Accessed 20 May 2026
2. Cimatti, A., Tian, C., Tonetta, S.: NuRV: A NUXMV Extension for Runtime Verification. In: Finkbeiner, B., Mariani, L. (eds.) RV 2019. LNCS, vol. 11757, pp. 382–392. Springer, Cham (2019). https://doi.org/10.1007/978-3-030-32079-9_23
3. Cremona, J., Civera, J., Kofman, E., Pire, T.: Gnss-stereo-inertial slam for arable farming. J. Field Robot. **41**(7), 2215–2225 (2023)
4. DeTone, D., Malisiewicz, T., Rabinovich, A.: Superpoint: Self-supervised interest point detection and description. In: CVPR Workshops, pp. 224–236 (2018)
5. Dosovitskiy, A., Ros, G., Codevilla, F., López, A.M., Koltun, V.: CARLA: an open urban driving simulator. In: CoRL, pp. 1–16. Proceedings of Machine Learning Research. PMLR (2017)
6. Goyal, S., Griggio, A., Tonetta, S.: System-level simulation-based verification of autonomous driving systems with the VIVAS framework and CARLA simulator. Sci. Comput. Program. **242**, 103253 (2025)
7. Grupp, M.: evo: Python package for the evaluation of odometry and slam (2017). https://github.com/MichaelGrupp/evo, Accessed 20 May 2026
8. Hutchinson, M., Tiwari, S., Zheng, Y., Moradi, R.: Carrier phase positioning techniques for mass market gnss receivers: enhancement of MSP3 precise point positioning (PPP) software. In: 7th International Colloquium on Scientific and Fundamental Aspects of GNSS (2019). conference abstract
9. Könighofer, B., Bloem, R., Jansen, N., Junges, S., Pranger, S.: Shields for safe reinforcement learning. Commun. ACM **68**(11), 80–90 (2025)

10. Lindenberger, P., Sarlin, P.E., Pollefeys, M.: Lightglue: Local feature matching at light speed. In: Proceedings of the IEEE/CVF International Conference on Computer Vision, pp. 17627–17638 (2023)
11. Liu, J., Gao, W., Hu, Z.: Optimization-based visual-inertial slam tightly coupled with raw gnss measurements. In: 2021 IEEE International Conference on Robotics and Automation (ICRA), pp. 11612–11618 (2021)
12. Macario Barros, A., Michel, M., Moline, Y., Corre, G., Carrel, F.: A comprehensive survey of visual slam algorithms. Robotics **11**(1), 24 (2022)
13. Machin, M., Guiochet, J., Waeselynck, H., Blanquart, J., Roy, M., Masson, L.: SMOF: a safety monitoring framework for autonomous systems. IEEE Trans. Syst. Man Cybern. Syst. **48**(5), 702–715 (2018)
14. Morelli, L., Ioli, F., Beber, R., Menna, F., Remondino, F., Vitti, A.: COLMAP-SLAM: A framework for visual odometry. Int. Arch. Photogramm. Rem. Sens. Spat. Inf. Sci. **48**, 317–324 (2023)
15. Morelli, L., Trybała, P., Razzino, A.V., Remondino, F.: Deep learning in visual odometry for autonomous driving. Int. Arch. Photogramm. Rem. Sens. Spat. Inf. Sci. **48**, 93–99 (2025)
16. Pereira, P.M.C., da Silva, H.D.M., Lima, C.M.G.S.: Advancements in multipath mitigation for gnss receivers: Review of channel estimation techniques. Space: Sci. Technol. **5**, 0278 (2025)
17. Pranger, S., Könighofer, B., Posch, L., Bloem, R.: TEMPEST - Synthesis Tool for Reactive Systems and Shields in Probabilistic Environments. In: Hou, Z., Ganesh, V. (eds.) ATVA 2021. LNCS, vol. 12971, pp. 222–228. Springer, Cham (2021). https://doi.org/10.1007/978-3-030-88885-5_15
18. Psiaki, M.L., Humphreys, T.E.: Gnss spoofing and detection. Proc. IEEE **104**(6), 1258–1270 (2016)
19. Revach, G., Shlezinger, N., Ni, X., Escoriza, A.L., Van Sloun, R.J., Eldar, Y.C.: Kalmannet: Neural network aided kalman filtering for partially known dynamics. IEEE Trans. Signal Process. **70**, 1532–1547 (2022)
20. Schonberger, J.L., Frahm, J.M.: Structure-from-motion revisited. In: CVPR Workshops, pp. 4104–4113 (2016)
21. Schreiber, M., Königshof, H., Hellmund, A.M., Stiller, C.: Vehicle localization with tightly coupled gnss and visual odometry. In: IEEE Intelligent Vehicles Symposium (IV), pp. 858–863. IEEE (2016)
22. Sha, L.: Using simplicity to control complexity. IEEE Softw. **18**(4), 20–28 (2001)
23. Skog, I., Handel, P.: In-car positioning and navigation technologies-a survey. IEEE Trans. Intell. Transp. Syst. **10**(1), 4–21 (2009)
24. Socha, K., Borg, M., Henriksson, J.: SMIRK: A machine learning-based pedestrian automatic emergency braking system with a complete safety case. Softw. Impacts **13**, 100352 (2022)
25. Sturm, J., Engelhard, N., Endres, F., Burgard, W., Cremers, D.: A benchmark for the evaluation of rgb-d slam systems. IROS 573–580. IEEE (2012)
26. Tao, L., Watanabe, Y., Yamada, S., Takada, H.: Comparative evaluation of kalman filters and motion models in vehicular state estimation and path prediction. J. Navig. **74**(5), 1142–1160 (2021)
27. Wisniewski, M., Chatzithanos, P., Guo, W., Tsourdos, A.: Benchmarking deep reinforcement learning for navigation in denied sensor environments. J. Intell. Rob. Syst. **111**(3), 103 (2025)

28. Zhao, X., Wu, X., Chen, W., Chen, P.C., Xu, Q., Li, Z.: Aliked: A lighter keypoint and descriptor extraction network via deformable transformation. IEEE Trans. Instrum. Meas. **72**, 1–16 (2023)
29. Zhu, N., Marais, J., Bétaille, D., Berbineau, M.: Gnss position integrity in urban environments: A review of literature. IEEE Trans. Intell. Transp. Syst. **19**(9), 2762–2778 (2018)

Cybersecurity and Mixed-Criticality in Safety-Critical Systems

Identifying Control and Monitoring Data for Forensics of Safety-Critical Attacks on Control Room Operations

Alexios Karagiozidis(✉) and Martin Gergeleit

University of Applied Sciences RheinMain, Wiesbaden 65195, Germany
alexios.karagiozidis@student.hs-rm.de, martin.gergeleit@hs-rm.de

Abstract. The increasing interconnection of Industrial Computer Systems heightens the risk of attacks on Industrial Control Systems (ICS). Remote attackers can interfere with the control room's ability to monitor and control a physical process, impairing operational safety. To assess or trace the impact of such attacks, digital forensics are essential, where data is identified and collected to generate evidence. Due to differences between office IT and operational technology (OT) used in ICS, established IT forensics cannot be applied directly, while existing approaches only cover the partial acquisition of control room related data. However, to integrate relevant data into acquisition or forensic methods, it must be identified where control room related data resides and which software processes and connections are of relevance. As ICS software, protocols, and data are often proprietary and vendor-specific, a vendor-independent identification approach is required. We propose a novel runtime-based reverse engineering procedure for identifying control room related data, paths, and software processes in Windows-based ICS environments. This enables one to determine which processes and connections are of forensic or safety relevance and need to be covered regarding acquisition or security-monitoring for control room operations. To establish our approach, we analyze how control and monitoring data is exchanged internally on an *ABB 800xA* Distributed Control System (DCS) and validate the derived approach by applying it to a *Siemens Process Control System (PCS7)* DCS/SCADA.

Keywords: Industrial Control Systems · Digital Forensics · Safety Critical Systems · Operational Safety

1 Introduction

ICS form the backbone of critical infrastructure, including energy, water, manufacturing, and transportation. As IT/OT convergence increases and these systems become increasingly connected to the internet, the risk of being attacked grows. In opposite to IT environments, security incidents in ICS can lead due to the physical process harm to humans, the environment and machinery, having

© The Author(s), under exclusive license to Springer Nature Switzerland AG 2027
E. Troubitsyna et al. (Eds.): SAFECOMP 2026, LNCS 16868, pp. 83–97, 2027.
https://doi.org/10.1007/978-3-032-34867-8_6

especially high risks regarding safety. Of particular concern is that an adversary with access to the ICS can:

1. **Manipulate or block the control room's view**: An adversary might alter displayed values including the indicated state of the physical process, preventing the detection of dangerous conditions (e.g., an overheating boiler).
2. **Block interventions by the control room or other systems**: If the control room or other systems detect a dangerous condition, their interventions can be prevented by blocking commands.
3. **Directly impact physical process machinery**: By sending unauthorized commands an attacker may cause a critical condition or even harm to human and damage to machinery.

To determine in a forensic investigation whether and to what extent the control room operations were impaired, exchanged control and monitoring data, including its associated host data must be identified and acquired. While this area remains largely unexplored, existing approaches focus on acquiring and examining data of already known identified control and monitoring paths (e.g. industrial protocols). However, how all relevant connections, including the software components responsible for control and monitoring can be identified, is not addressed. ICS systems can be complex, and due to proprietary software and protocols [1], the relevant components and communication paths are usually not readily apparent. Additionally, due to proprietary systems and the fundamental difference that ICS controls physical processes, standard IT forensics methods are insufficient [1]. To address this, we propose a novel runtime-based approach that identifies:

1. **All connections relevant for control and monitoring between ICS components, from the physical process to the control room.**
2. **All host-side data relevant for control and monitoring (e.g., responsible software processes and components, communication drivers, volatile data)**

The activity over these connections is essential, as it contains the values reported by the PLCs to the control room as well as commands sent in the opposite direction. Host-side data is required to be able to detect local manipulations or proof interventions in control and monitoring. For example, in another real-world ICS attack Stuxnet [2], a software component, specifically a DLL was manipulated so that the operator saw different indicated values than those actually transmitted. This example illustrates the importance of identifying where data must be collected in order to determine whether control and monitoring capabilities were affected during a security incident and to what extent.

In the next chapter, we introduce the ICS architecture within scope, followed by forensic fundamentals necessary for understanding. Afterwards we present the related work with similar existing approaches. In Sect. 4, we discuss safety related threats which are critical for control and monitoring capabilities. We use an ABB 800xA setup to examine how control and monitoring related data is available and how it can be identified by using Windows mechanisms as Inter-process-communication (IPC). Based on these findings, we derive a general procedure model that identifies the data path or involved software processes regarding control room operations, including their static and dynamic data. In Chap. 6, we apply our approach to a prototypical Siemens PCS7 testbed to determine whether our generalized procedure is vendor-independent.

2 ICS and Digital Forensics

This section introduces the ICS architecture and components relevant to our work including the essential digital forensics background.

2.1 Industrial Control Systems

ICS are based on the Purdue model architecture [7]. In Fig. 1, we present the ICS components in conjunction with the ABB 800xA system. Both ABB 800xA and Siemens PCS7 are in scope of our work, as they are widely deployed DCS systems across various safety-critical industries, including energy, railway, oil and gas, and chemical processing, where they manage large-scale distributed plant sites.

Level 3. In this level, all mixed IT/OT components are located, from which multiple field or plant sites are monitored and controlled. For this reason, it is often referred to as the control center. It consists of the following main components:

- **Engineering Workstation/Server (EWS)**: Utilizes project planning or engineering software to manage and configure field devices, including creating user control logic and downloading it to devices. Since it does not participate in the active control and monitoring via the control room, it is out of scope in this work.
- **Human Machine Interface (HMI)/Operator Client**: HMIs visualize the physical process and enable operators to monitor and control distributed sites from a centralized control room. They receive their data from the L1/2 components or the control server and forward commands by the operator.
- **Data Historian (DH)**: A centralized database that stores and archives operational data such as measured values or ICS events.
- **Aspect Server**: The aspect server is an 800xA specific component and centralizes the data including providing the configuration of the HMIs.
- **Data Gateway/Connectivity Server (CS)**: The connectivity server corresponds to the Data Gateway as it translates data between the L3 and L1/2. This is common for ICS as different sites may use different protocols, requiring a Data Gateway.

Level 1/2. This level consists of interconnected embedded controllers, typically PLCs, including safety controllers. PLCs primarily enable the control and monitoring of physical processes, while safety controllers consist of multiple fail-safe

PLCs that monitor safety-critical components. L1/L2 controllers have physical I/O connected to sensors and actuators, which are interfaced with the components of the physical components. Data from the PLCs is typically sent directly to the control server or HMI for monitoring and control.

Level 0. The final level is the physical process and its components. Depending on the ICS, this may include, for example, power generation and distribution, as well as chemical or oil processing.

2.2 Digital Forensics

In digital forensics, the goal is to determine after a security incident the answers to the questions, *what happened, when, why, where and how.* The process generally consists of the following phases:

Strategic and Operational Preparation. In the strategic preparation phase, forensic measures are set up in advance. This includes in particular, identification of available data sources and measures to acquire them in case of an incident. In the operational phase, triggered by an incident, a list of data to be acquired is established.

Acquisition and Examination. In this phase, the pre-determined data is acquired and subsequently examined to extract further data. This includes, for example, the acquisition of network traffic and the extraction of control and monitoring data, such as transmitted physical process values.

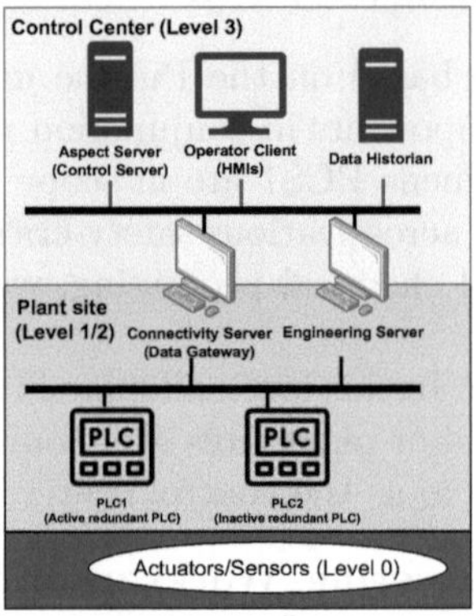

Fig. 1. ABB 800xA components/Purdue architecture.

Analysis. Here, the collected data is statistically analyzed to identify interrelations and gather additional evidence. The findings from this analysis are then used to address the forensic questions and reconstruct the incident.

This work addresses strategic preparation by identifying control and monitoring data paths and associated host data, enabling their integration into acquisition and examination measures. This ensures that, in case of an incident, it can be determined which data to acquire or whether and to what extent the control room was affected.

3 Related Work

In 2020, the MITRE ICS Attack catalog [8] was published, specifying known attack techniques against ICS and has since been updated multiple times. However, its main goal is more threat-detection and mitigation rather than tracing safety related impacts. The authors of [13] (2018) introduced *A Forensic Logging System for Siemens PLCs.* This system passively extracts physical process values and commands from known network traffic in a forensically sound manner, turning them into security events. More recently, the authors of [9] (2024) presented a reverse engineering approach capable of identifying the structure of proprietary PLC communication protocols. In 2025,the authors of *IFACT* [12] presented an approach, which acquires and extracts data from HMI software.

The latter three approaches cover important aspects for tracing incident impact. However, while network traffic is partially covered, the identification of which proprietary and undocumented connections and software processes are involved in control and monitoring data exchange remains uncovered. Similarly, although HMI application data is addressed, the associated data paths, further involved processes (forwarding the data to the HMI application), and static as well as dynamic host data related to control room operations are not considered. This work addresses these gaps by providing a vendor-independent approach to identify relevant processes, connections, and associated data in the first place.

4 Analysis of Threats and Data Regarding Control Room Operations

We first analyze which MITRE TTPs directly impact safety and control room operations, assessing why identifying control and monitoring data paths is essential. We then analyze how this data is exchanged internally in an ABB 800xA system, considering the involved Windows mechanisms, to derive a generalizable identification approach.

4.1 Safety Critical Threats with Impact on Safe Control Room Operations

The MITRE ICS TTP catalog includes 84 attack techniques, of which 36 are ICS-exclusive. We selected in Table 1 those directly impacting safe operations, categorized by their effect on either the physical process or control room visibility. Attacks directly affecting the physical process include sending unauthorized commands (T0855) to alter it, modifying parameters such as motor speeds (T0836), or restarting devices (T0816). Attacks targeting control room visibility include spoofing (T0856) or blocking (T0804) reported process values, suppressing alarms (T0878), or blocking commands (T0803). Some attacks affect both aspects, such as data destruction (T0809) that deletes critical software or service termination (T0814, T0881) of control and monitoring applications. Manipulating the I/O image (T0835) alters the PLC's internal process image, which can change physical I/O states (e.g., activating an output) or send false values to the control room.

These attacks can be combined, for instance, blocking control room visibility (T0878, T0804) by stopping or overloading relevant software and connections (T0809, T0814, T0881). Therefore, tracing and comprehending these attacks

Table 1. Potential security threats with direct impact on control room operations.

Direct impact on physical process				Direct impact on control rooms visibility				Impact on Both			
T0858 Change Operating Mode	T0855 Unauthorized Command Message	T0836 Modify Parameter	T0816 Device Restart/ Shutdown	T0856 Spoof Reporting Message	T0878 Alarm Suppression	T0804 Block Reporting Message	T0803 Block Command Message	T0809 Data Destruction	T0814 Denial of Service	T0881 Service Stop	T0835 Manipulate I/O Image

requires identifying the control and monitoring path including components. For example, it must be determined whether an executed command originated from the legitimate control room software or was injected by an attacker externally into the network or via a compromised HMI or other ICS component. The complete path with its associated static and dynamic data is also needed to trace manipulation of static files such as binaries of the control software. In the real-world Stuxnet attack [2], the s7otbxdx.dll used for PLC communication was replaced by a malicious wrapper that intercepted commands while replaying recorded process values to operators. The nature of these attacks demonstrates that the entire control and monitoring path, from HMIs to PLCs, including associated processes, connections, and host data (as executables or DLL files), must be identified to trace incidents and generate evidence.

4.2 Analyzing the ICS Software to Identify Data Flow and Inter Process Connections

To determine how control and monitoring paths including their data can be identified for a general approach, we use an ABB 800xA testbed, shown in Fig. 1 to analyze how data is exchanged. An initial analysis revealed that the HMI GUI does not run as a network process and must therefore obtain variables internally through other processes. The connectivity server, which forwards data between PLCs and HMI, employs two separate communication drivers. This indicates that identifying paths and data requires tracing all processes between the HMI application and the final L3 communication driver connected to the PLCs.
We tried two reverse engineering methods to derive a vendor-independent identification approach: First, implementation-level analysis of known control-related software components (e.g., HMI GUI), and second, external observation during system operation.

Implementation Level Analysis of Known Software for Control and Monitoring. We initially attempted to statically analyze known binary files of the HMI software and communication drivers through disassembly and decompilation [6] to determine process connections and accessed data. However, this approach proved ineffective: many libraries and connections to other processes are dynamically loaded and established. Furthermore, the control path includes multiple applications, and relevant runtime data cannot be obtained through static analysis alone. We further attempted live debugging of the binaries, where the HMI is paused to analyze its memory. This failed due to time-critical software and watchdog mechanisms causing constant system restarts, preventing such invasive analysis. It was revealed here, deep binary analysis is more suitable to comprehend low level vendor-dependent mechanisms [3], why we changed our strategy to external reverse engineering focusing on Windows mechanisms.

External Observation of Software Behavior and File Accesses. In a second attempt, we employed process analyzers, API monitors, memory dumpers, and file monitoring tools. Following [12], we searched the process memory for variables to identify processes that potentially handle control and monitoring data. While this revealed processes containing physical process variables, their specific role and involvement remained unclear. We further observed inter-process interactions by monitoring APIs responsible for IPCs including which IPC types are involved. This proved challenging at first: monitoring too many functions at once caused software crashes. By selecting and minimizing monitored APIs, we obtained valuable information, such as to which other processes the communication driver connects for processing and forwarding control and monitoring data. Additionally, we monitored all file system accesses of the known and potentially relevant processes, identifying dynamic and static data relating to control and monitoring functions, such as configuration files, operational databases (containing physical process values), temporary files, and software binaries. However, processes to be monitored must be identified beforehand, as the output of all accessed files is very extensive making analysis infeasible.

Since the HMI GUI receives control and monitoring data via intermediate processes, tracing these identifies host-internal data paths and forensically relevant processes for control room operations. Connected processes can be identified by examining shared IPC interfaces: while most IPCs (e.g., shared memory) are queryable via Windows functions, others require API monitoring. These findings form the basis for the generalized approach presented in the following section.

5 Procedure Model for Identifying Control and Monitoring Functions and Their Data

We present an approach that identifies the control and monitoring path from HMI applications through communication drivers to PLCs, including both static and dynamic data.

5.1 Procedure to Identify Control and Monitoring Path with Data

The identification procedure operates on a running ICS and proceeds as follows: First, vendor installer contents are extracted to identify industrial software binaries ($\mathcal{S}_{OT}$) and distinguish them from general IT or OS software. Next, variables K belonging to the supervised plant site F are identified. These serve as search patterns to identify if a given process contains monitoring and control data. Afterwards, the first network process P_1^{net} on host H_1 communicating with the PLCs is identified, establishing the starting point for tracing control and monitoring data flows (Steps 3–4).

1. Extract vendor installers: $\mathcal{S}_{OT} \leftarrow$ OT software, $\mathcal{S}_{IT} \leftarrow$ IT software
2. Identify PLC process P_X with variables K belonging to Plant Site F
3. Identify L3 host H_1 connected to PLC via P_X
4. Identify network process P_1^{net} on H_1 communicating with P_X
5. Initialize CM-DATA set: $R \leftarrow \{P_X, P_1^{net}\}$
6. For current process $P_i \in R$:
   (a) Get IPC objects (handles): $\mathcal{I}_i = H(P_i)$
   (b) Find connected processes: $\mathcal{C}_i = \bigcup_{h \in \mathcal{H}_i} \text{Owners}(I) \setminus R$
   (c) Check if $\mathcal{C}_i$ contains physical process variables
      - $R \leftarrow R \cup \{P \in \mathcal{C}_i \mid \text{Vars}(P) \cap K \neq \emptyset\}$ (CM-relevant)
      - $U \leftarrow U \cup \{P \in \mathcal{C}_i \mid \text{Vars}(P) \cap K = \emptyset\}$ (UNKNOWN)
7. Repeat step 6 until reaching L3 boundary (network process to next host H_{i+1})
8. Continue for hosts $H_2, \ldots, H_n$ until HMI host H_{HMI} (belonging to Plant Site F) with HMI GUI process P_{GUI} reached
9. For each $P \in R$, monitor filesystem accesses $\mathcal{F}_P$:
   (a) If $f \in \mathcal{S}_{OT}$: mark as STATIC, else DYNAMIC
   (b) If known format: mark READABLE + format, else UNKNOWN/BINARY
10. Return R, U, file classifications

The control and monitoring process set is initialized with P_1^{net} (Step 5). For each process $P_i \in R$, inter-process handles and interfaces $\mathcal{I}_i$ are determined to identify which other processes are connected as connected processes share the same handles (Step 6a-b). These connected processes are then classified based on whether they contain variables or alarms/events from K: relevant processes are added to R, while others are assigned to the unknown set U (Step 6c). The absence of variables does not necessarily indicate irrelevance, as processes may perform auxiliary functions such as data conversion.

This iterative procedure continues until reaching processes connected to the next host H_{i+1}, where it resumes from Step 6 with the first process connected to the previous host. The iteration terminates upon reaching the HMI application (Steps 7–8). Finally, file accesses of all identified processes in R are monitored to distinguish static data (identified by comparison with $\mathcal{S}_{OT}$) from dynamic data such as logs and databases (Step 9). Additionally, each file is classified as human-readable (e.g., ASCII, known formats) or binary, indicating whether further processing is required for analysis.

5.2 Windows IPCs (handles)

As established in our initial analysis, identifying how processes are connected and exchange data requires examining inter-process communication mechanisms. Windows systems provide OS-level IPC mechanisms [11] as examined in Step 6 of our procedure. While these mechanisms can serve various purposes such as synchronization or locking without data transfer and are essential for IT forensics, we focus only on those used for data transfer or transmitting physical process variables and control data. These mechanisms are introduced in the following:

1. **Shared memory (sections)**: Memory-mapped files and sections accessible by multiple processes.
2. **Named Pipes**: Bidirectional communication channels for local or network IPC. They can also base on TCP/IP.
3. **DCOM/RPC:** Distributed Component Object Model is a legacy application protocol for IPC data exchange. It remains widely deployed in industrial applications due to backward compatibility requirements and historical prevalence [4]. DCOM operates on Windows systems via Remote Procedure Calls (RPC) and was observed on our ABB 800xA.
4. **Host based TCP/IP**: Loopback connections (localhost/127.0.0.1) for inter-process networking

Except for DCOM, all IPC mechanisms can be identified by examining current endpoints and connections (sections, named pipes, loopback TCP/IP) through OS-based handle analysis. DCOM/RPC cannot be traced this way due to its dynamic port allocation and connection model. Without observing IPC interactions, one can only determine whether a process has an RPC port, but not its connection partners. Instead, communication partners must be identified via API monitoring. While *CoCreateInstance(process_name)* (DCOM) can be captured, it only occurs during process initialization. Our analysis found it more effective to monitor the *I_RpcBindingInqLocalClientPID* function, which executes before each remote function call and has the target process as parameter, enabling capture of both pure RPC and DCOM partners without recording the entire process communication from initialization. In practice, API monitoring can be implemented using frameworks such as Frida, while handle enumeration can be performed via the Windows Native API (NtQuerySystemInformation with SystemHandleInformation) or the Volatility 3 Python API.

6 Applying Our Approach to Identify Monitoring and Control Paths and Data to a Siemens ICS

To evaluate our model's applicability and vendor independence, we apply it to identify the data paths between HMI application and communication driver on a Siemens PCS7 system. All L3 components run as virtual machines.

6.1 Architecture of System for Evaluation (PCS7)

Figure 2 presents the main components of our Siemens PCS7 testbed, which follows the Purdue architecture as well.

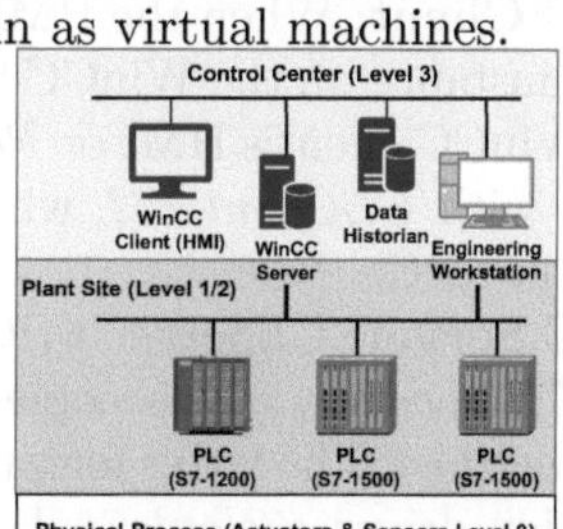

Fig. 2. Prototypical Siemens PCS7 testbed for evaluation.

The architecture is analogous to the ABB system, featuring two networks: one for HMIs and process control, and one for PLCs. A control server (*Windows Control Center, WinCC*) manages the HMIs, including their views and variables, while simultaneously handling communication with the PLCs, translating data between the HMIs (*WinCC Client*). In theory these are located in the control room, from which operators monitor and control the physical process. A single *WinCC Server* can centrally manage multiple plant sites (with different protocols) supporting centralized control and monitoring.

6.2 Identified Monitoring and Control Paths

Figure 3 shows the control and monitoring paths identified through our approach. All green boxes include control and monitoring data. As expected the same Windows IPCs as in ABB 800xA are used: Ethernet, DCOM, named pipe, and shared memory sections.

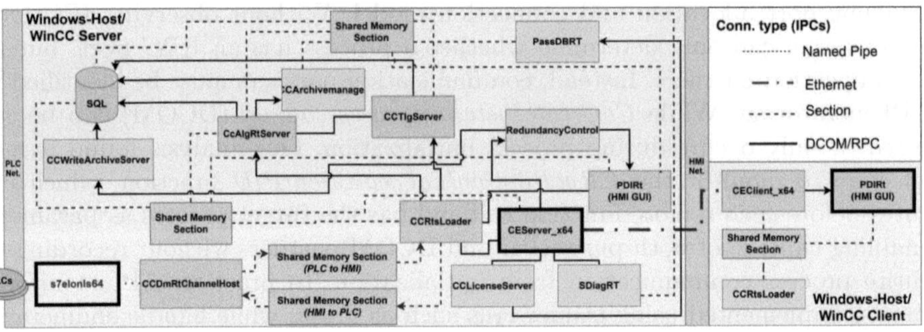

Fig. 3. Identified paths and processes responsible for control and monitoring data.

Identified Paths and Control/Monitoring Data Flow. To validate that these processes are related to control and monitoring, we monitored process APIs while performing HMI interactions (variable on/off) in addition to IPCs.

WinCC Client: When the HMI GUI (*PdlRt*) starts on the client, it also starts a *PdlRt* instance on the WinCC Server for redundancy. When a button is pushed on the WinCC client's HMI or *PdlRt*, it writes variables and values to the shared memory with *CEClient_x64*, which forwards them to *CEServer* on the WinCC Server via RPC.

WinCC Server: *CEServer* writes the received data to a shared memory section with *CCRtsLoader*, which copies the data to its shared memory with *CCDmRtChannelHost*. The latter forwards variables and values to the PLC via the communication driver *s7elonls64*. For the reverse direction (PLC value changes), a separate shared memory exists, both sections serving as ring buffers. Simultaneously, *CCRtsLoader* communicates with the local redundant HMI GUI (*PdlRt*) instance on the WinCC server to keep it synchronized with the WinCC Client HMI. Additional components (*CCTlServer*, *CcAlgRtServer*) handle archiving and logging of physical process data to the SQL database. If logging is disabled in WinCC settings (uncommon in real-world operations), the associated SQL-connected processes are not running.

The identified paths confirm that the same IPC mechanisms observed in the ABB 800xA are present in PCS7. Changes to their static data indicate potential manipulation, and attackers can spoof or intercept messages by targeting these processes directly. Moreover, monitoring these processes captures all control and monitoring activity, providing a complete record of control room operations regardless of where an attack occurred.

7 Evaluating the Models Applicability

To evaluate our approach, we discuss the identified data sources by our approach and their forensic relevance for investigating safety-critical attacks. We further assess how the approach allowed for systematic identification and address limitations that emerged during application.

7.1 Identified Processes and Software Complexity

The goal of our approach is to identify processes, paths, and their associated data for control room-critical attacks. Table 2 gives an overview of all distinct processes. Distinct means that multiple instances of the same process are counted as one. We provide the total number of running processes, specifying how many belong to PCS7 or the industrial software and how many to the Windows operating system. We apply the same categorization to network processes, with the final column indicating in how many process' memory we found physical process variables or alarms/events relevant for control and monitoring.

Table 2. Amount of processes divided in subcategories.

Processes/Component	Sum	PCS7	Windows	Windows Network	PCS7 Network	Cont./Monit. processes
WinCC Server	112	48	52	12	18	3
WinCC Client (HMI)	94	26	42	8	13	2

Analysis of the numbers indicates that fewer processes and connections require additional consideration. This allows one to concentrate efforts on the identified processes as reverse-engineering the relevant binaries to obtain vendor-specific details about implementation and communication. Furthermore, existing approaches as [9,10] can be now applied on the identified connections of network processes, which identify the fields of proprietary industrial protocols.

Table 3. Identified control room related data sources by file and network monitoring the identified software processes.

Located	Data Source	Extractable Artifacts	Relevant Information for Control Monitoring
HD	Industrial software	Static software components (*.exes, *.dll, *.ini) Config files;	Modification of control and monitoring software; Configured connections/ports/hosts;
HD	Operational Databases	Alarms/events; Variables/process image; System parameters;	Physical process events, commands, ph. process state; Set parameters, system configuration;
HD	Operational logs	Alarms/events, Command/parameter changes;	Physical process events, (control room) interactions;
HD	App./System logs	Software activity/diagnostics;	Software events (e.g. startup, crashes), error messages, system performance, integrity;
RAM	System memory	Software process state; Process connections/sockets; Registries/logon sessions;	Launched/running/terminated processes; Network processes; Existing/logged in users;
RAM	Process memory	Alarms/events,Variables/phys. process image; Listening/Open ports; Loaded files; Socket-tables; handle tables;	Indicated physical process state; Events; Active (control/monitoring) connections; Faceplate/supervised plant site; Loaded projects; Internal-process communication and interfaces; Accessed and shared files;
Network	Traffic	Connections/used ports; industrial protocol;	Active connections incl. payloads; Used protocol for exch. control and monitoring data;
Network	Protocol data;	Variables/reporting messages, payloads, commands, opened faceplates;	Transmitted physical process image, commands, control interaction/activities;

7.2 Identified Data of Identified Processes and Paths

When applying the last steps of our approach to the PCS7 system, we identified the control and monitoring data sources included in Table 3. We identified process data across three locations: persistent storage (HD), volatile memory (RAM), and network traffic. For each data source, the table specifies extractable artifacts and their forensic relevance regarding safety critical attacks. *Industrial software* binaries and *application logs* reveal manipulations, modifications, and software events (starts, crashes). Static components (e.g., executables or DLLs) are expected to remain unchanged during operation; deviations from a baseline established in Step 9 indicate potential manipulation of control room functions, as in the Stuxnet attack [2], where a DLL was manipulated to show different values to the operator than those transmitted by the PLCs. *Operational databases* and *operational logs* contain alarms/events, commands, and the process image (physical process state). The *process image* is transmitted by PLCs via network and displayed in HMI *process memory*, enabling consistency verification while also identifying it in the mentioned processes of Fig. 3. Consistency across this data is essential, as discrepancies can indicate spoofed or blocked messages (e.g. the attacker modifies only variables on the HMI, while variables of the communication driver to the PLC might include the actual values).

7.3 Advantages and Disadvantages

Finally, we discuss the benefits and drawbacks we determined during the application of our approach.

Advantages

- **Vendor independent:** We developed our approach through observation of an ABB 800xA system and successfully applied the procedure to a Siemens PCS7 system. As long as the ICS is Windows-based, our approach is vendor-independent.
- **Discovery of undocumented components, connections, and their purpose:** Examining Fig. 3, we identified multiple components involved in control and monitoring. Network processes related to control and monitoring such as *CEServer/CEClient* are particularly valuable when analyzing ICS traffic or implementing measures. Investigators and operators can determine the purpose of each process, including network connections and their relevance for control room operations. Through file monitoring of identified processes, we also discovered additional forensically relevant static and dynamic data, such as cached or temporary files of the HMI.
- **Suitable for production environments:** During our identification no crashes occurred and except for DCOM/RPC all interfaces were determined passively using existing Windows-Information. The only invasive operation is API hooking to identify DCOM/RPC partners while limited to monitoring of two single APIs or functions to keep interference at minimum.

Disadvantages

- **Only usable during runtime:** Since our approach examines active handles at the operating system level, it can only be applied during runtime of the system. The handles can be identified using existing windows-mechanisms.
- **Privileged rights required:** Examining available IPC handles of a process and monitoring APIs for DCOM initialization both require privileged rights. This can be challenging for production HMIs where user rights might be kept to a minimum. For safety reasons, we recommend applying the approach on virtual machines (VM) cloned from the production system during the strategic preparation.
- **Windows-only:** Our approach exclusively covers Windows systems and does not consider directly Unix/Linux-based ICS.

While our procedure represents a best-effort identification that does not formally guarantee completeness, we were able to identify control and monitoring processes relevant to control room operations. The approach was established on an ABB 800xA system and applied to a Siemens PCS7. By further inspecting these processes for file accesses, additional forensically relevant static and dynamic data was identified. The resulting data path map and identified processes can be integrated into security monitoring or forensic acquisition measures, or used as a basis for further reverse engineering of binaries. However, the approach is limited to Windows-based ICS due to its dependence on Windows-specific IPC mechanisms, and requires the system to be running at the time of application.

8 Future Work

Future work includes integrating our approach in existing analysis frameworks such as [5], where we present an analysis model to correlate or analyze specific ICS data (as control and monitoring artefacts) to determine MITRE TTPs, once specific data has been acquired or is available. This also includes executing control-room-relevant attacks and utilizing the data from the components identified in this work to trace and attribute such attacks. In this context detailed attack trace analysis and data correlation are directions for future work, as this work focused on establishing and demonstrating vendor-independent path and data identification for control room operations. Furthermore, the approach should be extended to Linux-based ICS environments by incorporating Linux-specific IPC mechanisms such as D-Bus or Unix sockets.

9 Conclusion

We presented a novel vendor-independent approach for identifying control and monitoring paths, data, processes, and connections in Windows-based ICS environments. By systematically tracing IPCs from PLCs to control-room HMIs, our method identifies relevant processes, connections, and software components without requiring vendor documentation or access to source code. Developed through analysis of an ABB 800xA system and validated on a Siemens PCS7 system, the approach demonstrates practical applicability across vendors. The identified components and data sources enable forensic investigators and operators to determine which processes and connections must be considered for forensic acquisition, security monitoring, and assessment of safety-critical attacks on control room operations. However, the approach currently requires the system to be running and represents a best-effort identification based on observable Windows IPC mechanisms rather than a formal completeness proof.

Disclosure of Interests. The authors have no competing interests to declare that are relevant to the content of this article.

References

1. Cook, M., Marnerides, A., Johnson, C., Pezaros, D.: A survey on industrial control system digital forensics: Challenges, advances and future directions. IEEE Commun. Surv. Tutorials **25**(3), 1705–1747 (2023)
2. Denning, D.: Stuxnet: What has changed? Fut. Internet **4**, 672–687 (2012)
3. Finck, C., Dohrmann, T.: A decade after stuxnet: How siemens s7 is still an attacker's heaven. Tech. rep., ENLYZE GmbH, Cologne, Germany (2023). https://colinfinck.de/files/EU-23-Finck-A-Decade-After-Stuxnet-How-Siemens-S7-is-Still-an-Attackers-Heaven-wp.pdf
4. González, I., Calderón, A., Figueiredo, J., Sousa, J.: A literature survey on open platform communications (opc) applied to advanced industrial environments. Electronics **8**, 510 (2019)

5. Karagiozidis, A., Gergeleit, M.: A forensic analysis framework for industrial control systems, pp. 780–790 (2025)
6. Keliris, A., Maniatakos, M.: ICSREF: A framework for automated reverse engineering of industrial control systems binaries. In: Network and Distributed System Security Symposium (NDSS) (2019)
7. Marcella, A.J.: Cyber Forensics: Examining Emerging and Hybrid Technologies, 5, vol. 4, chap. 6, pp. 211–267. CRC Press, Boca Raton, USA, 1 edn. (2021)
8. MITRE: Mitre att&ck for industrial control systems (2020). https://attack.mitre.org/tactics/ics/
9. Qin, Z., et al.: Reverse engineering industrial protocols driven by control fields, pp. 2408–2417 (2024)
10. Qu, Y., et. al, F.: Icepre: Ics protocol reverse engineering via data-driven concolic execution. Proc. ACM Softw. Eng. **2** (2025)
11. Russinovich, M.E., Solomon, D.A., Ionescu, A.: Windows Internals, Part 1: System Architecture, Processes, Threads, Memory Management, and More, 7th edn. Microsoft Press (2017)
12. Salmon, L., Baggili, I.: Out of control: Igniting scada investigations with an hmi forensics framework and the ignition forensics artifact carving tool (ifact) (2025)
13. Yau, K., Chow, K., Yiu, S.M.: A Forensic Logging System for Siemens Programmable Logic Controllers, pp. 331–349 (2018)

Remote Attestation for Secure Industrial Device Onboarding

Volodymyr Trykoz[1(✉)], Björn Leander[2], Saad Mubeen[1], and Mohammad Ashjaei[1]

[1] Mälardalen University, Västerås, Sweden
{volodymyr.trykoz,saad.mubeen,mohammad.ashjaei}@mdu.se
[2] ABB Process Automation, Process Control Platform, ABB, Västerås, Sweden
bjorn.leander@se.abb.com

Abstract. Security in industrial automation systems becomes more complicated as industry moves toward modern and flexible solutions such as modular automation and increased vendor diversity. Attack surfaces grow as more network interfaces are added and more devices become exposed to the internet and cloud. The rapid introduction and replacement of devices make it difficult to track vulnerabilities and ensure nothing malicious enters the system. Hence, trusting a device prior to its onboarding becomes critical for ensuring the overall security and safety of the system. This paper proposes a remote attestation mechanism for the onboarding process which follows the Remote ATtestation ProcedureS proposals. In order to provide evidence for the applicability of the proposed process, we provide a reference implementation. Moreover, we provide a security analysis of the proposed solution in the form of a threat model.

1 Introduction

As Industry 4.0 [3] becomes more and more mainstream, complex industrial systems are required to be flexible, scalable, and adaptable. This brings new challenges from the security perspective, as more attack surfaces open up due to support for flexibility and adaptability. The question of trust becomes imperative for secure operations of industrial automation systems, many of which may operate in safety-critical environments [4]. The systems must support flexibility by allowing users to quickly introduce new entities without compromising security. Without disrupting essential functionality, the system must be able to onboard a new device, provided it is deemed trustworthy. Modular automation [1] is an automation system design strategy allowing such dynamic behavior to support, e.g., scaling without interrupting operations.

In industrial automation systems, the Open Platform Communications Unified Architecture (OPC UA) is a widely used communication protocol in industrial automation and Industry 4.0 [21]. It has extensive security considerations and a part dedicated to secure onboarding of automation devices [10]. However,

© The Author(s), under exclusive license to Springer Nature Switzerland AG 2027
E. Troubitsyna et al. (Eds.): SAFECOMP 2026, LNCS 16868, pp. 98–113, 2027.
https://doi.org/10.1007/978-3-032-34867-8_7

the OPC UA onboarding process focuses on verifying identity, and does not provide any mechanism for validating device state. This limits the extent to which the onboarded device can be trusted. Even when the identity of the device is verified, its integrity cannot be fully assured. Consequently, it may be compromised by malware or firmware with critical vulnerabilities.

In the context of flexible industrial automation systems, remote attestation can be used for online state verification. During remote attestation, a device collects measurements about its software and configuration, and sends this evidence to a remote entity that can appraise it. Based on these measurements, the verifier determines whether the device is running permitted software or if deviations are present. The measurements are collected in a way that makes tampering extremely difficult without compromising the hardware root of trust, and they are cryptographically protected to ensure authenticity and integrity. In this regard, the Remote ATtestation ProcedureS (RATS) [5] working group defines a vendor-independent architecture, interfaces, and message types to enable secure remote attestation and related operations. It standardizes key actors and various processes used for attestation flows [7].

Previous works on integrating remote attestation with OPC UA focuses on isolated attestation and do not consider cross-protocol scenarios or the onboarding process [8,12]. In this paper, we provide a comprehensive OPC UA mechanism for performing remote attestation during secure onboarding, which hardens the system against compromised devices. The paper provides the following contributions.

1. It proposes an attestation-enhanced OPC UA onboarding process using the models and workflows proposed by RATS.
2. It presents a reference implementation to provide evidence on how this integration can be performed.
3. It provides a security analysis, based on threat modeling using the STRIDE classification, that identifies and evaluates relevant threats, upon which we subsequently propose mitigation strategies.

2 Related Work

There are very few works to date that have focused on integrating remote attestation into OPC UA. Birnstill *et al.* [8] investigate a basic model for retrieving Trusted Platform Module (TPM) quotes through OPC UA and verifying them. Their approach focuses exclusively on evidence transmission, and does not consider the broader onboarding workflow.

Another OPC UA–related attestation effort is presented by Juhola *et al.* [12]. They propose a simple reference attestation sequence in which a server provides evidence in response to a nonce-based request. The server aggregates evidence by interacting with its subsystems. Similar to [8], their work does not position attestation within the OPC UA onboarding process. They also do not adopt the RATS approaches and workflows, although they mention it as future work.

Matischek *et al.* [16] propose integrating a hardware-based secure element into OPC UA to offload certain cryptographic operations. Although their work does not address onboarding or remote attestation directly, it demonstrates how secure elements serve as essential building blocks for strong attestation guarantees, highlighting a complementary direction for strengthening OPC UA device trustworthiness.

With respect to reference implementations based on the RATS architecture, Usman and Asplund [20] provide a prototype implementation that integrates the RATS actors. Although the overall direction aligns with ours, their work does not address the OPC UA onboarding model, which is the focus of this paper.

In the context of OPC UA onboarding, Kohnhäuser *et al.* [14], later expanded in [17], analyze secure onboarding using DevIDs and hardware security modules. They focus on identity verification by leveraging DevIDs and associated cryptographic hardware. Our work builds further by assuming the identity check has already succeeded and extending the onboarding workflow with state verification via remote attestation.

3 Remote Attestation Process Proposal

The proposed onboarding model builds on two elements of trustworthiness: identity and state. To present the solution, we first describe the trust model and show how the current OPC UA onboarding process fits into it.

3.1 Secure Onboarding and Device Trustworthiness

Before a new device is introduced into an operational environment, its *trustworthiness* must first be established. In this work, two aspects of trust establishment is considered:

- *Identity*—the device must provide credentials for its identity. This authenticates the device, and protects against man in the middle attacks.
- *State*—the software on the device must be in an updated and uncompromised condition. This includes confirming that its software stack has not been tampered with (e.g., via rootkits), it is not running vulnerable or outdated components, and updates have been applied successfully.

When both aspects can be verified, the device is considered sufficiently *trustworthy* to join the system. At this point, the device can be provisioned with the necessary operational credentials, such as certificates, and with information regarding which entities it should trust. The onboarding workflow should be as streamlined as possible, with identity and state verification automated, remotely executable, and requiring minimal physical interaction with the device.

Using Device Identifiers (DevIDs) [11] is one mechanism for device identity verification. They are manufacturer-issued credentials that uniquely identify a specific device. A device owner or operator can authenticate these credentials by obtaining the corresponding verification material from the original manufacturer and comparing them to the credentials embedded inside the device. The specific validation procedures are outside the scope of this work; however, introducing the DevID concept is necessary for establishing the broader context in which our approach operates.

Remote attestation can be used for state verification, which is the process of verifying the integrity of a device or system (referred to as the *attester*) from a remote location. The attester generates *measurements* of its configuration (such as its boot software stack or other security-relevant components) and submits these measurements to a verifying entity known as the *verifier* [2]. The collected measurements constitute *evidence* that the system is in a specific state. The verifier compares the received evidence to a set of trusted reference values in order to determine whether the attester's configuration has been altered. Two common models exist for remote attestation: the *background-check* model, shown in Fig. 1(a) and the *passport* model, illustrated in Fig. 1(b) [18].

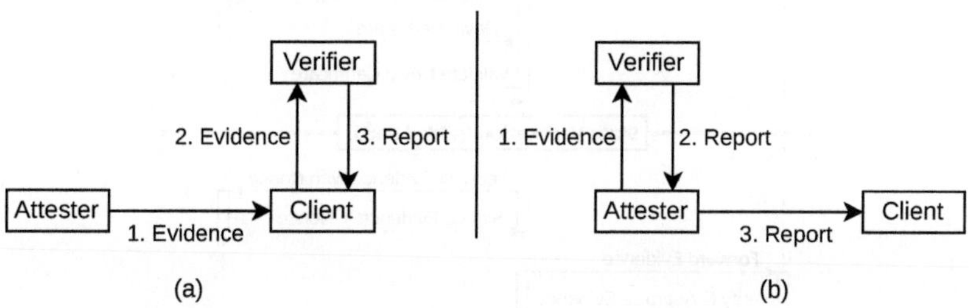

Fig. 1. Background check (a) and passport (b) models of remote attestation

In the background-check model, the attester sends measurements to the client (relying party), which forwards them to the verifier. The client relies on the attestation results to decide how to proceed. The verifier evaluates the evidence, produces an attestation result, and returns it to the client. In the passport model, the attester sends its measurements directly to the verifier and forwards the result to the client.

3.2 Aligning the OPC UA Onboarding Workflow with the Device Trustworthiness Model

OPC UA Part 21 [10] is focused on describing a secure onboarding model for new devices. Global Discovery Server (GDS) is one of the core entities in this process. Among its many roles, the GDS functions as a certificate authority, providing CA-signed certificates and trust lists to support PKI-based trust.

The current OPC UA onboarding process provides device authenticity, but lacks state validation. This means that although we can be reasonably confident that we are onboarding the correct device, we cannot say anything about the software or firmware it is running, whether it has been altered or corrupted, or whether it is running versions with known vulnerabilities. As a result, we risk provisioning a compromised device, which opens up a new attack surface for adversaries attempting to get entry into the system. Therefore, it does not fulfill the trust establishment requirements described in previous sections.

To address this limitation, we extend the procedure with a RATS-based attestation workflow to improve the trustworthiness assessment. The integration of the RATS interaction model into the onboarding process is illustrated in Fig. 2. The original steps remain unchanged, but a state-verification step is inserted into the middle of the workflow.

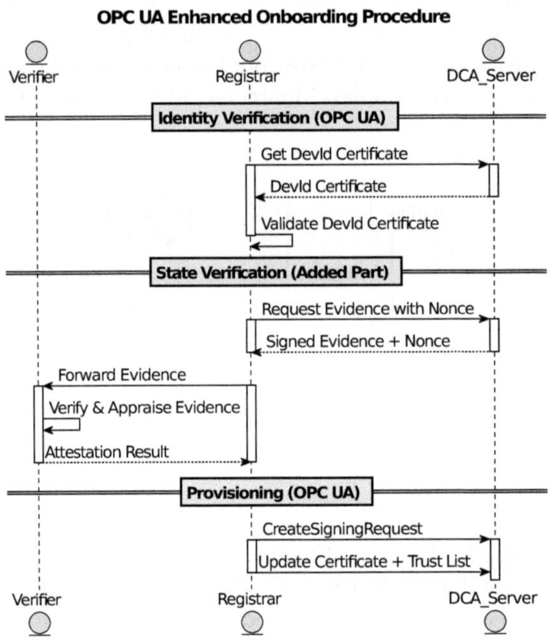

Fig. 2. Sequence diagram for the enhanced OPC UA onboarding model. The enhanced part is marked, and requires a new entity, the *verifier*, to be part of the onboarding.

The procedure for *state* verification is based on the RATS reference interaction models for remote attestation [7]. The registrar requests evidence that can be used to asses the state of the device. It sends a nonce that must be embedded into the evidence to prevent replay attacks. The device incorporates the nonce into the evidence, signs it, and returns it to the registrar. The registrar forwards the evidence to a verifier, which cryptographically validates its legitimacy by checking the signature and the included nonce. If the evidence is cryptographically valid, the verifier proceeds with appraisal and produces a report summarizing the device's state. This report is sent back to the registrar, which performs its own appraisal and decides whether to continue with provisioning. If the results indicate the device is in a corrupted or unhealthy state, the onboarding process is aborted. If the appraisal is successful, provisioning proceeds.

This workflow allows verification of both identity and state aspects of the trustworthiness, ensuring that only trusted devices are provisioned.

3.3 Interaction Model for Attested Onboarding

In order to make the interaction work, several adjustments are necessary to the core OPC UA entities in terms of their responsibilities and interactions. Figure 3 presents the high-level attestation system architecture, showing the involved entities and their interactions. The interactions are partially based on RATS interaction models and partially on the OPC UA onboarding model. The background-check model is used, as it fits the OPC UA worfklow, where there is an administrative client (Registrar) responsible for verifying and provisioning a device. The proposed approach naturally extends the client-centric workflow to also request and verify device evidence via the verifier.

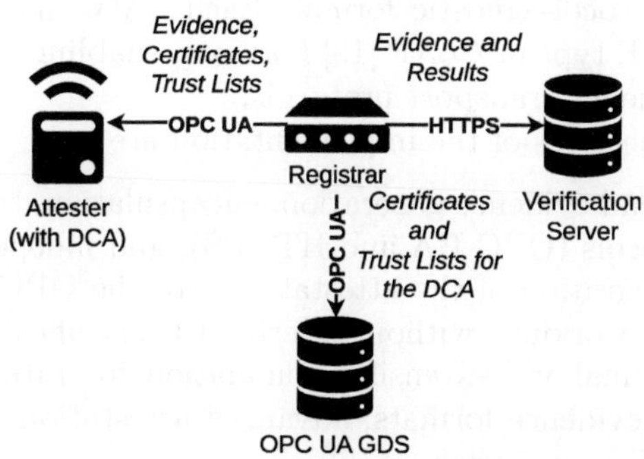

Fig. 3. Onboarding interaction model.

The following entities are part of the model.

- **Registrar** – initiates all interactions and acts as the central coordinator, requesting attestation evidence, forwarding it to the Verifier, and evaluating the result to ensure only devices with acceptable evidence receive operational certificates and trust lists. It maps to the Relying Party role in RATS.
- **Attester (DCA Server)** – generates verifiable evidence demonstrating its runtime integrity and allows the Registrar to update its operational certificate and trust list upon successful verification. It maps to the Attester role in RATS.
- **Verification Server** – evaluates evidence against device-specific appraisal policies and provides a verdict to assist the Registrar in determining whether the DCA Server can be safely provisioned. It maps to the Verifier role in RATS.

- **OPC UA GDS** – **G**lobal **D**iscovery **S**erver acts as the Certificate Authority (CA). It issues operational certificates and distributes trust lists post-attestation, enabling approved devices to join the network.

4 Reference Implementation

The reference implementation demonstrates the enhanced onboarding workflow. It is written in C# and available on GitHub[1]. Figure 3 shows the system architecture. Since our work targets OPC UA, all interactions between the Registrar, Attester, and GDS use OPC UA channels to align with the onboarding model [10].

To demonstrate a multi-protocol scenario, the Registrar communicates with the Verification Server over HTTPS. This highlights how the Registrar can interact with a Verifier that exposes an HTTPS endpoint for submitting evidence and receiving appraisal results. The OPC UA Foundation .NET stack is used for the OPC UA components, while ASP.NET is used for the HTTPS endpoint.

Conceptual Message Wrappers (CMWs [6]) transport attestation evidence and results in a protocol-agnostic format. Each CMW includes a media type identifier (e.g., MIME type or CoAP [19] Format), enabling interoperable decoding across heterogeneous transport protocols.

The primary objectives of the implementation are to:

- Validate end-to-end evidence generation, encapsulation, transport across heterogeneous protocols (OPC UA and HTTPS), and independent verification.
- Demonstrate integration of the attestation into the OPC UA push management onboarding workflow without altering the specification.
- Providing a minimal yet extensible foundation for future experimentation with alternative evidence formats, advanced attestation concepts, and more complex onboarding scenarios.

The prototype supports two different boot configurations for the attester. One configuration matches the reference values stored on the verification server, while the other intentionally deviates from them. This allows to demonstrate both a successful attestation flow and an unsuccessful one.

5 Threat Analysis

We want to ensure that our system does not expose any vulnerabilities that could allow attestation to be bypassed. We also want the system to remain available, avoid disclosing unnecessary information about itself, and maintain proper visibility into its own operations. We therefore perform threat modeling using the STRIDE [9] approach. STRIDE categorizes threats as **S**poofing, **T**ampering, **R**epudiation, **I**nformation Disclosure, **D**enial of Service, and **E**levation of Privilege. This approach enables a systematic security analysis of the system from the perspective of each threat category.

[1] https://github.com/vovatrykoz/UnifiedAttestation.

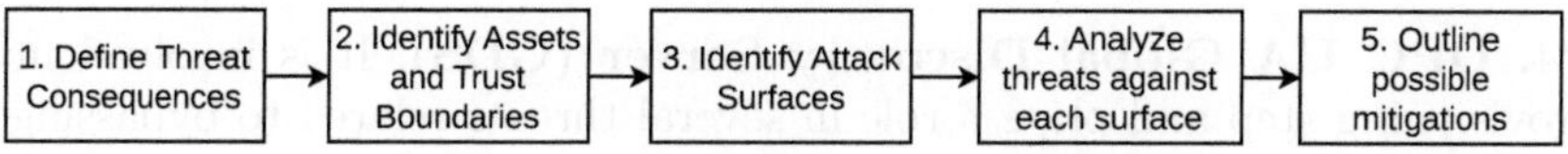

Fig. 4. The threat analysis methodology.

The methodology is inspired by [13,15] and summarized in Fig. 4. First the threat consequences are identified, then the system's assets and their trust boundaries are enumerated, followed by identification of the attack surfaces through which an attacker may attempt to achieve its goals. Each attack surface is analyzed according to STRIDE, and possible mitigations are outlined.

5.1 Threat Consequences

Each threat consequence is related to the attestation-enhanced onboarding workflow and may have severity on the scale: *Low*, *Moderate*, *High*, and *Severe*.

TC1 A device is onboarded while bypassing attestation. ⇒ *Severe*

TC2 Denial-of-Service (DoS) attack prevents a legitimate device from being onboarded. Threat to the availability of the device. ⇒ *Moderate*

TC3 DoS attack prevents the attestation system from performing its role. Threat to the availability of the onboarding system. ⇒ *High*

TC4 Attestation evidence or appraisal results are revealed to unauthorized parties. Threat to the confidentiality of attestation information. ⇒ *Low-Moderate*

TC5 The system denies having performed actions that actually happened or claims to have performed actions that never took place. Threat to non-repudiation of the system. ⇒ *Low-Moderate*

5.2 Assets and Trust Boundaries

The system assets along with their associated data-flows and trust boundaries are illustrated in Fig. 5. All the network connections and the Attester are seen as potential attack surfaces. We assume certain entities are already securely onboarded and trusted. Potential compromises of these entities are out of scope, as this work focuses on device onboarding.

- **P1: Attester**. Device to be onboarded. It interacts only with the Registrar over **DF1**.
- **P2: Registrar**. The Registrar is assumed to be uncompromised and trusted to perform its role as the attestation orchestrator. It therefore lies within a trust boundary. It initiates all interactions, and all other assets interact with it in response to its requests.
- **P3: Verification Server** is assumed to correctly perform evidence appraisal; therefore, it lies within a trust boundary. It interacts with the Registrar over data flow **DF2**

- **P4: OPC UA Global Discovery Server (GDS).** It is involved in the provisioning step and plays a role in several threats related to bypassing the attestation workflow. It interacts with the Registrar (**DF3**).

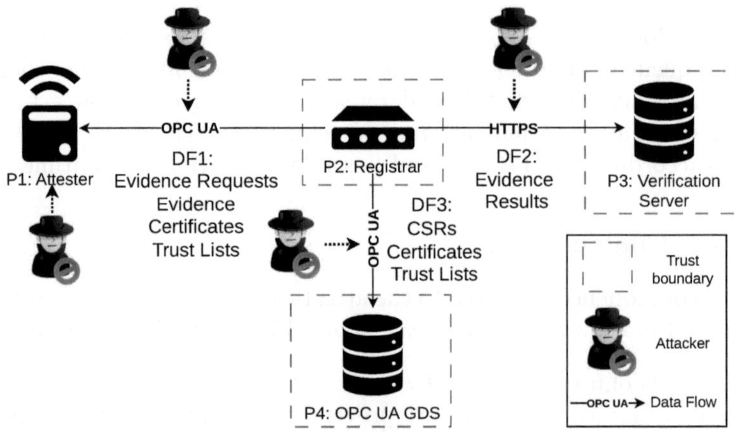

Fig. 5. Interaction diagram showing data flows and entities that can be manipulated by attackers under our threat model

5.3 Threats and Mitigations Enumeration

Threats are enumerated based on the STRIDE classification model, considering the defined attack surfaces, i.e., **P1** as well data-flows **DF1–3**. Additionally mitigations (M) are considered for each of the threats. The resulting threat analysis is summarized in Table 1, excluding non-applicable categories.

Attester Attack Surface (P1). An attacker may try to spoof a device's identity to appear authentic and be onboarded (Threat 1), leading to TC1. To mitigate this, identity verification must occur before attestation (M1). In OPC UA, the registrar verifies DevID certificates against reference credentials and rejects mismatches [10]. Because DevID private keys are stored in hardware security modules (e.g., TPMs), forging an identity is infeasible unless an attacker physically steals a genuine device. Requiring attestation evidence to be signed with a dedicated attestation key (M2) further increases the difficulty of compromising the device.

If an attacker obtains a legitimate device provisioned with an DevID and attestation keys, they may attempt to submit falsified evidence that does not reflect the software actually running on the device (Threat 2), again targeting TC1. This is a critical threat: an attacker in control of the device could manipulate the attestation process to measure or report only uncompromised components while omitting compromised ones, making the evidence appear healthy

Table 1. Threat analysis overview

Surface	Category	ID	Description	Consequences	Mitigations
P1: Attester	S	1	Attacker spoofs a legitimate device	TC1	M1, M2
	T	2	Attacker sends false evidence	TC1	M3, M4, M5
DF1: Registrar to Attester	S	3	Attacker spoofs the registrar	TC4	M6, M7, M8
	S	4	Attacker spoofs the attester	TC1	M1, M2, M9
	T	5	Attacker falsifies evidence	TC1, TC2	M2, M9
	T	6	Attacker replays older evidence	TC1, TC2	M10
	T	7	Attacker appends their certificate to a legitimate trust list	TC1, TC2	M9
	T	8	Attacker performs a relay attack	TC1	M1, M9
	R	9	Registrar fails to log events	TC5	M11
	I	10	Attacker intercepts evidence	TC4	M9
	D	11	Network-level DoS via packet dropping or oversized evidence payloads	TC2, TC3	M12, M13
DF2: Registrar to Verifier	S	12	Attacker spoofs the verifier	TC1, TC2	M9
	S	13	Attacker spoofs the registrar	TC4	M14
	T	14	Attacker falsifies evidence	TC1, TC2	M2, M9
	T	15	Attacker replays older evidence	TC1, TC2	M10
	T	16	Attacker replaces actual results with falsified results	TC1, TC2	M9, M15
	T	17	Attacker replays older results	TC1, TC2	M16
	R	18	Registrar or verifier fails to log events	TC15	M11
	I	19	Attacker intercepts evidence or results	TC6	M9
	D	20	Network-level DoS via packet dropping or oversized evidence/result payloads	TC2, TC3	M12, M13
	E	21	Unauthorized device sends evidence for verification	TC1, TC4	M14
DF3: Registrar to GDS	S	22	Attacker spoofs the registrar	TC1	M9
	S	23	GDS is spoofed	TC1	M9
	T	24	Attacker attempts to append their certificate to a legitimate trust list	TC1	M9
	R	25	Registrar or GDS fails to log events	TC5	M11
	I	26	Attacker intercepts certificates or trust lists	TC4	M9
	D	27	Network-level DoS via packet dropping, request flooding, or oversized payloads	TC2, TC3	M12, M13
	E	28	Unauthorized device requests certificates and trust lists from GDS	TC1	M14

despite a compromise. If the attacker accesses the device's signing credentials, they can generate evidence that is cryptographically valid but still misleading.

A hardware security module (HSM) (M3), such as a TPM, mitigates the risk of an attacker altering measurements, tampering with the attestation process, or accessing device credentials, thereby also strengthening M1 and M2. Additional hardening can be achieved through secure boot (M4) and runtime measurements (M5). Secure boot prevents unauthorized components from executing during startup, while runtime measurements monitor the device after boot to detect later modifications. A residual risk remains that a highly skilled attacker could bypass these protections, but doing so would require significant time and expertise and is likely to trigger at least one of the defenses.

Attester–Registrar Attack Surface (DF1). An attacker may attempt to spoof the Registrar to request attestation data from the Attester (Threat 3), targeting TC4 to learn more about the Attester/attestation process. This is mitigated by having the Attester keep the attestation endpoint hidden until it receives a preliminary trust list. After verifying the Attester's identity, the Registrar provides a limited trust list (M6), e.g., containing only the Registrar's certificate. The Attester then exposes its attestation endpoint but interacts only with entities included in this list. A residual risk remains that an attacker may discover and attempt to follow this sequence. However, doing so would also exclude the attacker from onboarding, which is a more severe consequence, while obtaining any attestation data would pose only a low to moderate risk, as it typically consists of hashes that require external metadata to be meaningful.

A more serious threat from a spoofed Registrar is that an attacker could attempt to inject its own certificate into the device's trust list to gain backdoor access. This can be mitigated by forcing a connection restart (M7) whenever the trust list is updated, making it practically infeasible for an attacker to capture the legitimate list, append its own certificate, and push an updated version in time. Limiting the Attester to a single simultaneous connection during onboarding (M8) further prevents such manipulation. Once onboarding is complete, the device trusts only legitimate system entities, reducing the attacker's opportunities. A residual risk remains that an attacker could provision a device by connecting first, but it would still lack access unless it also obtained a legitimate trust list from a genuine GDS and inserted its own certificate. Mitigations for this case are discussed in the DF3 section.

The reverse is also possible: an attacker may spoof the Attester to gain illegitimate access to the system (Threat 4), targeting TC1. This is mitigated through identity verification (M1) and the use of a secure, authenticated channel that ensures message-sender authenticity (M9). Additionally, attestation evidence should be signed using an attestation key (M2) that preferably resides in a HSM, providing an additional layer of identity protection. Combined, these measures strengthen identity trustworthiness. The residual risk is low, as an attacker would need to break the OPC UA security stack or compromise a TPM-protected key, which is considered practically infeasible.

An attacker may also replace legitimate evidence with falsified evidence in transit (Threat 5) or intercept and replay previously transmitted evidence (Threat 6). This may lead to TC1 if corrupted evidence is replaced with healthy evidence, or to TC2 if healthy evidence is replaced with corrupted-looking evidence. To prevent this, the evidence must be signed (M2), and the signed portion must include a nonce (M10). Signing ensures the evidence cannot be tampered with, while nonces prevent replay. Combined with M1, these measures assure that evidence cannot be forged over in transit over the network without physical access to a legitimate device.

An attacker may attempt to append their own certificate to a trust list in transit (Threat 7), achieving TC1. This is mitigated by using secure authenticated channels (M9), which provide message integrity and make tampering difficult. The residual risk considerations are similar to Threats 4 and 5.

An attacker may also perform a relay attack on the communication link (Threat 8), potentially inserting themselves into the system and bypassing attestation (TC1). DevID certificates help mitigate this (M1) by preventing unauthorized entities from initiating communication with the Registrar. However, a relay may still be possible through advanced network manipulation, allowing the attacker to insert themselves into an existing channel. This risk is further mitigated by using OPC UA secure authenticated channels (M9), which protect against meaningful message manipulation, such as tampering or replay, even if the attacker succeeds in relaying traffic. The residual risk is low and would require exploitable vulnerabilities in the OPC UA security stack implementation, which we consider unlikely.

If the Registrar does not keep logs of connections, such as device addresses and timestamps, troubleshooting becomes difficult (Threat 9), leading to TC5. This is easily mitigated by requiring core entities to maintain logs and, if needed, forward them to a centralized auditing system (M11). With the security protections provided by OPC UA, attacks on such logging systems become infeasible.

If the channel is unprotected, an attacker may intercept evidence in transit without spoofing the Registrar (Threat 10), resulting in unauthorized information disclosure (TC3). Using secure authenticated channels (M9) prevents this, as encryption ensures that message contents cannot be read by an attacker. As before, the residual risk is minimal and would require breaking the OPC UA security stack.

An attacker may drop packets to disrupt onboarding or inject oversized payloads to crash the Registrar or Attester (Threat 11), targeting TC2. Mitigations include message size limits (M12) and network hardening (e.g., firewalls, M13), which reduce the impact of DoS attacks, though some residual overload risk remains.

Registrar–Verifier Attack Surface (DF2). A major threat on this link is an attacker spoofing a Verifier, allowing them to approve or reject arbitrary evidence and thereby permit or block devices from joining the network (Threat 12), achieving TC1 or TC2. This is mitigated by using secure authenticated channels (M9): an HTTPS channel enforces client-side identity verification, making spoofing highly difficult. The residual risk is low and would require exploiting a vulnerability in the underlying Transport Layer Security (TLS) implementation.

The reverse is also possible: an attacker may spoof the Registrar and submit arbitrary evidence to learn about the Verifier's behavior or reverse engineer its verification logic (Threat 13), targeting TC4. Because API servers may not always authenticate clients, this can be mitigated by introducing an access-control mechanism (M14). The server would rely on tokens issued by an authorization service and grant or deny API access based on those tokens, which shifts authentication responsibility to an authorization entity. The residual risk depends on the security of the access-control system. Since this is a separate component, it is considered out of scope for this threat analysis.

Since the Registrar forwards attestation evidence to the Verifier, an attacker may replace valid evidence with falsified evidence (Threat 14) or replay older evidence (Threat 15), similar to DF1. These threats mirror Threats 5 and 6, so the same mitigations apply: signing (M2), secure authenticated channels (M9), and nonces (M10), along with the same residual-risk considerations.

An attacker may likewise replace (Threat 16) or replay (Threat 17) verification results on their return path to the Registrar, substituting negative results with positive ones to achieve TC1, or positive results with corrupted-looking ones to achieve TC2. As before, secure authenticated channels (M9) can protect the communication, while the results themselves should be signed (M15) and include a timestamp that defines their validity period (M16). This is analogous to Threats 5 and 6, with timestamps used instead of nonces. When properly implemented, and potentially combined with M1, these measures make manipulation infeasible unless the Verifier's security configuration is poor or the attacker gains physical access to it.

Troubleshooting this link is hindered if the Verifier does not keep logs of interactions and requests (Threat 18), leading to TC5. This is mitigated by M11, with residual-risk considerations similar to Threat 9.

An attacker may also intercept evidence or results in transit (Threat 19), leading to information disclosure and achieving TC4. This is mitigated by M9, with residual-risk considerations matching those of Threat 10.

A DoS attack may be launched by dropping packets or injecting oversized payloads to overwhelm either the Registrar or the Verifier (Threat 20), targeting TC2. Mitigations M12 and M13 apply here as well, with the same residual-risk considerations as in Threat 12.

Finally, an elevation of privilege may occur if an entity other than the Registrar communicates with the Verifier (Threat 21), potentially leading to TC4. In this case, the entity is not external but a previously onboarded, unauthorized one. This is mitigated by M14, with residual-risk considerations analogous to those discussed in Threat 13.

Registrar–GDS Attack Surface (DF3). An attacker on this link may attempt to spoof the Registrar and request certificates directly from the GDS, thereby bypassing attestation and obtaining credentials for a compromised device (Threat 22), achieving TC1. This is mitigated by using secure authenticated channels (M9), as the client is required to present a valid certificate, making Registrar spoofing difficult. The residual-risk considerations are similar to those in Threat 4.

A reverse scenario is also possible, where an attacker spoofs the GDS and performs its role by sending trust lists that include attacker-controlled certificates (Threat 23). This would grant the attacker backdoor access to the attester post-attestation, effectively allowing them to bypass attestation (TC1). This is mitigated by using secure authenticated channels (M9), as the Registrar must verify the GDS's certificate before establishing a connection, making spoofing significantly harder. The residual risk considerations mirror those of Threat 4.

An attacker may attempt to append their own certificate to a trust list in transit (Threat 24), achieving TC1. This is the same as Threat 7, just on a different link, so same mitigations and risk considerations apply.

Incident analysis on this link is hindered if the GDS keeps no record of interactions (Threat 25), leading to TC5. This is mitigated by M11, with residual risk considerations matching those of Threat 9.

An attacker may intercept certificates or trust lists in transit (Threat 26) to learn about the system configuration, thereby achieving TC4. Mitigation is provided by M9, with risk considerations similar to Threat 10.

As with other links, a DoS attack is possible: an attacker may drop packets or inject oversized payloads to disrupt communication and incapacitate either the Registrar or the GDS (Threat 27), targeting TC2 and TC3. Mitigations M12 and M13 apply here as well, with residual risk considerations parallel to those in Threat 11.

Finally, an elevation of privilege may occur if an internal entity other than the Registrar attempts to communicate with the GDS (Threat 28), potentially enabling device onboarding while bypassing attestation (TC1). This is mitigated by M14, with residual-risk considerations matching those of Threat 13.

To summarize, some threats can be mitigated by relying on the OPC UA security stack, such as DoS, certain tampering threats, and information-disclosure threats. However, it is not enough to cover everything: additional solutions are necessary for best protection. For best mitigation of the severe TC1, HSMs in combination with secure boot and runtime measurement are required. Careful configuration of *trust on first use* behavior on the attester and an access control system would further minimize the residual risks.

6 Conclusion and Future Work

In this paper, we proposed a method for integrating remote attestation into the OPC UA onboarding process. We began by defining the prerequisites a device must meet during onboarding, namely possessing the correct *identity* and being in a healthy *state*. If both aspects can be verified, the device can be considered *trustworthy*. We then examined how OPC UA currently handles onboarding and observed that it verifies identity, but not state. To address this gap, we extended the model to include a state-verification step via remote attestation. Our solution is based on the RATS reference interaction models [7], and we provided a reference implementation demonstrating how such an integration can be realized in practice. In addition, we conducted a threat analysis of the proposed attestation-enhanced onboarding workflow to identify potential vulnerabilities in the state-verification step and outlined several mitigations that improve the robustness of the overall process. We show that while the OPC UA security stack provides baseline protection, it is insufficient on its own to mitigate all identified threats.

Future work aims at exploring more complex attestation scenarios in which some devices are located behind isolated or segmented networks and cannot be reached directly. Moreover, the reference implementation is extended with a more realistic attesting environment, such as physical hardware with a TPM or an emulated TPM, as well as a more comprehensive system for managing reference values and endorsements within the onboarding workflow.

Acknowledgments. This work is supported by the Swedish Knowledge Foundation (KKS) via the SEINE and MARC projects and by the Swedish Governmental Agency for Innovation Systems (VINNOVA) via the FLEXATION and iSecure projects.

References

1. ABB: Modular automation - how it's being used and how you can take advantage of it. ABB (2024). https://new.abb.com/control-systems/modular-automation
2. Ambrosin, M., et al.: Collective remote attestation at the internet of things scale: State-of-the-art and future challenges. IEEE Commun. Surv. Tutorials **22**(4), 2447–2461 (2020)
3. Asim, M., et al.: SecT: A zero-trust framework for secure remote access in next-generation industrial networks. IEEE J. Sel. Areas Commun. **43**(6), 2293–2311 (2025)
4. Bhole, M., et al.: From manual to semi-automated safety and security requirements engineering: Ensuring compliance in Industry 4.0. In: 50th Annual Conference of the IEEE Industrial Electronics Society (2024)
5. Birkholz, H., et al.: Remote ATtestation procedureS (RATS) Architecture. RFC 9334 (2023). https://www.ietf.org/rfc/rfc9334.html
6. Birkholz, H., et al.: RATS Conceptual Messages Wrapper (CMW). Internet-Draft draft-ietf-rats-msg-wrap-21, IETF (2025). https://datatracker.ietf.org/doc/draft-ietf-rats-msg-wrap/21/
7. Birkholz, H., et al.: Reference Interaction Models for Remote Attestation Procedures. Internet-Draft draft-ietf-rats-reference-interaction-models-15, IETF (2025). https://datatracker.ietf.org/doc/draft-ietf-rats-reference-interaction-models/15/
8. Birnstill, P., et al.: Introducing remote attestation and hardware-based cryptography to OPC UA. In: 22nd IEEE International Conference on Emerging Technologies and Factory Automation (2017)
9. Howard, M., Lipner, S.: The Security Development Lifecycle. Microsoft Press (2006)
10. IEC: IEC 62541: OPC UA (2025). International Standard, multi-part
11. IEEE: IEEE Standard for Local and Metropolitan Area Networks - Secure Device Identity. IEEE Std 802.1AR-2018 (Revision of IEEE Std 802.1AR-2009) (2018)
12. Juhola, A., Kylänpää, M.: Experimental implementation of remote attestation over OPC UA protocol. In: 2022 International Conference on Networks, Communications and Information Technology (CNCIT), pp. 83–88 (2022)
13. Khan, R., et al.: Stride-based threat modeling for cyber-physical systems. In: IEEE PES Innovative Smart Grid Technologies Conference Europe (2017)
14. Kohnhäuser, F., et al.: Secure onboarding of IIoT devices using OPC UA. In: 27th IEEE International Conference on Emerging Technologies and Factory Automation (2022)
15. Leander, B., et al.: Redundancy link security analysis: An automation industry perspective. In: IEEE 29th International Conference on Emerging Technologies and Factory Automation (2024)
16. Matischek, R., Bara, B.: Application study of hardware-based security for future industrial IoT. In: 22nd Euromicro Conference on Digital System Design (2019)
17. Meier, D., et al.: Secure provisioning of OPC UA applications using the asset administration shell. In: 17th Conference on Industrial Electronics and Applications (2022)

18. Red Hat: Attestation in confidential computing. Red Hat Blog (2023). Accessed 12 Oct 2025
19. Shelby, Z., et al.: The constrained application protocol (coap). Request for Comments 7252, IETF (2014). https://www.rfc-editor.org/rfc/rfc7252
20. Usman, A.B., Asplund, M.: Remote attestation with software updates in embedded systems. In: IEEE Conference on Communications and Network Security (2024)
21. ZVEI—German Electrical and Electronic Manufacturers' Association: Process Industrie 4.0: The age of modular production. Tech. rep., ZVEI (2019)

Mixed-Criticality with Unsafe Operating Systems

Michael Peter[1], Radu Morus[2], George A. Ciusleanu[2], and Kai Lampka[1,2(✉)]

[1] RPTU Kaiserslautern, Kaiserslautern, Germany
{peter,lampka}@cs.uni-kl.de
[2] NXP Semiconductors, Eindhoven, USA
{Radu.Morus,George.Ciusleanu,Kai.Lampka}@nxp.com
https://disco.cs.rptu.de

Abstract. Open-source operating systems, e.g., FreeRTOS and Zephyr, are developed by distributed communities that typically follow non-standardized, ad-hoc software development practices. This makes their adoption in safety-critical contexts particularly challenging, especially when targeting the highest safety assurance levels. This work presents a software architecture designed to remove the operating system from the safety rationale. This is accomplished by leveraging the stage-2 memory protection unit (MPU) available in the hypervisor mode of modern real-time processors. In contrast to existing (hypervisor-based) approaches, the proposed concept provides finer granularity by enabling the protection of individual applications rather than entire virtual machines. Our evaluation on the latest generation of NXP embedded SoCs (S32N family) shows that the combined overhead of control-flow interception and the reprogramming of the stage-2 MPU for each transition into and out of a safety-critical application is only about 0.5 µs, which is comparable to typical interrupt path latencies. This indicates that the proposed scheme remains suitable even for low-latency, hard real-time applications.

Keywords: ISO 26262 · Safety-critical OS · Memory Isolation · Virtualization Mode

1 Introduction

1.1 Motivation and Contribution

Software development in accordance with a standard or norm requires the application of established best practices throughout the entire V-model lifecycle. The specific activities and rigor of these practices are determined by the assigned certification level.

For example, ISO 26262, the automotive standard for functional safety of software, distinguishes four Automotive Safety Integrity Levels (ASILs), ranging from ASIL-A as the lowest to ASIL-D as the highest safety level. When developing software targeting ASIL-B, which is the most commonly required safety level,

© The Author(s), under exclusive license to Springer Nature Switzerland AG 2027
E. Troubitsyna et al. (Eds.): SAFECOMP 2026, LNCS 16868, pp. 114–128, 2027.
https://doi.org/10.1007/978-3-032-34867-8_8

ISO 26262 recommends the implementation of the following quality-assurance measures:

1. Comprehensive testing to ensure full code coverage.
2. Static code analysis.
3. Compliance with established coding standards, such as MISRA [1].
4. Implementation of standardized software development processes.

At the highest safety level, ASIL-D, the standard further recommends the use of semi-formal design methods, such as sequence diagrams and state machines, the use of sophisticated integration testing and fault injection campaigns, e.g., see [14] for an overview, among other techniques that support correct design and implementation.

Consequently, compliance with ISO 26262 significantly increases software development and maintenance costs. Industrial experience suggests that the actual implementation effort is comparatively small relative to the overhead introduced by the required quality-assurance activities [16]. As a result, the post-development certification of pre-existing software can substantially increase project costs and typically requires full ownership of the safety-critical software. [1]

To reduce the certification burden for complex software stacks, ISO 26262 supports architectural decomposition. By partitioning the software into components with distinct safety responsibilities, only those components contributing to safety-related functionality must comply with the highest ASIL applicable to any feature they implement or support. The decomposition mandates that components, especially those residing at different ASILs, are independent, and fault propagation between them is prevented. This requirement, known as *freedom from interference* (FFI), must be justified through a safety analysis.

FFI comes in two dimensions: spatial and temporal FFI. This paper takes a focus on spatial FFI, i.e., memory isolation. How to ensure timing guarantees with an unqualified OS kernel is left to future work, e.g., along the lines of [8].

Systems that satisfy FFI requirements and contain components with different ASIL classifications are commonly referred to as mixed-criticality systems. Decomposing a safety-critical system into a mixed-criticality architecture, while ensuring spatial FFI between components, can be effectively achieved by leveraging Memory Protection Units (MPUs), which allow memory accesses originating from an application to be restricted to its designated private memory regions.

Monitoring memory accesses using an MPU can ensure that separation between independent applications is enforced. However, this protection depends on the correctness of the OS kernel, which is responsible for proper MPU management. A single defect in the OS kernel may corrupt the memory of safety-critical

[1] ISO 26262 also defines qualification approaches for pre-existing software components as an alternative to complete redevelopment. These approaches acknowledge that such components may not have been developed according to safety-compliant processes. To establish sufficient confidence, the standard prescribes verification activities tailored to the intended use of these components typically libraries used by safety applications rather than entire OS kernels.

applications or undermine existing protection mechanisms. For instance, incorrectly configured memory regions may unintentionally overlap, allowing applications to overwrite private code or data belonging to co-located safety-critical applications. As a result, full safety qualification must be applied to the entire OS kernel code base, and this qualification must comply with the highest ASIL found with any of the hosted safety applications.

Due to this all-or-nothing constraint, the post-production safety (or security) certification of an open-source OS kernel can become prohibitively expensive and is often limited to lower ASIL targets, such as ASIL-B, as reported in [3,9]. Moreover, each new OS release requires substantial re-investment, depending on the size and complexity of the respective update.

Addressing these challenges and enabling the efficient, cost-effective, and continuous integration of unqualified open-source OS kernels into industrial applications is the central objective of this work.

1.2 Organization

Section 2 presents related work. Section 3 introduces the processor architecture concepts relevant to the proposed scheme. Section 4 presents the proposed approach and its optimizations. sec. 5 presents and discusses the associated performance overheads. Section 6 concludes the paper.

2 Related Work

Second-Stage Memory Protection. Microsoft implemented page protection using second level address translation (SLAT) tables to reduce the attack surface of the NT kernel [11]. Since the OS kernel operates at a lower privilege level, neither its own software nor potentially malicious code can modify memory regions protected by the SLAT tables. This feature is known as Windows Kernel Data Protection (KDP). KDP is available in both static and dynamic variants. Static KDP protects read-only data sections within driver images, while dynamic KDP provides a secure memory pool where kernel modules can store sensitive information that is protected against tampering.

xMP [13] employs adapted SLAT mechanisms to safeguard both kernel and user-space memory. Standing for Selective Memory Protection for Kernel and User Space, xMP follows a similar approach to Microsoft's KDP by using dedicated memory regions that store critical data. Access to these regions is restricted and can only be performed via Intel-defined registers, provided an application-specific secret is available. The technique is limited to Intel processors that support selective memory protection features, such as Intel's Protection Keys for Userspace (PKU) or equivalent mechanisms.

This paper adapts SLAT-based memory protection for real-time processors, which generally lack Memory Management Units (MMUs) and are limited to MPUs. Unlike hypervisor-driven techniques, where protection applies to entire VMs and remains static during VM execution, protecting individual applications requires active coordination with the (guest) OS. This paper's contributions lie in implementing this coordination and evaluating the associated performance overhead.

Microkernel-Based Isolation. Microkernel-based OSs adhere to a design philosophy that moves almost all OS functionality into user-space programs, reducing the kernel itself to a minimal set of essential features. Because the kernel only implements basic functionality, its size is close to minimal, making it particularly appealing for certification purposes. A key milestone in this domain was the formal verification of the (open-source) seL4 microkernel [6], demonstrating that a functionally correct OS kernel is achievable–something previously considered infeasible [4].

While architecturally appealing, microkernel-based systems often do not align well with common industrial development practices. In practice, easy-to-use RTOSs such as FreeRTOS™ and Zephyr® are widely adopted despite their lack of certification. Migrating entire application stacks to microkernel-based architectures to address certification requirements represents a substantial development effort and can significantly disrupt initial project timelines. The approach proposed in this paper aims to increase the assurance level of open-source RTOS workloads while keeping the additional development effort low.

Embedded Hypervisor Technology. Over the past decade, hypervisor technology has been increasingly adopted in embedded devices. Arm® has added virtualization capabilities to its real-time profile, supporting hypervisors with the latest generations of Arm®Cortex®-R cores, primarily Arm Cortex-R52 and Arm Cortex-R82. This has been adopted in recent semiconductor offerings, as noted in Sect. 1. The adoption of virtualization-capable processors has also driven the development of a new wave of embedded hypervisors. Solutions range from commercial, safety-qualified products from GreenHills [15], ETAS [2,5], and others, to open-source hypervisors like L4Re Micro Hypervisor [7] and Bao [10,12].

Unlike traditional Virtual Machine (VM) based memory protection schemes, the presented scheme offers finer granularity, enabling the protection of individual applications within a VM rather than the entire VM. The scheme has been integrated into a minimalistic EL2 kernel currently under development at NXP Semiconductors. It leverages hardware-assisted virtualization features commonly found in Arm Cortex-R, allowing for seamless integration into typical embedded hypervisor architectures.

3 Background Material

Exception-level model of Arm® **v8-R.** Privilege levels–referred to as exception levels (EL) in Arm terminology–are essential for isolating software that shares a processor core. These levels define the access rights of the currently executing code to on-core resources, such as registers and the ability to execute specific instructions. Running in a higher privilege level allows system software to maintain control over application software, even if the latter behaves incorrectly or maliciously.

The Armv8-R architecture implements a strict privilege hierarchy where each successively lower privilege level (EL) imposes greater restrictions on CPU resource modification. At the lowest privilege tier (EL0), applications execute with minimal system access, while higher levels provide progressively greater control: EL1 is reserved for OS kernels, and EL2 serves as the hypervisor privilege level. Provoking a transition from a lower to a higher privilege level is commonly referred to as trapping. For example, one traps from a user-space program (EL0) to the underlying OS kernel (EL1) or from the OS kernel to the hypervisor kernel (EL2) (if used).

In general, privilege escalation occurs through specific mechanisms: explicit software instructions,[2] hardware interrupts, or exceptions (e.g., memory faults). The higher privilege level maintains strict control over these transitions by governing the entry vector (program counter), the interrupt masking state, and the active memory translation regime. Also, the superior privilege level retains complete visibility and restoration capability of the lower level's execution context.

Conversely, the architecture grants higher exception levels complete authority over resuming execution at lower privilege levels, including control over the full register state –notably the program counter and status registers– as well as processor mode configuration encompassing privilege level, interrupt masking state, and active memory region regime. This granular control over execution context enables robust multi-tasking implementations, allowing privileged software to perform secure context switches between isolated execution environments while enforcing system-wide isolation and resource allocation policies.

Hardware Enforced Spatial Isolation. The Armv8-R architecture implements hardware-enforced memory access control via its Memory Protection Unit (MPU). The MPU is a programmable hardware mechanism that can be configured by privileged software. Through this unit, system software can define precise access permissions for designated memory regions, enforcing protection policies at the hardware level. The MPU operates by continuously monitoring all processor transactions –instruction fetches and data accesses alike– as they are issued by the core. Upon detecting any access violation, whether unauthorized writes to read-only memory areas or attempted instruction execution from non-executable regions, the MPU triggers fault exceptions, which transfer control back to the privileged software.

In Armv8-R processors with virtualization support, the MPU implements a two-stage access control mechanism where execution levels partition control responsibilities. The stage-1 MPU configuration falls under EL1's domain, while the software running at EL2 manages the stage-2 MPU. A memory access is only granted upon successful completion of both checks.

[2] SVC instruction for EL0-to-EL1 and HVC instruction for EL1-to-EL2 transition.

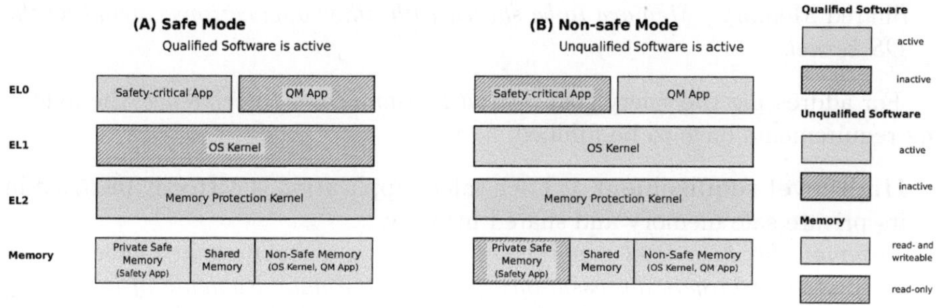

Fig. 1. Mixed-criticality addressed without safety qualified OS.

The two-stage MPU architecture enables comprehensive memory isolation across multiple levels and is commonly used as follows: the stage-1 MPU enforces protection between processes within individual virtual machines, while the stage-2 MPU ensures isolation between different virtual machines. The strength of these isolation guarantees depends on the level of trust placed in the software responsible for managing the respective MPU, namely the EL1 OS or the EL2-resident hypervisor kernel.

4 Memory Isolation for Safety Applications

This section begins by outlining the high-level safety requirements necessary to isolate and run a safety-critical application on top of an unqualified EL1 OS kernel. It then describes the Arm® 8-R features that enable efficient and transparent interception of context switches to and from any safety application, without impacting the EL1 OS kernel. These capabilities form the basis for a protocol that will be presented in the remainder of the section.

4.1 High-Level Requirements

Figure 1 illustrates an example use case for the proposed approach. In this scenario, the OS kernel hosts both safety-critical and quality-managed (QM) applications. The QM applications function independently of the safety-critical components, meaning they do not contribute–directly or indirectly–to any safety-related system functionality.[3]

The memory of each safety application is categorized as follows:

Private Safe Memory. Memory exclusively used by the safety-critical application under consideration. This region may also include safety-qualified device interfaces.

[3] As these components adhere "only" to best-practices of the SW life cycle, the terms *QM* and *non-safety critical* are often used interchangeably.

Shared Memory.] *Memory to be shared with other applications, including the OS kernel.*

For addressing the safety goal *Spatial Freedom from Interference*, the following requirements have to be fulfilled.

High-level requirement 1. Each safety application is correctly installed in its private safe memory and shared memory.
Rationale for high-level requirement 1. This is a requirement to the system integrator. It targets the assurance for code and data segments of any safety application for being initialized properly.[4]
High-level requirement 2. Any modification of a safety application's private memory by any entity other than the owning safety application must be strictly prohibited. Such attempts are considered safety violations.
Rationale for high-level requirement 2. This requirement forms the fundamental isolation barrier that prevents safety applications from being compromised either by faulty behavior from unqualified applications or by erroneous memory region configurations. Crucially, these access restrictions maintain their protective effect even against the OS kernel itself.
High-level Requirement 3. When a safety application is running, it must have unrestricted access solely to its own private and shared safe memory regions.
Rationale for high-level requirement 3. The rationale for this requirement is twofold. First, an application must have access to its private memory in order to execute. Second, safety applications must be protected from one another. Even if the safety applications have been fully qualified, misconfigured memory regions in the EL1 OS kernel cannot be entirely ruled out.

The requirements outlined above have a critical implication: Memory access permissions must be properly updated for every privilege level change involving a safety application. This requirement applies both to synchronous transitions, such as system calls initiated by the safety application, and to asynchronous preemptions, for example, those triggered by interrupts. Notably, control-flow transitions between EL0-resident applications always pass through the EL1 OS kernel.

4.2 Interception of Control-Flow Transitions

Fulfilling the above requirements necessitates that the EL2 kernel intercepts control transfers involving a safety application. On these occurrences, it dynamically updates the stage-2 MPU configurations, thereby tightening or lifting memory access restrictions.

[4] Commonly, this is achieved by using runtime checks for validating the integrity of the loaded image, e.g., as part of a (secured) bootchain. Thereby, we ensure that only intact safety applications are started at the correct memory locations. For brevity, any (sub-)requirements concerning integrity at system start are assumed to be satisfied and will not be examined in greater detail.

One may identify different hardware features of interest for intercepting control-flow transitions. The mechanisms are as follows.

Stage-2 MPU Exceptions. Upon accesses to non-congured addresses, the MPU generates an exception to be handled at EL2. Decoding the exceptions enables the EL2 kernel to determine the address of the faulting instruction and there by to identify the source of the access request. This approach can be used for the interception of both EL0-to-EL1 and EL1-to-EL0 transitions.

Trap General Exceptions (TGE). To interpose control flow transitions from EL0 to EL1, the EL2 kernel can leverage a mechanism that forces EL0-exceptions to trap directly into the EL2 exception vector.[5]

Virtual Interrupt Interception. The Arm® v8 virtual interrupt interception is a function that allows a hypervisor to control and manage the execution of physical interrupts in a virtual environment. To do so, a virtual interrupt is marked as pending by the hardware and forwarded to the EL2-resident exception handlers at the earliest possible delivery point. In systems where (unsafe) OS kernel code (EL1) always executes with interrupts disabled—a condition that, while common in many systems, cannot be universally guaranteed - this mechanism allows to interpose on EL1-to-EL0 transitions.

Explicit Hypervisor Invocation. The Hypervisor Call (HVC) instruction, available on Arm processors with virtualization extensions, enables a control transfer from EL1 to EL2. By inserting HVC instructions into the entry and exit paths of the guest kernel, both EL0-to-EL1 and EL1-to-EL0 transitions can be intercepted. Additional measures, such as write-protection, may be required to ensure the integrity of these control flow paths.

4.3 Protected Execution of Safety Applications

We assume the following structure of any safety application:

1. Initialization Phase. The safety application is loaded and it sets up necessary components.
2. Execution Phase. The safety application repeatedly executes its core function. The execution phase is usually implemented as an infinite (control) loop, often referred to as the main control loop or server loop.

[5] This feature is influenced by the TGE-bit in the Hypervisor Configuration Register (HCR_EL2). Once set, the EL2 kernel can seamlessly intercept control transfers from safety applications to the EL1-rooted OS kernel.

The pseudo-C code shown in Algorithm 1 illustrates this structure. During initialization, the safety application is not yet provided with any spatial protection by the EL2-resident kernel. Once initialization is complete, and before entering its main loop (line 5), the safety application must synchronize with the EL2 kernel to activate memory protection.

As part of this process, the EL2-resident kernel verifies the integrity of the safety application, for example by computing hash values over memory regions containing executable code. The details of the integrity verification process are platform-dependent. On some platforms, the isolation monitor itself may be responsible for performing the integrity assessment. Alternatively, the resulting integrity measurements may serve as evidence to an external health supervisor that the safety application has been properly prepared for execution. After successful integrity verification, the EL2-resident kernel transitions into the execution phase.

During this phase, it must ensure that the high-level requirements outlined above are consistently satisfied. To that end, the EL2-resident monitor interposes on every control-flow transition into and out of the safety application. This is achieved using one of the methods described in the previous section. At each transition, the execution environment is adjusted accordingly. Memory access restrictions for regions owned by the safety application are either relaxed (upon entry into the safety application) or enforced (upon exit from it).

Depending on the mechanism used to interpose on privilege-level transitions, additional adjustments may be necessary. For example, when using TGE, exceptions must be redirected to the EL2-resident kernel when entering the safety application and routed back to EL1 when leaving it. This behavior is controlled by setting or clearing the corresponding flag in the HCR_EL2 register.

The mode of operation of the scheme presented so far is illustrated in Fig. 1. Part (A) illustrates the case a safety application is running which requires the lifting of the memory protection. Part (B) illustrates the case in which a safety application is inactive, requiring write protection of its private safe memory.

Algorithm 1: Structure of a safety application

```
1 // Initialize safety app
2 do_initialization();
3 activate_protection();
4 // Entering the main control loop
5 while (true) do
6     wait(..);
7     do_something();
8 return;
```

4.4 Register Integrity

The EL1-OS is responsible for managing application context switches, which includes preserving and restoring CPU register states. When relying on an unqualified OS, the correctness of context switching operations for safety-critical applications cannot be assumed. Therefore, protection of safety applications must extend beyond the application's safe private memory and encompass the CPU state when context switching to a safety application.

To do so, we add a register state verification step before resuming the execution of any EL0-resident safety application. This check ensures that the system context is valid and consistent prior to the safety application's execution.[6]
Register state validation must be carried out before resuming a safety application. This could be achieved with a mechanism analogous to the TGE intercept (see Sect. 4.2, p.7), but one that permits execution in EL1 and triggers traps to EL2 upon a transition to EL0. However, the architecture does not offer a built-in mechanism for intercepting EL1-to-EL0 privilege transitions.

Three alternative approaches can be employed:

1. Remove execution privileges from the safety application's code when preempting it. This causes an access fault whenever the safety application is resumed.
2. Implicit interception of EL1-to-EL0 control flow transitions using the Virtual Interrupt Intercept feature.
3. Explicit interception of EL1-to-EL0 control flow transitions via HVC (Hypervisor Call) instructions.

Details on the architecture support for these mechanisms have already been given in Sect. 4.2, p. 7. In the following we briefly discuss the applicability of these methods.

Fault-on-Execute for Intercepting EL1-to-EL0 Transitions. Such a strategy conflicts with scenarios where the safety application shares code with other tasks, a common setup found in memory-restricted embedded systems. As a result, the interception of EL1-to-EL0 transitions cannot rely on fault-on-execute.

Virtual Interrupts for Intercepting EL1-to-EL0 Transitions. On an EL0-to-EL1 transition, the EL2 isolation kernel would pend a virtual interrupt. When the EL1 OS resumes an application, this interrupt triggers a transition into EL2. If the application being resumed is a safety application, the EL2 isolation kernel lifts the protection before allowing it to execute. For a regular application, it resumes execution while keeping the memory protection of any safety application

[6] The extent to which the register state must be preserved depends on the situation: for a system call, some registers are modified by the EL1 OS to convey return values to the safety application, whereas for asynchronous preemption, such as a timer interrupt, execution must resume with an identical register state.

enforced. In both cases, the virtual interrupt is cleared and the TGE flag is set, which allows for EL0 execution and reroutes traps bound for EL1 into EL2.

This approach relies on a critical assumption: that EL1 OS kernel code always executes with interrupts disabled. While this condition is common in many systems, it cannot be universally guaranteed. For OSs that allow interrupts, that method is not applicable.

While this approach does not require modifications to the EL1 OS, it introduces significant overhead because all EL1-to-EL0 transitions are intercepted. By explicitly intercepting only the EL1-to-EL0 transitions of safety applications (as described below), this overhead can be avoided, albeit at the cost of making some changes to the EL1 OS.

HVC-Based Interception of EL1-to-EL0 Transitions. Although this approach requires modifications to the guest OS, by inserting HVC instructions into the exit path of the EL1 OS kernel, we consider it preferable due to the limitations of virtual interrupt interceptions as indicated above. The advantages –such as a simplified EL2-kernel architecture and the preservation of CPU state integrity– outweigh the relatively minor changes needed in the EL1 OS.

By tracking the safety status of a task, e.g. through a dedicated flag in the Process Control Block (PCB), the EL1 OS can determine whether a task can be resumed normally (for non-safety tasks) or whether the resumption must be routed through the isolation kernel in EL2. Importantly, the protection guarantees do not rely on correct behavior of the EL1 OS. If the isolation kernel is invoked during the resumption of a non-safety application, it will ignore any request to lift write protection on private safe memory. Conversely, if a safety application is resumed directly by the EL1 OS without proper coordination, the violation will be detected when the application executes its next write operation.

5 Evaluation

5.1 Setup: Isolation Kernel

To minimize development effort, we integrated the proposed protection scheme into an experimental real-time hypervisor. This kernel was designed to support the evaluation and profiling of relevant hardware, particularly for multicore processors built on Arm Cortex-R52, e.g., the latest generation of NXP® platforms. The isolation kernel specifically targets processors built upon Arm Cortex-R52 and features a minimal codebase, comprising approximately 5.5 kSLOC of C code and 150 SLOC of assembly.

The kernel itself resides at EL2 and provides virtual machines as its sole execution environment, fully leveraging the Arm Cortex-R52 hardware-assisted virtualization features. This ensures that core functionalities–including CPU privilege levels, interrupt handling (via the GIC CPU interface), memory management (through the MPU), and timers–behave as they would on a non-virtualized platform. For selected components, such as interrupt configuration via the GIC distributor, we adopted a para-virtualized design, exposing the shared device through hypercalls rather than fully virtualizing it using a trap-and-emulate approach. As a result, existing OSs can be reused with relatively little effort, particularly when compared to the complexity of porting them to the abstractions of traditional microkernels, such as threads and IPC.

5.2 Basic Profiling of the S32N5 Processor

This section presents measurements obtained from running workloads on top of our isolation kernel on the NXP S32N55 platform without using the functionality to protect individual applications. The S32N55 is a real-time processor designed to consolidate multiple vehicle functions onto a single chip. It features 16 Arm Cortex-R52 cores as well as two pairs of Arm®Cortex®-M7 cores, the latter of which were not considered in our evaluation.

The measurements presented in this section quantify the costs of typical real-time workload operations, with a particular focus on interrupt path latencies. These results provide a baseline for comparison with the overhead introduced by activating and deactivating the memory protection of regions owned by safety applications, as examined in the following section (Table 1).

Table 1. EL2-mode profiling of the S32N5

Measurement	Time (μs)	Comment
IRQ latency.	1.03	Highest priority VM, and the highest priority interrupt
IRQ latency	1.29	Second RTOS VM Prio-1 *Includes Second RTOS latency
SGI latency (same VM)	0.5	Second RTOS VM Prio-1
IRQ to current VM	0.42	No VM switch
VM to WFI	0.743	VM transitions to WFI (save, no restore)
VM wake up from WFI due to IRQ	1.028	PHY CPU in WFI, IRQ to VM occurs (restore only)
IRQ to other than current VM	0.803	includes a VM switch
High prio VM (WFI) → low prio VM	1.182	Highprio VM calls WFI, IRQ arrives meanwhile, switch to low-prio VM
Boot time	55	Time from reset to first instruction in first VM

5.3 Profiling the Protection Scheme

Building on the understanding of the Arm Cortex-R52 characteristics from the previous section, we evaluated the overhead introduced by enabling and disabling memory protection for safety applications. In our setup, we selected TGE as the interception mechanism for EL0-to-EL1 transitions and triggered the activation and deactivation directly from the safety application in EL0 using HVC instructions.

In a real-world deployment, these operations would not be initiated by the safety application itself. However, the resulting overhead would remain the same. Therefore, the measured results are representative of the expected overhead when individually protecting safety applications.

Figure 2b shows a screenshot from a debugging session, where the safety application is currently executing ($vTask10$). Executing the HVC instruction triggers a cascade of service routines and related function calls. In this sequence, the function *El2mp_HyperTrap* is responsible for reprogramming the stage-2 MPU. In the evaluated scenario, this operation temporarily lifts the memory access restrictions for the safety application's private safe memory.

The measured time for this operation is approximately 250 processor cycles, which corresponds to about 217 ns at a clock frequency of 1.2 GHz. The reverse operation, which restores the protection of the private safe memory, requires 324 cycles, corresponding to approximately 270 ns.

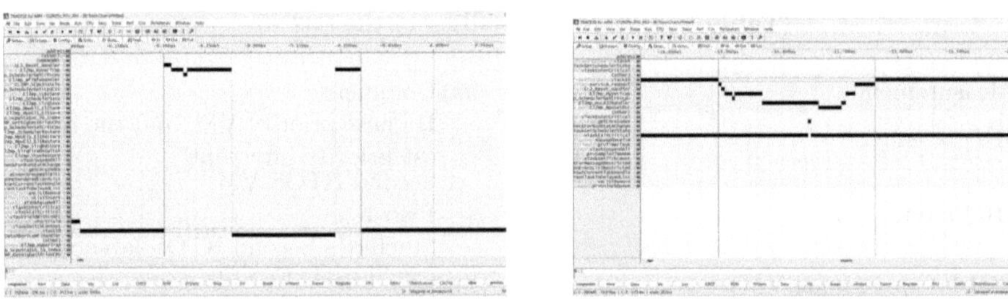

(a) Deactivate protection (b) Activate protection

Fig. 2. Protection change.

For context, the average interrupt latency is approximately 0.42 μs when no VM switch is required and about 0.8 μs when a VM switch occurs. This comparison shows that the overhead introduced by activating and deactivating the protection mechanism, while noticeable, is not prohibitive. These results indicate that the proposed approach remains practical even for low-latency, hard real-time applications.

6 Conclusion

This paper presented a novel memory protection scheme that we implemented in an experimental EL2-resident isolation kernel. The scheme is designed to manage access rights to memory regions that are private to individual safety applications. It ensures that only the currently executing safety application has the appropriate write access to its private (safe) memory, while all other software components–including the EL1 OS kernel–are prevented from modifying these regions.

In systems where safety-critical and quality-managed applications coexist on top of an unqualified EL1 OS kernel, these guarantees are essential to prevent latent corruption of safety-critical applications caused by unidentified defects in the unqualified OS. The measured round-trip times of approximately 0.5 μs for activating and deactivating the protection for a safety application indicate that the proposed scheme is sufficiently efficient for use in low-latency, hard real-time applications.

These results suggest that the presented approach may pave the way for incorporating unqualified open-source OS, such as FreeRTOS and Zephyr, into safety-critical environments without requiring a development process that is fully compliant with standards such as ISO 26262.

NXP is a registered trademark of NXP B.V. All other product or service names are the property of their respective owners.
Arm, Arm Cortex, Arm Cortex-R, Arm Cortex-R52, Arm Cortex-R82, Arm Cortex-M7, Armv8-A, Armv8-R are trademarks and/or registered trademarks of Arm Limited (or its subsidiaries or affiliates) in the US and/or elsewhere. The related technology may be protected by any or all of patents, copyrights, designs and trade secrets. All rights reserved.
FreeRTOS is a trademark of Amazon.com, Inc. or its affiliates.

References

1. Motor Industry Software Reliability Association. Misra-c. https://misra.org.uk/
2. ETAS. Etas webpage. https://www.etas.com/
3. Red Hat. https://www.redhat.com/en/about/press-releases/red-hat-reaches-key-milestone-push-functional-safety-certification-red-hat-vehicle-operating-system Red Hat Press Release 6'th January 2025
4. Heiser, G.: The seL4 Microkernel: An Introduction (2025). White paper
5. SYSGO Embedding Innovations. Sysgo - embedding innovations webpage. https://www.sysgo.com/
6. Klein, G., et al.: SEL4: formal verification of an OS kernel. In: Proceedings of the 22nd ACM SIGOPS Symposium on Operating Systems Principles (SOSP), pp. 207–220 (2009)
7. Lackorzynski, A., Kloetzke, J.: The L4RE micro hypervisor. Whitepaper available at: https://www.kernkonzept.com/kk_white_paper/whitepaper-l4re-micro-hypervisor
8. Lackorzyński, A., Warg, A., Völp, M., Härtig, H.: Flattening hierarchical scheduling. In: EMSOFT '12, pp. 93–102, New York, NY, USA (2012). Association for Computing Machinery
9. Lampka, K., Thurlby, J., Lackorzynski, A., Hähnel, M.: Safety certification with the open source microkernel-based operating system l4re. In: Trapp, M., Saglietti, F., Spisländer, M., Bitsch, F., (eds.), Computer Safety, Reliability, and Security - 41st International Conference, SAFECOMP 2022, Munich, Germany, September 6-9, 2022, Proceedings, vol. 13414 of LNCS, pp. 31–45. Springer (2022)
10. Martins, J., Tavares, A., Solieri, M., Bertogna, M., Pinto, S.: BAO: a lightweight static partitioning hypervisor for modern multi-core embedded systems. In: Bertogna, M., Terraneo, F., (eds.), Workshop on Next Generation Real-Time Embedded Systems (NG-RES 2020)

11. Microsoft: (2020). Microsoft kdp
12. Bao project. Bao, a lightweight static partitioning hypervisor webpage. http://www.bao-project.org/
13. Proskurin, S., Momeu, M., Ghavamnia, S., Kemerlis, V.P., Polychronakis, M.: XMP: selective memory protection for kernel and user space. In: 2020 IEEE Symposium on Security and Privacy (SP), pp. 563–577 (2020)
14. Schirmeier, H., Hoffmann, M., Kapitza, R., Lohmann, D., Spinczyk, O.: Fail*: towards a versatile fault-injection experiment framework. In ARCS **2012**, 1–5 (2012)
15. Green Hills Software. Green hills software webpage. https://www.ghs.com/
16. Stucki, A.: Functional safety. what is the development effort?

Safety of AI-Based Systems

Formal Target Aware Training of Machine Learning Models

Nicolas Valot[1(✉)], Louis Fabre[1], Benjamin Lesage[2], Ammar Mechouche[1], and Claire Pagetti[2]

[1] Airbus Helicopters, Marignane, France
nicolas.valot@airbus.com
[2] ONERA, Toulouse, France

Abstract. Integrating Machine Learning models within certified safety critical avionics systems faces several challenges among which we focus on the difficulty to demonstrate that a trained Machine Learning Model meets its requirements when deployed on the target system. Indeed, a *semantic gap* exists between training framework models and deployed inference models on target hardware. Certification standards require the rigorous implementation of ML model descriptions with design assurance, as well as a demonstration that properties verified during training (such as accuracy and robustness) are preserved on the target. To that end, we propose the *Target Aware Training* (TAT) approach, that ensures semantic preservation by construction. Practically, the training framework's native inference engine is substituted with a compiled shared library derived from design-assured code for the target hardware. Doing so, we achieve bit-accurate replication and eliminate the need for error-prone transformations to and from intermediate representations like ONNX. The approach also allows the specification, control, and formal verification of activation and accumulator bounds. We validate our approach on the ResNet18 architectures, guaranteeing the absence of numerical overflows even for unseen data through a generative verification method. This dual contribution provides a robust path toward meeting DO-178C design assurance objectives for airborne ML applications.

1 Introduction

Context. Machine Learning (ML) algorithms offer appealing opportunities for advanced functional capabilities in rotorcraft avionics. Practical applications include real-time helicopter weight estimation [18], predictive maintenance based on structural hardware loads [10], and automated obstacle detection [1]. However, to be embedded in airborne systems, ML-based systems must comply with stringent certification requirements to guarantee safe operation. To this end, regulatory bodies have established specific guidance–notably the EASA concept paper [7] and the forthcoming EUROCAE/SAE ED-324/ARP-6983 standard [8]. These documents address supervised, off-line trained, ML models. Both

© The Author(s), under exclusive license to Springer Nature Switzerland AG 2027
E. Troubitsyna et al. (Eds.): SAFECOMP 2026, LNCS 16868, pp. 131–145, 2027.
https://doi.org/10.1007/978-3-032-34867-8_9

approaches delineate high-level objectives to 1) assess the ML model achieves its *intended function* and 2) demonstrate that the *properties* verified during training are maintained during execution on the target hardware. For that purpose, they promote a W-shaped development life-cycle (outlined in Fig. 1).

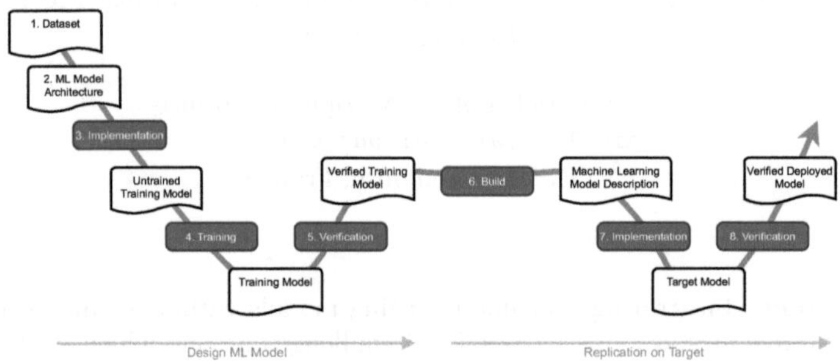

Fig. 1. ED-324/ARP-6983 W-shape Development Life Cycle.

The 1st-V focuses on the design of ML model and involves: 1. 2. defining a data set split into a training and a test data sets; 3. selecting a *ML model architecture* (e.g. a Resnet18 [13]); 4. implementing the *Untrained Training Model* in a training framework (e.g. PyTorch); 5. learning the parameters of the Training Model (TrM) with the framework (e.g. PyTorch); and 6. verifying the *properties* (e.g. performance, robustness, stability, generalization) of the TrM with corresponding metrics (e.g. MSE, accuracy, mAP) . 7. specifying a ML Model Description (MLMD) from the trained and verified ML model, that is the starting point of the implementation of the *Target Model* (TaM) in the 2nd-V; 8. implementing the TaM following the DO-178C guidance [20]; 9. verifying the TrM properties are preserved on the target platform (e.g. avionics processors), i.e. on the TaM, per ML guidance.

Proposed Approach. We propose a *Target Aware Training* (TAT) method identified in the development life-cycle in Fig. 2, to reduce the gap between training on host and replication on target. The training framework directly interacts with the *Target Model Source Code* to update the parameters of the model (1st-V, step 4). This simplifies the replication activities (2nd-V, step 7): the implementation of the TaM is the compilation of the MLMD, defined as the Target Model Source Code with the trained weights. On-target TAT requires no additional step, while on-host TAT still benefits from the proximity of the TrM, MLMD, and TaM, as the only model transformations are compilation steps. TAT further offers shortcuts towards TaM verification (Step 8) using Bit-accurate preservation, semantic replication as per [25], and falling back to the traditional approach if both fail.

We further propose a method to specify, enforce, and verify the input and output ranges of neural network-typed ML models. In accordance with design

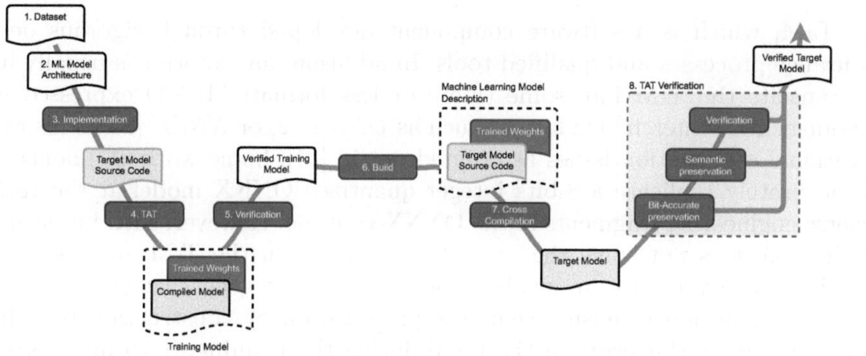

Fig. 2. Proposed Development Life Cycle.

assurance processes, the goal is to ensure that the interface of the model is well-defined, with strict range constraints, such that intermediate computations do not cause out-of-range errors. Our approach combines experimental and formal (abstract interval propagation) methods to compute sound and tight bounds on the input and output ranges of selected operations. The model architecture further enforces the computed range while reducing the complexity of the analysis, by clamping operations' input to their range.

In this work, we consider medium-sized convolutional models that can be implemented in using C code and deployed on avionic multi-core processors.

Outline. In Sect. 2, we describe the semantics preservation problem and how TAT addresses the problem. Section 3 points out the neural network layer range generalisation challenge and proposes a formal approach to tackle it. Finally Sect. 4 summarizes an experimental evaluation of the TAT and the range generalisation methods. We evaluate the training time overhead of our TAT method, and demonstrate formally bounded ranges of the ResNet18 model layers, using PYTORCH for the training, and C code for the target.

2 TAT: an Answer to the Semantic Preservation Problem

To address the replication problem, that is ensuring TrM and TaM satisfy the same properties, we rely on the concept of *semantic preservation* [25]. *Semantic preservation* is 1) the preservation of the TrM properties by the TaM; and 2) the implementation of the MLMD with design assurance.

2.1 Problem Statement and Existing Approaches

The W-shaped development lifecycle highlights a critical challenge in bridging the semantic gap between 1) the (untrained) TrM defined by a proprietary, high-level internal representation of a specific framework (e.g., PYTORCH), and 2) the TaM, which is a software component developed through rigorous design assurance processes and qualified tools. In addition, an exporter is usually used to translate the TrM into some (more or less formal) MLMD expressed in a vendor-neutral interchange format such as *ONNX* [2] or *NNEF* [24]. This raises additional certification issues as pointed out in [11]. Some work [9] manages to bit-accurately replicate a 8-bits integer quantized ONNX model at the cost of reverse engineering fragments of the ONNX-runtime. However, such bit accurate replication does not solve the semantic gap between the TrM and its ONNX MLMD. Moreover, such an involved method does not generalize well.

A first solution to ensure semantic preservation by construction is to integrate the target Hardware in the Loop during the training. Such an approach, referred to as *on-device training* [17], was motivated by dataset confidentiality and end-user customization. ML models are directly trained on a target device, entailing that TrM = MLMD = TaM. However, there are two main drawbacks. First, the design assurance forbids to keep dead code on the target, and the single TrM/TaM contains the loss, gradient, and weight update code. Second, the training uses the target hardware and faces a significant memory wall [14].

A second solution [25] proposes a methodology to check whether a TaM preserves the properties of the TrM. Practically, the approach mixes 1) a mathematical relation between the verification metrics (e.g. MSE, accuracy) and the difference between the TrM and TaM outputs, 2) a test based strategy to assess said difference to check whether they fit within some pre-defined margin. This requires 1) the definition acceptable metrics margins for each property and 2) a design modification when the execution discrepancy is greater than the margin.

Those solutions address the semantic preservation problem but still face strong limitations. We would like to ensure the design is correct ahead of time, like on-device training, while keeping training framework efficiency and capabilities.

2.2 Target Aware Training Method

We propose a *Target Aware Training (TAT)* method to integrate the TaM within the training environment. Practically, we substitute the *forward* inference function of the training environment with the one developed for the target system with design assurance. Thus, there would be no distinction between the description, i.e. the source code, used to implement the TrM and the TaM. The differences between the two would boil down to the compilation and execution environments.

As depicted in Fig. 2, we are in essence specialising the W-lifecycle steps by: developing with design assurance the Target Model Source Code first, and compiling them as the TrM (Step 3); training the TrM parameters (Step 4), verifying the TrM properties (Step 5); and finally cross compiling the TaM source code for the target system (Step 7). The training is *target aware*, because the training process exercises the exact same source code as the airborne target. This simplifies the verification of the replication between TrM and TaM (Step 8), i.e. ensuring those environments are identical (on-target TAT) or produce identical results (bit-accurate replication), to take credit of verification artifacts from the TrM for the TaM.

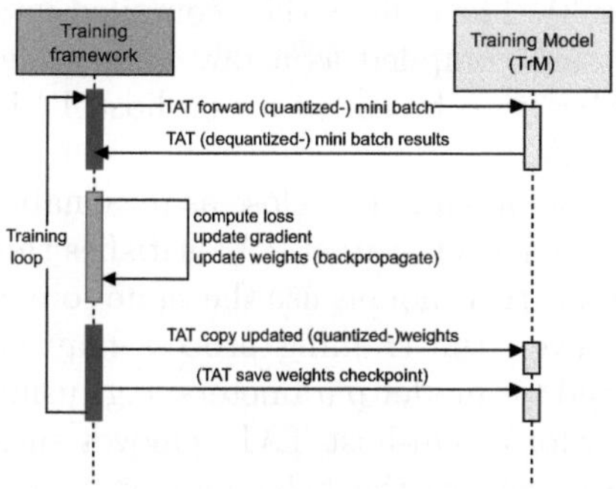

Fig. 3. Target Aware Training insertion points in the training loop.

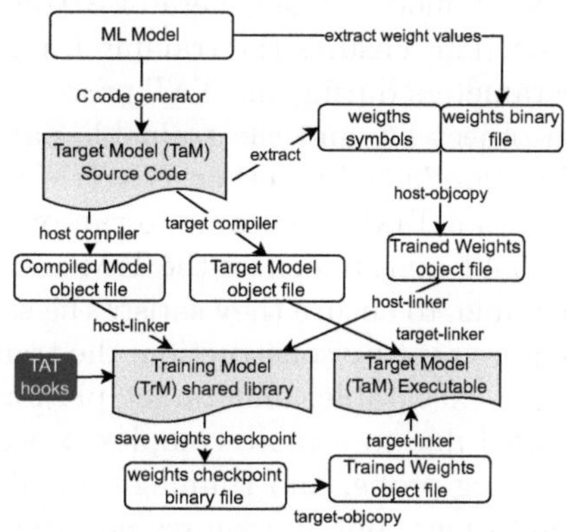

Fig. 4. Training and Target Model implementation process.

Figure 3 captures the interaction between the training framework and the Training Model (TrM) library:

- **Forward Inference:** For each mini-batch, the training framework (e.g., PYTORCH) calls its native inference engine for gradient update and discards its output, then calls the TrM's forward pass.
- **Gradient Computation:** The TrM's results are read back into the framework to compute the loss and perform backpropagation. The framework updates the weights in the Python model using legacy optimization features.
- **Weight Synchronization:** Batch normalization layer weights are fused. Layer weights are copied back into the TrM's memory space for the next iteration.
- **Integrated Checkpointing:** The framework can trigger the TrM to dump model parameters and integrity checksums into a raw memory image. This file is intended to be loaded as-is on the target system, ensuring bit-level consistency between TrM and TaM.

Applying TAT on-target, the TrM can be trained on target using the same executable, as it contains only the implementation of the model forward function. On-host, TAT requires compiling the TaM source code for the target, and linked together with the weight parameters checkpoint gathered during the training. TAT can be applied to quantized models as well.

2.3 Training and Target Model Implementation

We now describe our application of TAT, per the process outlined in Fig. 4. We choose to implement the TaM source code in C code compatible with design assurance process. Such a choice ensures adequacy with current Airborne practices, at the cost of being solely applied to medium size models. A C-code generator [22] produces the TaM source code from a ML Model, e.g. a PYTORCH or ONNX model. The generator extracts the trainable model parameters from the code. This ensures the training framework can identify and update the required parameters during the TAT process (see Fig. 3). The code is then compiled into an object file and the trainable parameters are compiled as a raw binary file. Both are linked together as a TrM shared library to be called for on-host TAT, or into a TaM executable to run on-target TAT.

TAT aims to ensure the TrM and TaM are identical, or as close as reasonably possible, to ensure they satisfy the same properties. On-target TAT satisfies this requirement by construction: the training and target models use the same source code, executable, and target platform. However, the training process may be slowed due to a more complex process to update model parameters, e.g. using a debug probe, and running on a slower platform. On-host TAT removes such limitations, but it requires additional care to ensure the behaviour on target will be the same (i.e. bit-accurate replication) than during training. The main concern is that numerical data representations and operations match during training and on target:

- to ensure a common floating-point semantics at the object code level for both the host and target systems. Indeed, non IEEE 754 options (e.g. fast math) can introduce differences in floating-point calculations.
- to get rid of possible compiler effects that could alter the number encoding, e.g. litteral numbers rounding. For that, the model parameter values are saved as a checkpoint raw binary file, which is converted to an Executable and Linkable Format (ELF) object file with GNU *objcopy*, using model parameters symbol names.

3 Range Generalization Challenge

Under rigorous design assurance process (e.g. DO-178/ED-12 [20]), software engineering necessitates well-defined interfaces with strict input/output constraints. When it comes to neural networks, a subclass of ML models, this entails that the ranges of the input tensor, the output tensor, and the intermediate tensors must be specified and verified for the Trained Model.

3.1 Problem Statement

During training, a neural network's operations are *exercised* within specific tensor value ranges determined by the training dataset. We remind that a neural network is a graph where the nodes are the layers (or operations), and the edges capture the dataflow between layers. The data exchanged are *tensors* that are represented and encoded as vectors.

Definition 1 (Interface of a Node). *Consider a neural network defined by the graph $G = (V, E)$. For each node $v \in V$, we define its interfaces by $v.i$ its input interface, and $v.o$ its output interface by:*

- $v.i.dim \in \mathbb{N}^*$ *its input dimension meaning that an input x contains $v.i.dim$ elements $x = (x_1, \ldots, x_{v.i.dim})$, $x \in \mathbb{R}^{v.i.dim}$,*
- $v.i.range = [\underline{v.i}, \overline{v.i}]$ *its input range such that $\forall k \in [1, v.i.dim], \underline{v.i} \leq x_k \leq \overline{v.i}$,*
- $v.o.dim \in \mathbb{N}^*$ *its output dimension,*
- $v.o.range = [\underline{v.o}, \overline{v.o}]$ *its output range.*

Example 1 (LeNet5 Model). Figure 5 shows the LeNet5 graph with explicit layer interfaces. For instance, the first convolutional *conv1* has an input dimension of $conv1.i.dim = 28 \times 28 \times 1 = 784$ and an output dimension $conv1.o.dim = 24 \times 24 \times 6 = 3456$.

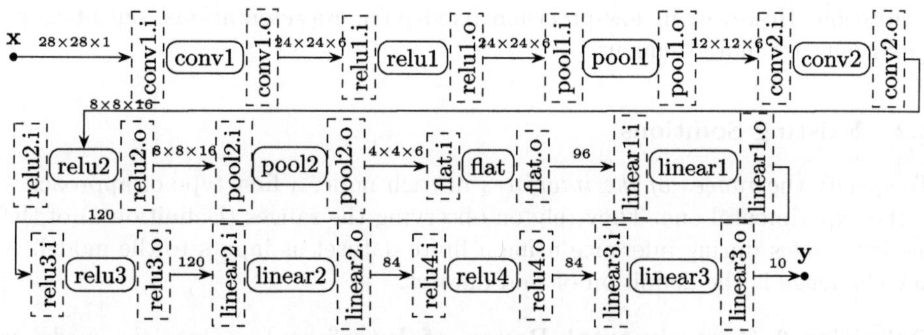

Fig. 5. LeNet5 and its interface

Property 1 (Connecting interfaces and associated ranges). Let v a node and $Pred(v) = \{v' | (v', v) \in E\}$ its predecessors. The interface of v is defined by that of its predecessors,

$$v.i.dim = \sum_{v' \in Pred(v)} (v'.o.dim)$$

Similarly the range of $v.i$ is exactly the one obtained from its predecessors,

$$v.i.range = [min_{v' \in Pred(v)}(\underline{v'.o}), max_{v' \in Pred(v)}(\overline{v'.o})]$$

The computation of $v.o.dim$ and $v.o.range$ depends on $v.i$ and the operation performed by the node.

Back to the design assurance approach, the interfaces must be identified, the ranges specified and verified. From a design assurance perspective, the model must remain within its verified operational envelope and any *out-of-range* behavior is classified as a mis-prediction. This requirement is of interest to ensure the *functional safety* of the system.

Definition 2 (Out-of-range behavior). *For a given trained model, an* out-of-range *behavior occurs when the model produces intermediate inference values outside the ranges which were observed during training.*

A special case of *out-of-Range* behavior is the *out-of-domain* behavior. This consists in having a value exceeding the type representation (e.g. 32-bit floating point - FP32). This situation is extremely detrimental. When the model is computed with low precision data type, this inevitably leads to semantic faults due to modular arithmetic wrap around; while the computed value should exceed the type representation, and thus the computing node range, the actual value wraps around and could result in a value that may fall back into said range. In FP32 formats, this issue is often hidden by a vast dynamic range and IEEE-standard saturation. However, in low-precision fixed-point representations, out-of-range conditions are more frequent.

3.2 Existing Solutions

To specify the *ranges of the interfaces* of each node, a first type of approaches is the experimental one. They rely on observing the ranges (Definition 1) of the model's nodes during inference using a finite dataset as inputs to the model. A key challenge is the definition of said dataset.

Definition 3 (Experimental Range of Interfaces). *Once the model is trained, we execute the inference for all the training dataset, and store the minimal and maximal values observed on each interface. We denote this input (respectively output) experimental range for each node v by $v.i.obs$ (resp. $v.o.obs$).*

Example 2 (Experimental range of MobileNetv2). Let us consider the first convolution of MobileNetv2 [21] trained for the dataset Imagenet [6]. The experimental range for *conv1* is $conv1.i.obs = [0, 1]$ and

$$conv1.o.obs = \left[\min_{x \in IMAGENET\_train}(conv1(x)), \max_{x \in IMAGENET\_train}(conv1(x))\right] \tag{1}$$

Computing the interface range solely with the training dataset is rarely sufficient to compute trustworthy bounds, as will be discussed in Example 3. Adversarial attacks such as FGSM [12] and PGD [16] instead generate synthetic data to assess properties on the model, e.g. assess robustness by targeting an output decision change, with an imperceptible change in the input image. Therefore, we adapt the PGD algorithm to induce out-of-range outputs at each model node.

Experimental methods have a limited coverage due to the discrete nature of the dataset. A second approach is thus to rely on formal methods which assess whether properties hold for the full range of data, i.e. for any input. These are supported by abstract interpretation tools used for safety critical systems design assurance (e.g. Astre, Frama-C [5]). Due to their complexity, the analysis of neural networks benefits from dedicated tools, able to compute the output of each node over an abstract interval domain, i.e. a sound bound of the set of possible values. The challenge is then to balance the accuracy of the selected domain, with the complexity of propagating it through a model. Simpler approaches such as Interval Bound Propagation (IBP) may result in large overestimates.

A third, complementary, approach is to clamp the activation (outputs of nodes) to enforce specific ranges. It was introduced by MobileNetv2 [21] with the use of $\text{ReLU6}(\mathbf{x}) = \text{Clamp}(\text{ReLU}(\mathbf{x}), 6)$ which limits the upper bound to the value 6 after convolutions. The rationale for choosing the value 6 is not detailed more than mentioning algorithm stability in low precision. This architectural work-around leads to a better average accuracy than with unbounded ReLU, but it does not preclude out-of-range events.

While MobileNetv2 and PACT [3] propose to clamp activation function at convolution *output*, we propose to specify the bounds with the experimental calibration method, and use them as a clamp parameter for parameterized layers' *input*. Similarly to PACT, we can further move the bound through the training process in order to optimize accuracy. We refine in next section how this bound is related to the parameterized layer's weight $\mathbf{w}$.

Example 3 (Range propagation). To illustrate the second and third approaches, we performed an IBP analysis using the Auto-Lirpa [27] tool[1]. Figure 6 presents the output ranges computed through IBP for the convolutional layers of State-of-the-Art pre-trained models. The x-axis represents the convolutional layers in topological order from model input to output. The y-axis is the range magnitude in log scale. Applying IBP on convolutions leads to a pessimistic over-approximation which in turn leads to an exponential growth of ranges during model traversal. We observe that using clamped activation function *ReLU6* stops the exponential growth of the ranges.

3.3 Approach to Specify the Interfaces of Each Layer

We want to guarantee at operation that unseen data do not produce values exceeding interface ranges in the TaM. To that end, we study intermediate layer bounds and propose a method to specify, enforce, and verify these bounds. While PACT [3] and A2Q [4] methods, address the specific context of quantized models, we generalize to unquantized ones. Our contribution impacts requirements on the *ML Model Architecture* (see Fig. 2), which are defined following 3 steps.

[1] Experimentally-derived ranges were omitted due to their comparatively limited range.

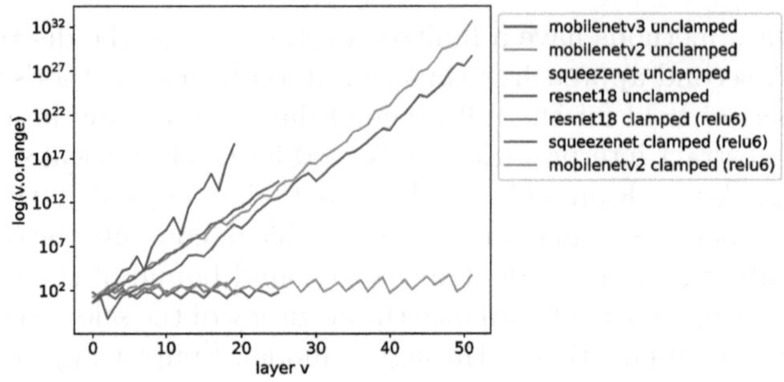

Fig. 6. Exponential growth of IBP computed ranges across convolutional layers of famous image classifier models with unbounded (unclamped: ReLU) activations, and some bounded ones (clamped: relu6).

Step 1: Specification of Input Ranges. To specify the input ranges, we compute the experimental range of interfaces, per Definition 3, with a calibration dataset (a subset of the training set) for pre-trained models. Untrained models have to rely on an arbitrary bound, similar to ReLU6, or a value derived from the weight initialization method for untrained models. We thus compute $v.i.obs$ for all v in the model.

Step 2: Enforcing Input Range Specification. We add before each layer, a clamp to the $v.i.obs$ values computed in Step 1. Figure 7 illustrates the resulting architecture. Doing so, we enforce the respect of the input interface specification, $v.i.range = v.i.obs$. The clamp interval C_v is initialized to $v.i.obs$,

$$\text{clamp}(x, C_v) = \max(-C_v, \min(C_v, x))$$

Note that we only clamp *parameterized* nodes which are the ones responsible for making the $v.o.range$ range grow rapidly through the model. This is because parameterized layers, namely *Linear* and *Convolution*, build upon the fundamental **scalar product** $\mathbf{x} \cdot \mathbf{w} + \mathbf{b}$ operations, with weight $\mathbf{w}$ and bias $\mathbf{b}$ parameters.

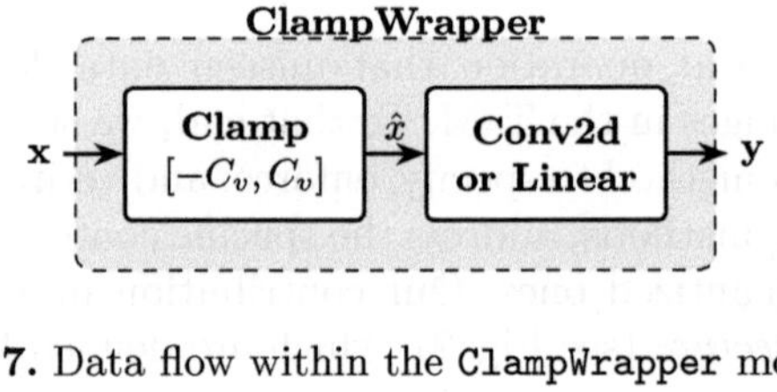

Fig. 7. Data flow within the `ClampWrapper` module.

Step 3: Specification of the Output Ranges. The specification of the output range *v.o.range* relies on formal methods, namely IBP. As inputs are bounded and enforced by the clamp, IBP can be applied on a per node basis, as opposed to the whole neural network.

Our approach thus combines different methods to specify and enforce input and output bounds of each node in a neural network. Step 1 provides tight but possibly optimistic estimates based on a known dataset where available (pre-trained models), or arbitrary values (untrained models). Those bounds are then enforced as *v.i.range* by Step 2[2]. The use of formal methods in Step 3 provides sounds estimates for *v.o.range*. The combination of Step 1 and 2 reduces the likelihood of over-pessimistic output bounds, as observed in Example 3, and supports the independent analysis of each node.

In the next section, we propose to use our proposed TAT and formal bounds methods to design, implement and verify neural network models with specified node ranges.

4 Bounded TAT Verification

Having introduced our approaches to TAT and bounding layer ranges, we propose a workflow to support bounded TAT, its application, and verification. The workflow maps onto the first 3 steps of the W-lifecycle illustrated in Fig. 2. We choose a pre-trained ResNet18 [13] as a starting point to evaluate the method.

First in the *ML model architecture* phase, we pre-process the model in order to wrap the inputs of convolution and linear layers with a trainable clamping parameter C_v (Sect. 4.1). Second, in the *Implementation* and *Training* phases, we generate the code of the target ML model with design assurance and perform a target aware training (TAT) (Sect. 4.2). We further consider changes to the loss function to finetune C_v during training while ensuring that the output of each node v remains below the specific limit *v.o.range*. Third, in the *Verification* phase, we employ a generative method to maximize activation tensor ranges and experimentally verify that the property (limit *v.o.range*) holds for adversarial valid input (Sect. 4.3).

Overall, the workflow ensures that 1. the source code semantic of the target is exercised and validated in the training framework, 2. that this model runs in a specified and controlled range, 3. that the controlled range property is verified by robustness test values beyond what can be achieved with the dataset.

4.1 Preprocessing

The considered ResNet18 neural network is preprocessed per Steps 1, 2, and 3. We compute an input range *v.i.range* for each layer v experimentally using a calibration dataset. We systematically clamp multiplication intensive layers

[2] Note that clamping a node's inputs to its observed range does not warrant retraining pre-trained models as said range is derived from the training dataset.

(i.e. convolution and linear layer), as described in Fig. 7, given they are the main cause of unbounded output ranges. And we finally apply IBP to specify *v.o.range* per layer and per channel.

4.2 TAT Experiment

The source code is generated using ACETONE framework[3], with channels last format (for CPU-SIMD) and Winograd 2D-convolution 3×3 kernels. We generated the shared library according to the process in Fig. 4 and loaded the library into the PYTORCH training loop. The C code for target performs forward call sample by sample (optimized for latency). In addition, we extended ACETONE to generate a batch loop function using OpenMP multithread; this function is only required and compiled for training, it is not compiled nor embedded in the target. For big models, to improve the training loop duration, the C forward function might be used only for validation at the end of epochs.

As the weights norm tends to grow during training as described in [15] (Fig. 2, Eq. 2) and [23], we also define a loss function including a range-barrier regularization term $\mathcal{L}^{(v)}_{\text{barrier}}$ which constrains the weights $\mathbf{w}_v$ and bias $\mathbf{b}_v$ to force node v's output to stay below the specified *v.o.range*. The range-barrier treats the sum of the dot-product bound and the bias as the constrained variable. μ is the "hardness" of the barrier, and γ a guard for strictness of specified *v.o.range*.

$$\mathcal{L}^{(v)}_{\text{barrier}} = \max\left(\left(C_v \cdot \|\mathbf{w}_v\|_1 + |b_v| - v.o.range + \gamma\right), 0\right)^2$$

$$\mathcal{L}_{\text{task}} : \text{the classification or regression loss to the ground truth}$$

$$\mathcal{L}_{\text{total}} = \mathcal{L}_{\text{task}} + \mu \sum_{v \in L} \mathcal{L}^{(v)}_{barrier}$$

By optimizing $C_v, \mathbf{w}_v$, and b_v under this barrier, the model learns a balanced resource allocation. If a high bias is necessary for accuracy, the penalty will force a reduction in C_v or $\|\mathbf{w}_v\|_1$ to create the required numerical headroom, ensuring $|y_s| < v.o.range$ at all times.

Experiment Setup: We measure a training loop of 8 steps with mini-batches of size 32. We vary the ratio of (evenly-spaced) steps with callbacks to the TAT TrM, following the process in Fig. 3, from a training loop without TAT (0/8, the baseline) to calling TAT for each training step (8/8). For verification, the baseline is the PYTORCH evaluation loop of the model (w/o backpropagation nor TAT). We use for host and target specific conservative IEEE-745 compiler options for floating-point.

Experiment Result: Figure 8 presents the overhead of TAT on the training duration over the baseline. The x axis represents increasing ratio of {1/8, 1/4, 1/2, 1/1} TAT hook calls in the training loop. The batch normalization fusing during the weight copy represents 20% of the 51% overhead. The y axis represents the overhead (in percentage) of the baseline. The measured overhead for the verification of the model properties using the TrM is 30%. We further verified that the cross-compiled TaM bit-accurately[4]replicates TrM inferences.

---

[3] https://github.com/onera/acetone.

[4] TAT also demonstrated bit-accuracy and similar training overhead on smaller-scale industrial case studies, e.g. a model to predict helicopter gross weight [19,26].

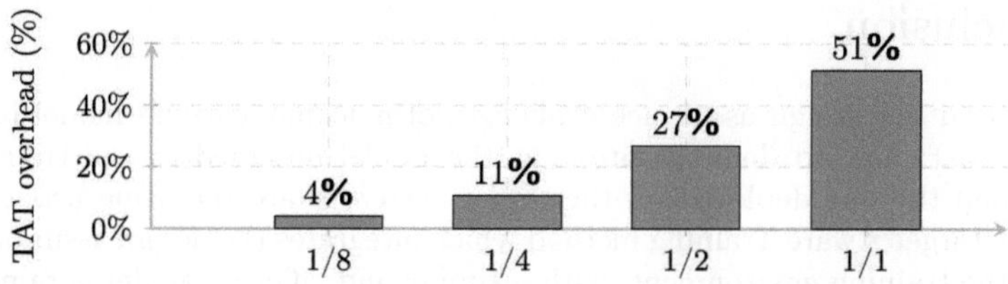

Fig. 8. Analysis of Target Aware Training (TAT) overhead across different hook call frequencies in the training loop.

4.3 Verification of Out of Bounds Robustness: Generative Adversarial Method

In order to verify that our model with a clipping range *v.i.range* does not overflow the specified *v.o.range*, we need to go beyond the available datasets to capture unseen data possibilities.

To generalise to unseen image data, and any node v in the model's graph, we use the PGD principle [16] with the objective to maximize *v.o.obs* in the direction of *v.o.range* for any $\mathbf{x} \in v.i.range$.

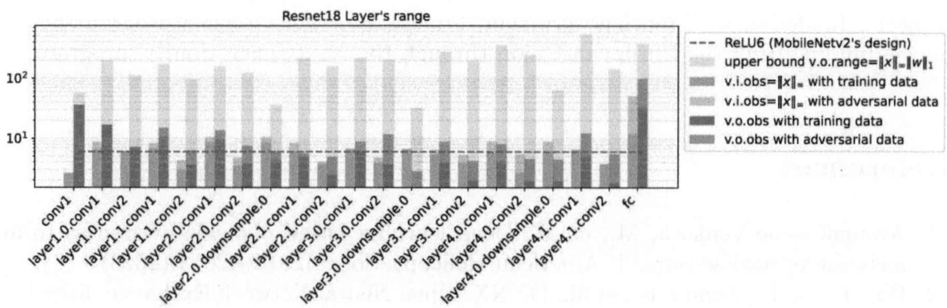

Fig. 9. Resnet18 model, preprocessed according to Sect. 4.1.

Figure 9 captures the computed and observed ranges for the considered ResNet18 model. Computed upper bounds, $\|x\|_\infty \|w\|_1$, are reported per layer in gray. The color of the bar distinguishes between input and output observed ranges, respectively in orange (*v.i.obs*) and purple (*v.o.obs*). The shade of the color identifies the dataset. The ranges obtained with the training data, i.e. with ImageNet samples, use a darker shade. The ranges obtained with the generative adversarial method are lighter.

Compared to the unbounded model in Fig. 6, the proposed input Clamp wrapping does reduce the upper bound computed through IBP for ResNet18. As opposed to ReLU6, the enforced bounds stem from actual training data and requirements. Observations do validate the soundness of the computed bounds: the observed ranges are always lower than the computed bounds. However, those remain over-pessimistic compared to what can be achieved using the generative adversarial method. Observations with training data consistently result in smaller ranges than with said adversarial methods.

5 Conclusion

We introduced the design assurance challenges of machine learning model in a safety critical context: how to demonstrate that the model designed in the training environment, and the one deployed on the target system share the same properties. We proposed a Target Aware Training method which integrates the design assurance source code into the training environment, with a simple and efficient model parameters layout. With some intent on compiler options requirements, our Target Aware Training method enables bit-accurate replication of models from training to target environment, and is free of error-prone intermediate representations (ONNX protobuf, TFlite flatbuffer).

In addition, we proposed a method to specify, control and verify neural networks layer bounds combining empirical, formal, and generative techniques. Nevertheless, we observed that the safe bounds *v.o.range* are of high magnitude compared to empirical ones *v.o.obs*. In a quantizing scheme where scaling factors are computed out of *v.o.range*, data resolution could be reduced, and rounding error magnitude increases. The weight norms values are driving layer bounds and therefore the tensors resolution for low precision implementation. To increase resolution, we need to reduce weight norms.

Acknowledgement. The work presented in this paper partially relies on the PATO project supported by the Directorate General of Civil Aviation (DGAC). The PATO project is funded by the French government through the France Relance program, based on the funding from the European Union through the NextGenerationEU program.

References

1. Ammalladene-Venkata, M., et al.: Deep learning based obstacle awareness from airborne optical sensors. J. American Helicopt. Soc., 42012–42024 (2023)
2. Bai, J., Lu, F., Zhang, K., et al.: ONNX: Open Neural Network Exchange. https://onnx.ai/ (2019)
3. Choi, J., Wang, Z., Venkataramani, S., Chuang, P.I., Srinivasan, V., Gopalakrishnan, K.: PACT: parameterized clipping activation for quantized neural networks (2018). CoRR abs/1805.06085
4. Colbert, I., Grob, F., Franco, G., Zhang, J., Saab, R.: Accumulator-aware post-training quantization (2024). ArXiv abs/2409.17092
5. Cuoq, P., Kirchner, F., Kosmatov, N., Prevosto, V., Signoles, J., Yakobowski, B.: Frama-C: a software analysis perspective. In: International Conference on Software Engineering and Formal Methods, pp. 233–247. Springer (2012)
6. Deng, J., Dong, W., Socher, R., Li, L.J., Li, K., Fei-Fei, L.: ImageNet: a large-scale hierarchical image database. In: 2009 IEEE Conference on Computer Vision and Pattern Recognition, pp. 248–255 (2009)
7. EASA: Concept Paper: First usable guidance for Level 1 machine learning applications - Proposed Issue 02 (2024)
8. EUROCAE WG-114/SAE joint group: ED-324/ ARP-6983 REV.DRAFT 5B Process Standard for Development and Certification/Approval of Aeronautical Safety-Related Products Implementing AI (2024). on going standardization

9. Gabreau, C., Teulieres, M.C., Jenn, E., Lemesle, A., Butucaru, D.P., Thiant, F., Fischer, L., Turki, M.: A study of an ACAS-xu exact implementation. using ed-324/arp6983 (2024)
10. Gallimard, C.D.C., et al.: Harmonic decomposition to estimate periodic signals using machine learning algorithms: application to helicopter loads. In: International Joint Conference on Neural Networks (IJCNN), pp. 1–8. IEEE (2022)
11. Gauffriau, A., De Albuquerque Silva, I., Pagetti, C.: Formal description of ML models for unambiguous implementation. In: 12th European Congress on Embedded Real Time Software and Systems (ERTS 2024) (2024)
12. Goodfellow, I.J., Shlens, J., Szegedy, C.: Explaining and harnessing adversarial examples. In: 3rd International Conference on Learning Representations (2015)
13. He, K., Zhang, X., Ren, S., Sun, J.: Deep residual learning for image recognition. In: IEEE Conference on Computer Vision and Pattern Recognition. CVPR (2016)
14. Li, S., Tian, C., Tam, K., Ma, R., Li, L.: Breaking on-device training memory wall: A systematic survey. CoRR **abs/2306.10388** (2023)
15. Lobacheva, E., Kodryan, M., Chirkova, N., Malinin, A., Vetrov, D.P.: On the periodic behavior of neural network training with batch normalization and weight decay. Adv. Neural. Inf. Process. Syst. **34**, 21545–21556 (2021)
16. Madry, A., Makelov, A., Schmidt, L., Tsipras, D., Vladu, A.: Towards deep learning models resistant to adversarial attacks. In: CoRR (2017)
17. McMahan, B., Moore, E., Ramage, D., Hampson, S., y Arcas, B.A.: Communication-efficient learning of deep networks from decentralized data. In: Proceedings of the 20th International Conference on Artificial Intelligence and Statistics, AISTATS (2017)
18. Mechouche, A., Houles, M., Gallimard, C.D.C., Maisonneuve, P.L.: On the trustworthiness of machine learning models in health and usage monitoring of in-service helicopters. In: 81th Vertical Flight Society Forum (VFS) (2025)
19. Mechouche, A., Valot, N., Fabre, L.: Towards learning assurance for in-flight machine learning-based helicopter weight estimator. Vert. Flight Soc. Forum **82** (2026). West Palm Beach
20. RTCA/EUROCAE: DO-178C/ED-12C - Software Considerations in Airborne Systems and Equipment Certification (2011)
21. Sandler, M., Howard, A.G., Zhu, M., Zhmoginov, A., Chen, L.: MobileNetV 2: inverted residuals and linear bottlenecks. In: 2018 IEEE Conference on Computer Vision and Pattern Recognition, CVPR, pp. 4510–4520 (2018)
22. Silva, I.D.A., Carle, T., Gauffriau, A., Pagetti, C.: ACETONE predictable programming framework for ML applications in safety-critical systems (Artifact). Dagstuhl Artif. Ser., 06:1–06:2 (2022)
23. Soudry, D., Hoffer, E., Nacson, M.S., Gunasekar, S., Srebro, N.: The implicit bias of gradient descent on separable data. J. Mach. Learn. Res. **19**(70) (2018)
24. The Khronos NNEF Working Group: Neural Network Exchange Format (2018)
25. Valot, N., Fabre, L., Lesage, B., Mechouche, A., Pagetti, C.: Implementation of airborne ML models with semantics preservation. In: 44th Digital Avionics Systems Conference (DASC) (2025)
26. Valot, N., Mechouche, A., Lesage, B., Pagetti, C., Fabre, L.: Towards on-board implementation of ML-based helicopter weight estimator. Vert. Flight Soc. Forum **82** (2026). West Palm Beach
27. Xu, K., et al.: Automatic perturbation analysis for scalable certified robustness and beyond. Adv. Neural. Inf. Process. Syst. **33** (2020)

From High-Dimensional Spaces to Verifiable ODD Coverage for Safety-Critical AI-Based Systems

Thomas Stefani(✉), Johann Maximilian Christensen, Elena Hoemann, Frank Köster, and Sven Hallerbach

Institute for AI Safety and Security, German Aerospace Center (DLR), Sankt Augustin, Ulm, Germany
thomas.stefani@dlr.de

Abstract. While Artificial Intelligence (AI) offers transformative potential for operational performance, its deployment in safety-critical domains such as aviation requires strict adherence to rigorous certification standards. Current EASA guidelines mandate demonstrating complete coverage of the AI/ML constituent's Operational Design Domain (ODD)—a requirement that demands proof that no critical gaps exist within defined operational boundaries. However, as systems operate within high-dimensional parameter spaces, existing methods struggle to provide the scalability and formal grounding necessary to satisfy the completeness criterion, and no standardized engineering method exists to bridge the gap between abstract ODD definitions and verifiable evidence. This paper addresses that gap with a structured, multi-step method that integrates parameter grouping, a dependency analysis distinguishing deterministic from stochastic relationships, physically grounded constraint definition, and criticality-based discretization into a single ODD coverage verification process. Applied to a case study on AI-based near mid-air collision avoidance with the VerticalCAS system, the method reduces the verification space by roughly 60% while preserving all safety-relevant combinations and yields an explicit, iterable list of uncovered regions, providing a repeatable process for demonstrating the completeness EASA requires. Designed to scale toward higher-dimensional ODDs, the method advances a Safety-by-Design approach to the certification of future AI-based flight systems.

Keywords: ODD Coverage · AI Engineering · Safety-by-Design

1 Introduction

The integration of Artificial Intelligence (AI) and Machine Learning (ML) into safety-critical applications in aviation represents a transformative shift in aerospace engineering [14]. While neural networks offer unprecedented capabilities for processing complex sensor data and optimizing real-time decision-making, such as in the next-generation Airborne Collision Avoidance Systems

© The Author(s), under exclusive license to Springer Nature Switzerland AG 2027
E. Troubitsyna et al. (Eds.): SAFECOMP 2026, LNCS 16868, pp. 146–160, 2027.
https://doi.org/10.1007/978-3-032-34867-8_10

(ACAS X), their *black-box* nature directly conflicts with traditional deterministic safety assurance [2,3,12]. In this highly regulated domain, the transition from *experimental* to *certified* systems depends not on performance alone, but on verifiable trust. The European Union Aviation Safety Agency (EASA) has addressed this challenge by outlining a roadmap for AI trustworthiness [6], emphasizing the necessity of Operational Design Domain (ODD) coverage, since ODDs define the area where the AI-based system is intended to operate. Current EASA guidelines mandate that for an AI/ML constituent to be certified, the applicant must demonstrate "completeness" of the ODD. This implies proving that the system has been verified across all relevant operating conditions it might encounter, ensuring no safety gaps exist within the defined boundaries [7]. However, a fundamental engineering gap exists between the abstract definition of an ODD, which often comprises a mere list, and the practical demonstration of its coverage. Safety-critical systems typically operate in high-dimensional parameter spaces, where variables such as relative velocity, closure rate, bearing, and environmental factors interact. As dimensionality increases, the state space expands exponentially. Traditional sampling or brute-force simulation cannot provide the formal completeness guarantees required by certification authorities, as they either leave vast regions of the high-dimensional space unexplored or prove to be computationally infeasible [16]. Furthermore, current engineering frameworks lack a standardized method to bridge the gap between an abstract ODD definition and a demonstration of coverage. While formal methods provide exactness, they often struggle to meet the scalability requirements of deep neural networks. Conversely, simulation-based testing offers scalability but often at the cost of high computational effort needed to claim *complete coverage* that satisfies certification authorities. There is a pressing need for a systematic approach to identify safety-critical regions within the ODD and focus verification efforts there.

This paper addresses this challenge by proposing a structured method for verifiable ODD coverage that translates abstract ODD definitions into quantifiable, verifiable metrics. To ensure computational viability in high-dimensional spaces, we introduce a specialized method for criticality-based dimension reduction. Later, the method is validated through an empirical study of AI-based near mid-air collision (NMAC) avoidance, providing a blueprint for the rigorous evidence required to satisfy EASA's stringent ODD completeness standards. By bridging high-dimensional data with regulatory compliance, this work advances a method that can be integrated into a Safety-by-Design methodology for the certification of future AI-based flight systems.

The remainder of this paper is organized as follows: The Sect. 2 reviews the regulatory context and existing approaches for parameter space coverage. Next, Sect. 3 details the proposed multi-step method for demonstrating ODD completeness through discretization and constraint definition. In Sect. 4, this method is applied to a case study on the VerticalCAS airborne collision avoidance system. Finally, Sect. 5 discusses the engineering implications and scalability of the approach, followed by a conclusion and future outlook in Sect. 6.

2 Related Work

The safety assurance of AI-based systems in aviation is situated at the intersection of upcoming regulatory requirements and the novel engineering challenges of demonstrating compliance for AI-based systems. This section reviews the current state of the art in both the regulatory context and ODD coverage to contextualize the proposed method.

2.1 Regulatory Context

The EASA Concept Paper, *Guidance for Level 1 & 2 Machine Learning Applications*, establishes a framework for AI trustworthiness in aviation, centering on Learning Assurance [7]. A critical requirement within this framework is the rigorous definition and verification of the ODD, as detailed in multiple objectives, including DM-08 and LM-16. These guidelines mandate that applicants demonstrate not only the representativeness of data but also the completeness of the verification process across all operational dimensions. EASA specifically requires evidence that no safety gaps exist within the ODD, suggesting quantitative approaches, such as subdividing the parameter space into hypercubes, to assess data density and model stability. However, as the ODD encompasses increasingly complex, high-dimensional spaces, providing the evidence of completeness demanded by these objectives becomes a significant engineering hurdle that this paper addresses.

2.2 Parameter Space Coverage

Quantifying coverage in high-dimensional parameter spaces has become a fundamental requirement for evaluating the verification and operational safety of AI-based systems. Weissensteiner [20] established a high-level ODD coverage process for automated driving systems, using an adaptive k-means clustering algorithm under predefined boundary conditions to reduce the number of required scenarios without compromising verification quality. While this provides essential evidence for safety argumentation, the method does not account for complex interactions between parameters and therefore remains vulnerable to the *curse of dimensionality.*

A complementary line of work addresses parameter dependence rather than scenario count. Höhndorf et al. [10] characterize dependence structures among operational parameters using vine copulas with subset simulation, optimizing failure-probability estimation by pruning implausible or low-likelihood combinations, and focusing verification on realistic regions. Copulas, however, do not reduce the nominal dimensionality of the ODD: they impose a dependence structure on the same space, and each pairwise copula adds correlation parameters that may require non-trivial measurement. Their cost thus grows with dimensionality, leaving open the need for a framework that scales to real-world ODDs while satisfying EASA's exhaustive completeness criteria.

A third group of approaches defines coverage geometrically. Hirschle et al. [8] evaluate methods such as the convex hull of the sampled data, which gives a clear geometric boundary but is insensitive to internal density and scales poorly, and the union of hyperspheres around samples, which is highly sensitive to the chosen radius and can overestimate coverage beyond unity. They propose kernel density estimation as a more scalable alternative for identifying low-density regions.

The idea of subdividing an ODD to make coverage tractable has recently been pursued at the ODD level by Schäfer and Cichon [15], who partition an ODD into micro-ODDs (μODDs) with generic-volume object representations, building a bottom-up completeness argument from manageable sub-domains rather than enumerating explicit instances. The allocation of verification effort across such partitions has a longer history in software reliability engineering, where two limitations of a structural criterion are well established. First, uniform coverage may misallocate effort: Cotroneo et al. [4] instead sample inputs according to an operational profile in their RELAI scheme, contrasting with the strict, likelihood-agnostic completeness this work pursues. Second, exercising a partition does not guarantee that faults within it surface, as captured by Voas's propagationinfectionexecution (PIE) model [19], in which fault detection requires not only that a location be executed but also that the execution infect the program state and that this state propagate to an observable output. Covering a bin, in our sense, satisfies only the first condition–underscoring that coverage is a structural prerequisite for, rather than a guarantee of, correct behavior.

On the regulatory side, the ForMuLA report by EASA and Collins Aerospace [5] proposes the closest analog to a completeness check: dividing the ML constituent ODD into subspaces at a required granularity and verifying that a sufficient number of examples populate each subspace, supported by Design-of-Experiment and Latin-hypercube sampling. In their case study, however, this check is demonstrated only on a low-dimensional categorical cross-product (mission pattern against operating environment) and is assessed by visual inspection of the resulting distribution rather than by a quantitative metric. The subspace granularity is left unspecified, and the sampling-based nature of the approach provides coverage in expectation rather than the exhaustive guarantee implied by EASA's completeness wording.

Thus, despite these advances in quantifying ODD sub-spaces, a distinct void remains: there is no systematic engineering process that bridges these individual metrics with EASA's overarching demand for ODD completeness, a prerequisite for the formal certification of AI-based systems [7]. This work addresses that void by turning the subspace idea into a quantitative, exhaustive completeness criterion over the discretized parameter space, equipped with methodical dimension reduction to keep it computationally feasible.

3 Method for Demonstrating Completeness of the ODD

In EASA's Concept paper for AI learning assurance, the anticipated Means of Compliance (MOC) DM-08 requires *"... coverage of the whole ODD by the test data set, with the necessary level of completeness and representativeness."* [7]. However, the process of demonstrating the necessary level of completeness remains undetermined. This work provides a comprehensive method that engineers can follow. It is part of a domain-agnostic effort to establish a Safety-by-Design AI engineering pipeline in which demonstrating adequate coverage is one component of the safety argumentation [9]. Figure 1 illustrates the workflow for assessing coverage completeness in a set of scenarios. It begins with the defined ODD, which serves as input to the first step of parameter grouping. Subsequently,

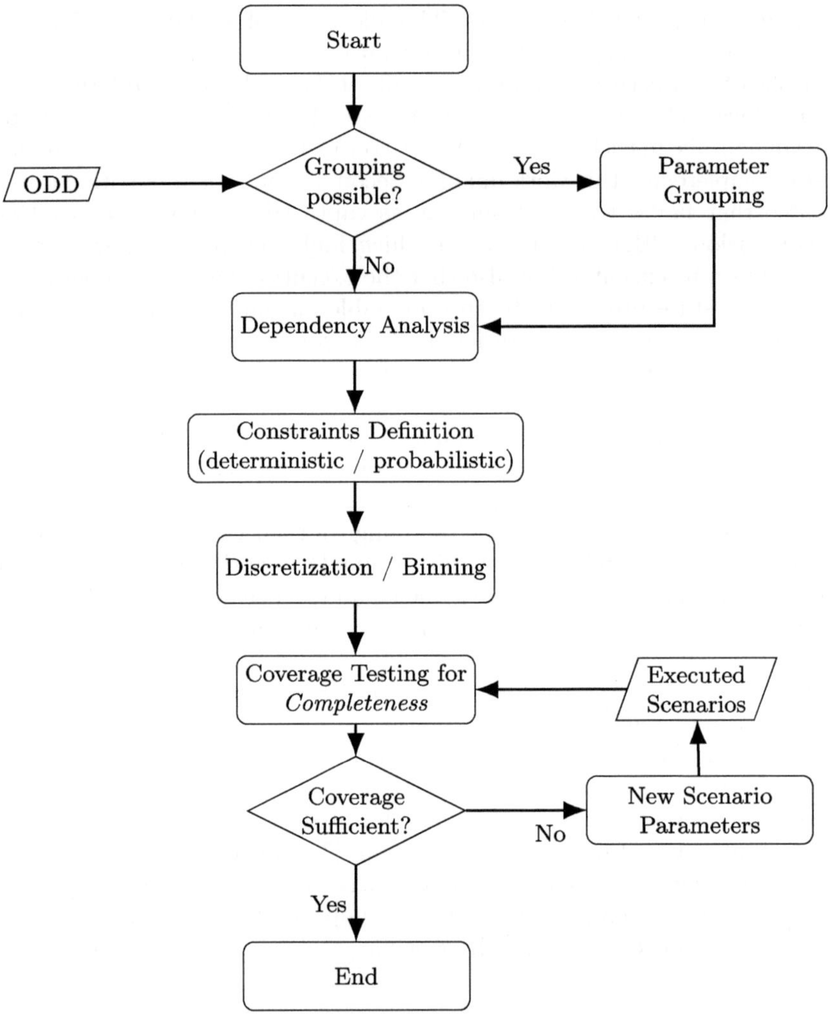

Fig. 1. Process flow to comply with the coverage requirement of completeness by reducing high-dimensional spaces to a verifiable ODD contributing to a Safety-by-Design approach for AI-based Systems.

the reduced parameter space feeds into the dependency analysis, constraint definition, and discretization steps before coverage testing is performed against the database of executed scenarios.

A decision step first checks whether *Parameter Grouping* is possible. If possible, parameters may be grouped to reduce the dimensionality of the coverage space before any further analysis is performed. For example, multiple ODD parameters describing similar conditions, such as precipitation type and precipitation intensity, could be combined into a single precipitation condition parameter. Such grouping reduces the number of parameter combinations to consider during coverage testing. However, it is important to ensure that the grouping does not mask safety-relevant distinctions and therefore does not reduce the effectiveness of the coverage analysis. To this end, grouped parameters should be validated, e.g., through sensitivity analysis or by temporarily refining the grouped ranges, to confirm that no additional safety-critical behaviors appear. This step operates at the schema level, and should therefore precede all subsequent steps, since grouping changes which parameters exist and thus which dependencies, constraints, and bin sizes need to be defined.

In the *Dependency Analysis* step, the relationships between the remaining parameters are characterized before any space reduction is performed. This is a prerequisite for the subsequent constraint definition step, as the nature of the dependency—whether deterministic or stochastic—determines the appropriate reduction method. Deterministic dependencies arise when the relationship between parameters can be derived analytically from physical laws or system kinematics. In such cases, the dependency structure is fully known and can be expressed as an exact analytical constraint. Stochastic dependencies, by contrast, arise when no closed-form relationship exists and the joint parameter distribution must be estimated from data. In these cases, statistical methods such as vine copulas, Bayesian networks, or kernel density estimation may be employed to characterize the joint probability distribution of ODD parameters [10]. When relying on such estimated dependencies, it is essential to distinguish correlation from causation, as a statistical association between parameters does not necessarily reflect a causal relationship and may yield misleading constraints if treated as one. It is important to note, however, that such statistical methods do not reduce the dimensionality of the ODD itself; they impose a joint probability distribution over the full parameter space rather than eliminating dimensions. Consequently, if the completeness criterion demands that all bin combinations be covered regardless of their likelihood, statistical dependency models provide no direct benefit to the coverage metric. Their utility lies instead in informing the subsequent constraint definition step, where low-likelihood regions can be filtered, or in prioritizing scenario generation toward high-likelihood uncovered regions.

In the *Constraints Definition* step, the insights from the dependency analysis are translated into concrete reductions of the parameter space. For deterministic dependencies, analytical constraint functions are derived directly from the known physical relationships, removing provably unreachable or irrelevant bin combinations from the required coverage set. For stochastic dependencies, likelihood-based filters derived from the estimated joint distribution can be used to exclude parameter combinations with negligible probability of occurrence, provided this probabilistic relaxation of the completeness criterion is accepted by the relevant certification authority. Beyond dependency-driven constraints, this step also identifies combinations that are physically implausible, operationally irrelevant, or of lower safety criticality. Such combinations can be excluded or tested more sparsely without significantly affecting the assessment of safety-relevant coverage. The result of this step is a reduced parameter space containing only those combinations that are both physically plausible and safety-relevant.

In the *Discretization and Binning* step, the reduced parameter space is divided into a finite number of bins per parameter, since continuous parameters make exhaustive coverage of the parameter space infeasible. Performing discretization after grouping, dependency analysis, and constraint definition ensures that binning operates on the already-reduced relevant parameter space rather than the full ODD, avoiding unnecessary computational overhead. The bin size is determined by the criticality c of each parameter, which must be defined in advance and can be specified either uniformly or as a function of c. This step is crucial, as it directly determines the number of bins and the corresponding bin combinations to be evaluated during coverage analysis. If bins are chosen too large, relevant regions of the parameter space may remain insufficiently explored, potentially leaving safety-critical gaps undetected. Conversely, selecting overly small bins significantly increases the number of combinations and can make the analysis computationally expensive. Therefore, defining an appropriate criticality measure and deriving suitable bin sizes from it is an important step in balancing computational feasibility with sufficient resolution of the parameter space. The method applies to ODD parameters that are either continuous or can be meaningfully mapped to a continuous or ordinal representation, such as translating ‘time of day’ categories into illuminance ranges. Parameters that are purely categorical without such a mapping are outside the current scope and require separate treatment.

In the *Coverage Testing* step, the discretized and reduced parameter space is systematically analyzed to determine whether the executed scenarios sufficiently cover the ODD. Each parameter is divided into a finite number of bins, and coverage is evaluated across all combinations of these bins. Let B_i denote the set of bins for parameter p_i after grouping and discretization. The full discretized parameter space is given by the Cartesian product

$$\mathcal{B} = B_1 \times B_2 \times \cdots \times B_n. \tag{1}$$

The preceding grouping, dependency analysis, and constraint-definition steps jointly yield a set of admissibility conditions $\mathcal{C}$. Applying these conditions reduces the full space to the *reduced* parameter space

$$\mathcal{B}_{\text{reduced}} = \{b \in \mathcal{B} \mid b \text{ satisfies all conditions in } \mathcal{C}\} \subseteq \mathcal{B}. \tag{2}$$

The reduction is intentionally conservative: a bin combination is removed only if it is provably infeasible or non-critical, while combinations of uncertain relevance are retained. As a result, $\mathcal{B}_{\text{reduced}}$ is an over-approximation of the truly relevant space and may still contain less-relevant bins, but no safety-relevant combination is discarded. A bin combination $b \in \mathcal{B}$ is considered *covered* if at least one data point from an executed scenario lies within the corresponding parameter ranges of b. Coverage r_{cov} is considered sufficient only if all bin combinations in the reduced parameter space are covered, i.e.,

$$r_{\text{cov}} = \frac{|\mathcal{B}_{\text{reduced,covered}}|}{|\mathcal{B}_{\text{reduced}}|} = 1. \tag{3}$$

If uncovered regions remain, the analysis produces a list of missing bin combinations. These uncovered combinations can be translated into new scenario parameters, which are then executed and added to the existing scenario set. The coverage analysis is then repeated, forming an iterative process that continues until the required coverage of the reduced parameter space is achieved.

4 Case Study: AI-Based Collision Avoidance System

The Vertical Collision Avoidance System (VerticalCAS) is an advanced aircraft collision avoidance system currently under development. It is designed as a successor to the widely deployed *Traffic Collision Avoidance System II*, with the goal of improving vertical separation advisories and enhancing overall airborne safety. VerticalCAS provides automated guidance to pilots by issuing resolution advisories based on the relative vertical motion and position of nearby aircraft, taking into account updated sensing, prediction, and advisory logic compared to its predecessor [13]. The corresponding advisories are listed in [11]. Table 1 lists the key state variables of the VerticalCAS system [11]. It includes relative altitude (h), own and intruder vertical rates ($\dot{h}_{\text{own}}, \dot{h}_{\text{int}}$), the time to closest point of approach (CPA) (τ), and the previous advisory indicator (s_{adv}), along with their corresponding ranges of values. These variables define the state space and are used as the ODD for VerticalCAS. The values are oriented on Kochenderfer et al. [11] but differ slightly since in prior work [1,3,17,18], other scenario settings were used, and the main purpose of this work is to demonstrate the process flow rather than focus on the actual numbers.

Table 1. VerticalCAS State Variables and the parameter ranges, oriented on the GitHub release https://github.com/sisl/VerticalCAS

Variable	Description	Values	Bins	Bins adj.
h	Relative intruder altitude	[−1500ft, 1500ft]	100	100
$\dot{h}_{\text{own}}$	Ownship vertical rate	[−3200ft min^{-1},3200ft min^{-1}]	32	32
$\dot{h}_{\text{int}}$	Intruder vertical rate	[−3200ft min^{-1}, 3200ft min^{-1}]	32	1
τ	Time to CPA	[0 s, 60 s]	60	60
s_{adv}	Previous advisory	see [11]	9	1

Data from our prior work was used to populate the database for the scenarios executed. For this, the data was aggregated and stored in a unified CSV file with 1.97 million rows of VerticalCAS state variables.

Parameter Grouping. In this work, parameter grouping is not applied. The VerticalCAS state space under consideration already comprises a small set of well-defined parameters, making additional grouping unnecessary. However, this step is intended for larger ODDs with many parameters, where multiple related variables can be combined into higher-level parameters, reducing the dimensionality of the scenario space and the number of bin combinations to be analyzed.

Dependency Analysis. Before any reduction is performed, the relationships among the VerticalCAS state variables are characterized to determine their nature. Two dependencies dominate the relevant encounter dynamics. First, the safety-relevant range of the relative altitude h depends on the time to CPA τ: for large τ, the full altitude range can still evolve into a conflict, whereas for small τ only small relative altitudes can. Second, the criticality of an encounter depends on the joint configuration of h and the relative vertical rate $\dot{h}_{\text{rel}}$, since their relative sign determines whether the aircraft are converging or diverging. Both relationships follow directly from the vertical kinematics of the encounter and are therefore *deterministic*: they admit a closed-form description rather than requiring estimation from data, and are accordingly handled in the next step through exact analytical constraints. Stochastic dependency models such as vine copulas, Bayesian networks, or kernel density estimation, by contrast, are designed to characterize dependence structures that have no closed-form description and must be inferred from observed data [10]. As the governing dependencies in this use case are deterministic and the parameter space is low-dimensional, there is no stochastic dependence structure to estimate, so such methods would add modeling cost without refining the reduction. They are therefore not applied here, but remain the appropriate tool for larger ODDs in which parameters exhibit genuine stochastic interactions.

Constraint Definition. The two deterministic dependencies identified above are now translated into explicit admissibility conditions that remove provably infeasible or non-critical bin combinations. The first constraint operationalizes the h–τ coupling. The bound is derived directly from the vertical kinematics of the encounter. A conflict at the CPA requires the relative altitude to be reduced below the NMAC threshold h_{NMAC} within the remaining time τ. In the considered scenarios, the intruder maintains level flight, so the relative vertical motion is governed by the ownship alone, whose vertical rate is limited to$\pm$3200ft min^{-}1, see Table 1. The relative vertical closing rate is therefore bounded in the worst case by $\dot{h}_{\text{rel,max}} =$ 3200ft min^{-}1, attained when the gap is closed at the maximum rate; for encounters in which the intruder may also maneuver vertically, the same construction applies with $\dot{h}_{\text{rel,max}}$ given by the sum of both aircraft's maximum rates. The largest relative altitude from which co-altitude is still reachable within τ is therefore

$$h_{\max}(\tau) = \min\left(h_{\text{ODD}}, h_{\text{NMAC}} + \dot{h}_{\text{rel,max}}\,\tau\right), \tag{4}$$

with $h_{\text{ODD}} = 1500\,\text{ft}$ and $h_{\text{NMAC}} = 100\,\text{ft}$. States above this bound are highly unlikely to reach a conflict within the available time and are excluded from the coverage analysis. Because the exclusion rests on kinematic reachability rather than on an assumed likelihood, no state capable of producing a conflict is discarded. The bound grows linearly with τ and reaches the full ODD range at $\tau = (h_{\text{ODD}} - h_{\text{NMAC}})/\dot{h}_{\text{rel,max}} \approx 26\,\text{s}$, beyond which the entire altitude range remains safety-relevant. It should be noted that deriving such constraints generally requires domain knowledge of the system's dynamics. The resulting envelope is shown in Fig. 2, where h is plotted over τ with a heatmap of the per-bin datapoint count; as the underlying scenarios were not designed to demonstrate coverage, substantial uncovered area remains.

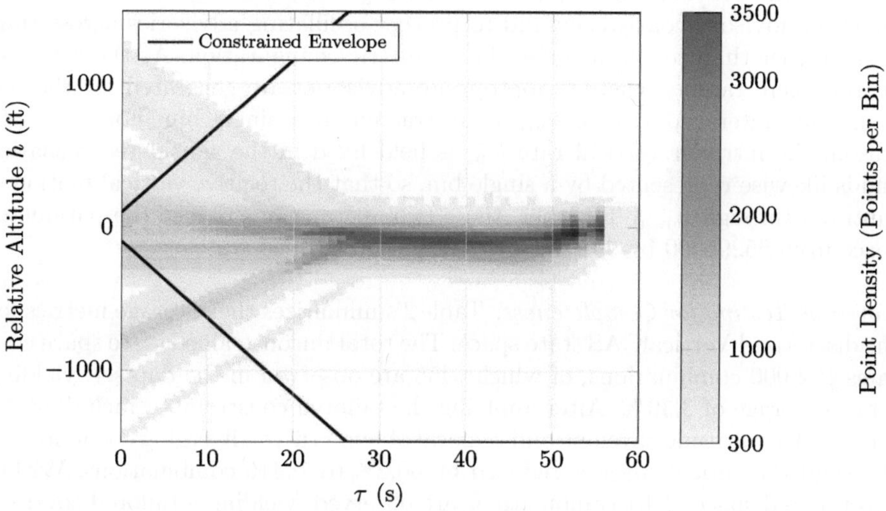

Fig. 2. Example visualization of h over τ. The heatmap indicates the number of points contained in each bin in the dataset. The black line indicates the applied constrained envelope.

The second constraint operationalizes the geometric coupling between h and $\dot{h}_{\text{own}}$. Encounters in which the aircraft are vertically diverging—ownship below the intruder ($h > 0$) while descending ($\dot{h}_{\text{own}} < 0$), or above the intruder ($h < 0$) while climbing ($\dot{h}_{\text{own}} > 0$)—produce increasing vertical separation over time and are therefore non-critical. These states are removed, retaining only converging or level-flight configurations for the coverage analysis.

Discretization and Binning. The bin size of each parameter is governed by its criticality c: the more a region of a parameter influences the safety of the advisory, the finer it must be resolved. Here, criticality is assessed from expert knowledge, although a more detailed process could derive its distribution from simulation-based analyses. Criticality may not be constant across a parameter range. For τ, for instance, an advisory issued with enough time to react is far more valuable than one issued a fraction of a second before a potential collision, which would suggest a finer resolution at reasonable larger τ. For the purpose of this demonstration, however, we assume a constant criticality distribution, which yields constant bin sizes for all parameters. The resulting bin counts are listed in Table 1. For h, $\dot{h}_{\text{own}}$, and $\dot{h}_{\text{int}}$, slightly different bin sizes than in [11] are used, while for τ a resolution of one bin per second is retained. The bin size for h is set slightly below the height of an A320[1] to provide an adequate resolution for the advisory prediction. Taken together, the full discretization yields 55 296 000 states. This space is reduced by abstracting two variables that do not describe the current encounter geometry. The state variable s_{adv} encodes the advisory issued in the previous time step and originally takes nine values. Its role is to maintain advisory consistency and to prevent oscillating advisories across time steps, rather than to characterize the geometric configuration. As the coverage analysis here targets whether appropriate advisories are generated for the relevant encounter geometries, s_{adv} is abstracted to a single bin. For the same reason, the intruder vertical rate $\dot{h}_{\text{int}}$ is held fixed in the considered scenarios and is likewise represented by a single bin, so that the relative vertical motion is captured through $\dot{h}_{\text{own}}$. Together, these two abstractions reduce the parameter space from 55296000 by 99.7% to 192000 unconstrained states.

Coverage Testing for Completeness. Table 2 summarizes the coverage metrics for the discretized VerticalCAS state space. The total unconstrained state space contains 192 000 combinations, of which 6455 are observed in the dataset, yielding a raw coverage of 3.36 %. After applying the tailored constraints—including the τ-dependent altitude envelope and removal of non-critical diverging encounters—the valid theoretical space is reduced by 60.2% to 76416 combinations. Within this tailored space, 2001 combinations are observed, yielding a tailored coverage of 2.62%. While the constrained coverage is lower in this example, it demonstrates that the effort required to achieve full coverage with newly generated scenarios is reduced because only physically meaningful and relevant combinations are left to consider.

In addition, the script used for this coverage analysis generates a list of missing combinations in the theoretical state space. These missing combinations correspond to regions of the parameter space that have not yet been observed in the dataset. By translating these combinations back into physical values, new scenarios can be created and added to the dataset. Iteratively generating

[1] https://www.airbus.com/en/products-services/commercial-aircraft/passenger-aircraft/a320-family.

Table 2. Coverage report for VerticalCAS: Unconstrained vs. Constrained

Metric	Unconstrained	Constrained
Combinations	192 000	76 416
Combinations Covered	6455	2001
Coverage	3.36%	2.62%

and executing such scenarios allows coverage to be systematically increased, ultimately achieving *complete* coverage of the constrained operational envelope.

5 Discussion

The transition of AI into safety-critical aviation relies on verifiable trust. Our method bridges abstract ODD definitions and EASA's completeness requirement, turning the loosely defined notion of covering *the whole ODD* into a quantitative, exhaustive criterion over a discretized parameter space—a systematic, end-to-end process where none previously existed. By evaluating completeness across the Cartesian product of discretized bins rather than parameter-wise, it verifies that all parameter ranges are covered across all others, exposing safety gaps in interdependent regions of the ODD that independent parameter testing would overlook, even in high-dimensional state spaces. Two design choices carry implications beyond the case study. The first is how dependencies are treated: statistical models such as vine copulas impose a joint distribution but do not reduce dimensionality, so under a strict criterion requiring every bin combination regardless of likelihood, they offer no direct benefit to the coverage metric. Their role is to inform constraint definition and prioritize scenario generation; genuine reduction instead stems from grouping, the abstraction of non-geometric variables, and constraints that provably remove unreachable or non-critical combinations. The second is how those constraints are derived: grounding the τ-dependent altitude bound directly in the encounter's vertical kinematics makes exclusion rest on physical reachability rather than on an assumed likelihood. The reduction is therefore conservative—a combination is removed only if provably infeasible or non-critical—so the reduced space over-approximates the truly relevant one and discards no safety-relevant combination. Performing binning last, on the already-reduced space, keeps the process tractable. Applied to existing simulation data, the method yielded relatively low coverage in the VerticalCAS case study—expected, since the original scenarios were not designed for ODD completeness; the result's value lies in pinpointing untested operational regions and translating them into new scenario parameters. Coverage here is an exercising criterion: a bin counts as covered once a single scenario falls within it, demonstrating that the region was tested, but not that the model behaves correctly there. It is thus necessary but not sufficient for safety—a structural complement to behavioral verification rather than a substitute. Two limitations frame the contribution. First, coverage is defined over the ODD's input parameter space, not the ML constituent's learned latent representation. Input-space completeness ensures the operational conditions an applicant must justify are systematically exercised, but does not guarantee uniform coverage of the internal feature space, where semantically distinct inputs may collapse to nearby representations or nearby inputs diverge. The method thus targets EASA's operational completeness rather than AI-specific failure modes such as latent-space clustering or distribution-shift sensitivity; we regard parameter and latent-space coverage as complementary. Second, weighting all bins equally can direct effort toward operationally improbable regions; the optional likelihood-based relaxation, filtering low-probability combinations subject to certification-authority acceptance, is the intended mechanism for this trade-off.

Finally, the VerticalCAS study was kept compact so the method's mechanics remain transparent; it is therefore a proof of concept rather than an operational-scale demonstration. For high-dimensional ODDs, where the unconstrained space grows exponentially, the advanced stages—criticality-based bin sizing, parameter grouping, and stochastic dependency modeling for constraint definition—are what keep verification tractable while preserving safety-relevant coverage. Demonstrating the method at that scale is the principal direction for future work.

6 Conclusion and Outlook

This paper addresses the engineering gap between abstract ODD definitions and the *completeness* evidence required by EASA for AI/ML certification. We proposed a structured, end-to-end method that reduces the parameter space through grouping, dependency analysis, and constraint definition before testing coverage exhaustively over the Cartesian product of the discretized bins. A central element is constraint-based dimension reduction, which restricts verification to physically plausible and safety-relevant regions and thereby keeps the completeness check computationally feasible in high-dimensional spaces. Applied to the VerticalCAS collision-avoidance use case, the method reduced the unconstrained state space by approximately 60% (from 192 000 to 76 416 combinations) while preserving all safety-relevant combinations, and the same analysis yields an explicit list of uncovered bins that can be translated into new scenario parameters. This iterative gap-closing loop turns EASA's qualitative completeness requirement into a quantifiable, repeatable engineering procedure. It contributes a building block toward a Safety-by-Design approach for future AI-based systems. Several directions remain open. Because the case study deliberately used a compact parameter set, the parameter-grouping step—intended to combine related variables and curb combinatorial growth in larger ODDs—was not exercised, and demonstrating it on a higher-dimensional ODD with many interrelated parameters is a primary next step. In the same setting, stochastic dependency modeling, e.g., through vine copulas, Bayesian networks, or kernel density estimation, can inform constraint definition where relationships are not analytically known. Further work will replace the uniform criticality assumption with empirically or simulation-derived criticality distributions, yielding non-uniform bin resolutions that concentrate verification effort where it matters most. Finally, extending the coverage criterion from the input parameter space toward the model's learned representation, and coupling it with behavioral assurance evidence, would broaden the method from operational completeness toward AI-specific safety guarantees. As a concrete step in that direction, and because covering a bin only exercises it rather than verifying behavior, coupling the criterion with fault-injection or mutation-based analysis in the spirit of the PIE model would yield stronger per-region evidence.

Disclosure of Interests. The authors have no competing interests to declare that are relevant to the content of this article.

References

1. Christensen, J.M., et al.: Advancing the AI-based realization of ACAs X towards real-world application. In: 36th IEEE International Conference on Tools with Artificial Intelligence (ICTAI), pp. 57–64. Herdon, VA, USA (2024). https://doi.org/10.1109/ICTAI62512.2024.00017
2. Christensen, J.M., et al.: Formulating an engineering framework for future ai certification in aviation. Aerospace **12**(6) (2025). https://doi.org/10.3390/aerospace12060482, https://www.mdpi.com/2226-4310/12/6/482
3. Christensen, J.M., et al.: Towards certifiable AI in aviation: a framework for neural network assurance using advanced visualization and safety nets. In: 43rd Digital Avionics Systems Conference (DASC). IEEE, San Diego (2024). https://doi.org/10.1109/dasc62030.2024.10749321
4. Cotroneo, D., Pietrantuono, R., Russo, S.: RELAI Testing: a technique to assess and improve software reliability. IEEE Trans. Softw. Eng. **42**(5), 452–475 (2016). https://doi.org/10.1109/TSE.2015.2491931
5. EASA and Collins Aerospace: Formal Methods Use for Learning Assurance (Formula). Tech. rep, European Union Aviation Safety Agency (EASA) (2023)
6. European Union Aviation Safety Agency (EASA): Artificial intelligence roadmap 2.0. techreport, European Union Aviation Safety Agency (EASA), Postfach 10 12 53, 50452 Cologne, Germany (2023). https://www.easa.europa.eu/en/document-library/general-publications/easa-artificial-intelligence-roadmap-20
7. European Union Aviation Safety Agency (EASA): Easa concept paper: Guidance for level 1 & 2 machine learning applications. techreport, European Union Aviation Safety Agency (EASA), Postfach 10 12 53, 50452 Cologne, Germany (2024). https://www.easa.europa.eu/en/document-library/general-publications/easa-artificial-intelligence-concept-paper-issue-2
8. Hirschle, M., Kirov, D., Aievola, R., Adamy, J.: Coverage-driven synthetic data generation for machine learning assurance. In: 2024 AIAA DATC/IEEE 43rd Digital Avionics Systems Conference (DASC), pp. 1–10. IEEE (2024). https://doi.org/10.1109/dasc62030.2024.10748692
9. Hoemann, E., et al.: Towards a domain-agnostic safety-by-design AI engineering pipeline. In: Proceedings of the 1st International Workshop on Engineering Autonomous Systems Intelligence (EASI 2026), co-located with the 38th International Conference on Advanced Information Systems Engineering (CAiSE 2026) (2026)

10. Höhndorf, L., Dmitriev, K., Vasudevan, J.K., Subedi, S., Klarmann, N., Holzapfel, F.: Artificial intelligence verification based on operational design domain (odd) characterizations utilizing subset simulation. In: 2024 AIAA DATC/IEEE 43rd Digital Avionics Systems Conference (DASC), pp. 1–10. IEEE (2024). https://doi.org/10.1109/dasc62030.2024.10749586
11. Julian, K.D., Kochenderfer, M.J.: Guaranteeing safety for neural network-based aircraft collision avoidance systems. In: 2019 IEEE/AIAA 38th Digital Avionics Systems Conference (DASC), pp. 1–10. IEEE, San Diego, CA, USA (2019). https://doi.org/10.1109/dasc43569.2019.9081748
12. Julian, K.D., Kochenderfer, M.J., Owen, M.P.: Deep neural network compression for aircraft collision avoidance systems. J. Guid. Control. Dyn. **42**(3), 598–608 (2019) https://doi.org/10.2514/1.G003724
13. Kochenderfer, M.J., Holland, J.E., Chryssanthacopoulos, J.P.: Next-generation airborne collision avoidance system. Lincoln Laborat. J. **19**, 17–33 (2012). https://api.semanticscholar.org/CorpusID:2372044
14. Precedence Research: Artificial intelligence in aviation market size, share, and trends 2024 to 2034. resreport 1748, Precedence Research (2022). https://www.precedenceresearch.com/artificial-intelligence-in-aviation-market
15. Schäfer, S., Cichon, M.: Incremental validation of automated driving functions using generic volumes in micro- operational design domains (2025). https://arxiv.org/abs/2512.11351
16. Späth, H., Daw, Z.: Towards detecting unintended behaviors in machine learning algorithms. In: 2024 AIAA DATC/IEEE 43rd Digital Avionics Systems Conference (DASC). IEEE (2024). https://doi.org/10.1109/dasc62030.2024.10749477
17. Stefani, T., et al.: Automated scenario generation from operational design domain model for testing ai-based systems in aviation. CEAS Aeronaut. J., 1–19 (2024). https://doi.org/10.1007/s13272-024-00772-4
18. Stefani, T., et al.: Applying model-based system engineering and devops on the implementation of an AI-based collision avoidance system. In: 34th Congress of the International Council of the Aeronautical Sciences (ICAS). CEAS, Florence, Italy (2024). https://www.icas.org/icas_archive/icas2024/data/papers/icas2024_0869_paper.pdf
19. Voas, J.M., Miller, K.W.: Software testability: The new verification. IEEE Softw. **12**(3), 17–28 (1995). https://doi.org/10.1109/52.382180
20. Weissensteiner, P., Stettinger, G., Khastgir, S., Watzenig, D.: Operational design domain-driven coverage for the safety argumentation of automated vehicles. IEEE Access **11**, 12263–12284 (2023). https://doi.org/10.1109/ACCESS.2023.3242127, https://ieeexplore.ieee.org/document/10036064

From Concept to Capability: Evaluating 3D Gaussian Splatting for Synthetic Scene Editing in Autonomous Driving

Ali Nouri[1,4(✉)], Yifei Zhang[2,3], Yifan Zhang[3], Tayssir Bouraffa[1], Zhennan Fei[1,4], Zijian Han[2], Håkan Sivencrona[1], and Anders Heyden[3]

[1] Volvo Cars, Gothenburg, Sweden
{zhennan.fei,hakan.sivencrona}@volvocars.com
[2] Zenseact AB, Gothenburg, Sweden
[3] Lund University, Lund, Sweden
[4] Chalmers University of Technology, Gothenburg, Sweden
ali.nouri@volvocars.com

Abstract. The perception of an Autonomous Driving System (ADS) critically depends on relevant, comprehensive, and diverse datasets to ensure its safety while operating in the environment (i.e., Operational Design Domain). Field data collection lacks completeness with respect to the list of rare but still possible safety-related scenarios needed for the development, verification, and validation of the ADS. 3D Gaussian Splatting (3DGS) has shown promising capabilities for the reconstruction and editing of scenes based on data collected by cameras and LiDAR sensors. However, the industrial fidelity evaluation of reconstructions is underexplored, which is crucial when employing such methods in safety-related systems, especially for ADS. This becomes more challenging as ADS operates in a dynamic, uncontrolled environment with limited viewpoints and often partially occluded objects. This paper addresses this gap by proposing and implementing a framework to systematically analyze the capabilities and limitations of 3DGS for use in the reconstruction of safety-related scenes. It focuses on the quality of reconstruction for vehicles and pedestrians, which are the two critical object classes for safe ADS. Our findings provide industry insights into the fidelity degradation of reconstructions from multiple novel viewpoints, both lateral and longitudinal, enabling the integration of these methods into real-world industrial AD software development and testing pipelines.

Keywords: Autonomous Driving System · Verification · Simulation · 3DGS

1 Introduction

Verification and validation (V&V) serve as a crucial safety net to catch design mistakes and insufficiencies, especially for ADS perception systems, which largely

© The Author(s), under exclusive license to Springer Nature Switzerland AG 2027
E. Troubitsyna et al. (Eds.): SAFECOMP 2026, LNCS 16868, pp. 161–175, 2027.
https://doi.org/10.1007/978-3-032-34867-8_11

rely on ML-based software. Virtual environments are a common approach for performing Software-in-the-Loop (SiL) testing and are also highlighted in safety reports of most ADS players [19]. They are particularly useful for testing scenarios that are rarely encountered in the real world, such as animals on the road, while others may be too dangerous to collect, such as emergency collision data involving children or high-speed impacts.

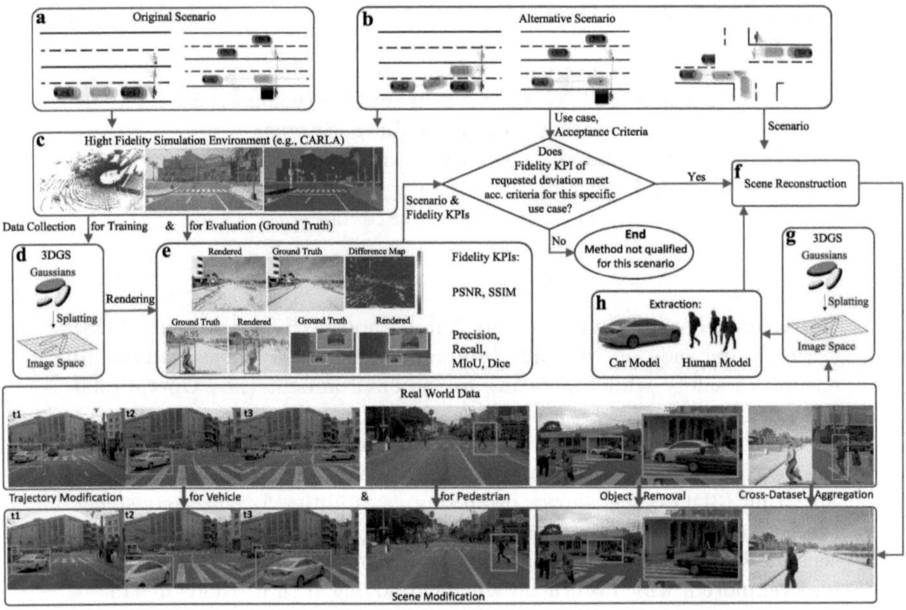

Fig. 1. Presents the Dynamic Fidelity Evaluation framework for scene reconstruction: Initially, within the simulation environment (c), training data from the original scenario (a) and evaluation data from an alternative scenario (b) are collected. Reconstruction (d) is then performed based on the original view, and then a novel viewpoint is rendered. The rendered data is compared (e) against the evaluation data, fidelity KPIs are extracted. Finally, if the KPIs meet the acceptance criteria for each use case (e.g., AEB testing "Avoid Collision" during "Cutting-in"), the reconstruction (g, h, and f) is considered valid. The bottom of the flow presents multiple examples showcasing the capabilities of 3DGS in constructing alternative scenarios using real-world datasets.

With a realistic simulation, all of these cases can be safely and massively generated within a reasonably short time frame. Moreover, a wide range of scenarios can be applied to ensure sufficient data diversity (Fig. 1). These variations extend beyond environmental conditions like lighting, weather, or background, and also include differences in vehicle trajectories and viewpoints. Moreover, the complexity of end-to-end transformer-based self-driving algorithms brings high requirements for data quality and volume [16]. Satisfying these requirements is challenging using only real-world vehicle-collected data, since the data collection process using test vehicles or fleets is slow and costly. However, the benefits of synthetic data in terms of efficiency and cost have shown great potential to address these issues. Data diversity is another crucial aspect to consider, since covering as many scenarios as possible during training is key to assure safety and robustness. Using a game engine like Unity or Unreal Engine to generate synthetic data for training or testing can be seen as one potential solution. However, the purely game-engine-based method requires a huge amount of human effort for asset creation and environment tuning. It is also very challenging to ensure that manually created assets and environment settings are realistic enough for neural networks. As the need for different environments and fidelity levels increases, this method is very limited in terms of automation and scalability. Other data enrichment methods, such as GANs, can efficiently generate a large variety of new data but lack precise control over the data content.

Recent work in neural rendering methods, such as Neural Radiance Fields (NeRF) and 3D Gaussian Splatting (3DGS), underscores promising potential to address the aforementioned demand. 3DGS, in particular, has gained attention due to its efficient reconstruction from camera and LiDAR data as well as real-time rendering performance, which is crucial for the training and V&V of ADS software stacks. Despite these advancements, most studies evaluate 3DGS in specific scenarios without shedding light on its capabilities and limitations that must be considered when using it in a real industrial project. This is crucial for safety-related use cases, as the fidelity of the method's output must be considered, and arguments around this aspect are often the missing part of the process. Hence, this paper aims to critically assess the capabilities and limitations of 3DGS and proposes a framework to systematically define the boundaries based on acceptable fidelity metrics. In this work, we mainly focus on scenarios involving vehicles (high exposure) and pedestrians (high exposure in urban and high severity). Additionally, we integrate multiple methods to improve the reconstruction of pedestrians within the framework. By rigorously identifying the fidelity boundaries of 3DGS, this research aims to serve as an industrial guideline for stakeholders, highlighting the capabilities and limitations of synthetic data generation and providing clear guidance for the effective integration of 3DGS into ADS development and V&V.

2 Related Work

Data Gap and Simulation: Despite the significant increase in the quality and variety of autonomous driving datasets in recent years, real-world data often proves insufficient and not inclusive of all unknown or complex scenarios [16]. Moreover, some scenarios are the result of analysis methods, such as Hazard Analysis and Risk Assessment [20], which lead to requirements and then to implemented algorithms. However, when moving toward end-to-end approaches, it becomes the role of data to represent these scenarios, which are difficult to capture because they are hypothetical.

Synthetic data generation by employing computer graphics and simulations is proposed as a potential cost-effective and flexible approach to improve diversity in datasets and construct difficult or unsafe scenarios [22]. Several simulation environments have been developed for ADS, such as LGSVL [21] and Car Learning to Act (CARLA) [5]. For instance, CARLA is an Unreal Engine-based simulation environment focused on urban environments with customizable weather conditions. However, data-driven methods are gaining attention, as assets in simulation environments are not easily scalable or realistic enough [4], especially when it comes to tackling previously rare or unseen objects.

NeRF vs. 3DGS: Prior work on neural scene reconstruction for view synthesis in driving environments largely follows two families: volumetric, NeRF-style methods and explicit, 3D Gaussian Splatting (3DGS). NeRF represents a scene as an implicit radiance field and achieves high-quality novel views from sparse, real-world imagery [18]. However, the combination of per-scene optimization and volumetric ray marching leads to comparatively high compute cost for both training and inference, which becomes a practical limitation for autonomous driving where large volumes of re-constructed data are needed for development, verification, and validation [8]. Even with accelerated variants, end-to-end reconstruction typically remains minutes to days per scene depending on hardware, resolution, and view coverage, and city-scale deployments commonly partition space into many sub-models, increasing operational overhead. 3DGS addresses these efficiency constraints by replacing volumetric integration with explicit anisotropic 3D Gaussians rendered via differentiable splatting/rasterization, while learning per-primitive appearance and opacity [15]. Under common settings, this formulation yields interactive to real-time rendering and shorter time-to-usable reconstructions, with competitive perceptual quality relative to NeRF-style renderers [8], [15]. For AD/ADAS use cases, 3DGS's runtime profile is better aligned with industrial needs, since perception-in-the-loop evaluation demands high frame throughput, frequent re-rendering from novel viewpoints, and coverage of large outdoor environments. Beyond camera-only setups, recent 3DGS pipelines also integrate multi-sensor inputs (e.g., LiDAR) to stabilize geometry and reduce view-dependent artifacts, which is valuable for safety-critical analysis and editing workflows [8,15].

Within the 3DGS family tailored to road scenes, StreetGS [27] is a specialized implementation of the 3D Gaussian Splatting approach to reconstruct both static and dynamic elements of street scenes from multiple camera views. To improve the fidelity of specific road users, such as pedestrians and cyclists, and to capture details of human posture and movement, methods including OmniRe [4], SplatAD [8], and DistillNeRF [25] have been introduced. Taken together, 3DGS currently offers a more practical balance of fidelity, efficiency, and ecosystem compatibility for industrial-scale AD/ADAS pipelines, which motivates our focus on 3DGS-based benchmarking in this work.

Pedestrian Gesture Modelling: As pedestrians are involved in scenarios with the highest risk (i.e., high severity in the event of an incident and high exposure in driving scenarios), specialized human modeling and generation methods are required to improve the fidelity of the reconstructed scenes. Human body gestures are an important input for perception, as they affect pedestrian trajectory prediction, which can be used in the precautionary behavior of both

AD and ADAS functions. The complex motion characteristics and appearances of the human body require specialized modeling techniques, which impose limitations on the aforementioned methods. Skinned Multi-Person Linear Model (SMPL) [17] improves human reconstruction by incorporating the known human body structure, which is further enhanced by NeuMan [14].

Safety Standards: Reconstruction techniques such as 3DGS are relatively new and fidelity guidelines are still missing in standards such as ISO 26262 [10], 21448 [9], 8800 [11] and 4804 [12]. For instance, ISO/PAS 8800 (Road vehicles - Safety and artificial intelligence) [11] proposes the use of synthetic datasets to address insufficiencies in training and testing data, but does not prescribe any specific method or process for ensuring the quality or validity of such datasets.

Research Gap: Existing scene reconstruction methods mainly demonstrate reconstruction quality or editing capability from specific viewpoints. However, for AD applications, the key open question is not only whether the method can reconstruct handpicked scenarios, mostly under longitudinal deviations and specific conditions, but also whether the method can remain reliable when employed for novel viewpoints in different directions that were not sufficiently covered by the original sensor viewpoints and can be evaluated against ground truth. This is particularly crucial for employing these methods in real industrial use cases, where vehicle-mounted cameras and LiDAR sensors have viewpoint limitations, especially for dynamic actors such as vehicles and pedestrians. Therefore, the fidelity boundaries of 3DGS reconstructions must be identified before such reconstructed datasets can be employed in AD development, verification, or validation, which will be addressed in this study.

Research Questions: Hence, this study addresses the following research questions.
RQ1- How reliable is 3DGS for reconstructing novel-view traffic scenarios from real-world sensor data in AD?
RQ2- How can the fidelity of reconstructed novel-view traffic scenarios from real-world sensor data be dynamically evaluated and assessed for AD?

Contributions: To answer these questions, this paper proposes a systematic and dynamic fidelity evaluation framework for assessing 3DGS-based scene reconstruction and editing under controlled lateral, longitudinal, and elevation viewpoint deviations (*RQ2*). Employing the framework, multiple 3DGS methods are also compared for vehicles and pedestrians separately, for both computer vision and AD perception metrics. Based on these results, the paper identifies and reports fidelity boundaries for reconstructed novel views in different directions (*RQ1*).

3 Methodology

This section describes the methodological framework designed to identify the capabilities and limitations of 3DGS by evaluating its fidelity and applicability under different spatial deviations from the original viewpoint, as illustrated in Fig. 1. Our approach includes systematic data acquisition for both the original viewpoint and the ground truth of novel viewpoints to accurately evaluate reconstruction performance. We first compared state-of-the-art methods, then focused on the best-performing model, assessing its capabilities across different object classes. Finally, we integrated complementary methods to address its identified limitations.

3D Gaussian Splatting Reconstruction: We implemented a pipeline employing 3DGS, and among various candidate models, we selected DeformGS, StreetGS, PVG, and OmniRe. We then focused further on OmniRe, as it outperforms the others. Our pipeline uses OmniRe, a dynamic 3DGS method, to reconstruct scenes from multi-modal simulation data. OmniRe organizes the scene as a hierarchical Gaussian Scene Graph, modeling static backgrounds, vehicles, and pedestrians with distinct sets of Gaussians. Inputs include calibrated RGB images and LiDAR point clouds from CARLA, with preprocessing to extract camera parameters and segment dynamic objects. During training, OmniRe optimizes the spatial, visual, and transparency attributes of the Gaussians to minimize rendering error. By integrating LiDAR depth, it enhances convergence speed and spatial precision, particularly around object boundaries. We also introduce training augmentations like camera perturbations and spatial regularization to improve generalization and prevent overfitting to training trajectories. Due to the complexity and non-rigid behavior of humans, these methods lack adequate performance in modeling pedestrians. Given their criticality, we subsequently employed more specialized models for human body reconstruction [17]. Our framework leverages the Gaussian scene representation to enable real-time, training-free edits of reconstructed environments. By directly manipulating the Gaussian nodes, we can remove or translate rigid objects, simply omitting a vehicle's Gaussians to delete it, or applying a rigid-body transform like translation and rotation to reposition it along a new trajectory. These operations let us synthesize safety-related scenarios, such as a jaywalking pedestrian suddenly dashing into the road or an oncoming car veering into the ego-lane, while preserving photorealism and physical consistency. Finally, a lightweight trajectory module, such as straight interpolation, S-curve, lane-change, produces novel-view camera paths across the edited scenes, yielding the tailored synthetic sequences for downstream evaluation.

Data Acquisition from Simulation: CARLA (Car Learning to Act), a high-fidelity simulation environment, is used to generate a diverse dataset by altering only one factor per scenario. CARLA is an open-source simulation platform built on Unreal Engine 4, capable of recreating various weather conditions (e.g., clear day, rain, sunset). It supports multiple sensor modalities, such as cameras and LiDAR, and automatically provides detections and ground truth data, including semantic segmentation. CARLA offers several key advantages that make it an ideal for evaluating reconstruction methods:

1- Controlled Environment: CARLA enables precise control over agents in the scene, lighting conditions, and environmental factors, which is essential for eliminating confounding variables.
2- Novel Viewpoint Sampling: Unlike real-world datasets, which are restricted to sensors mounted on the vehicle, CARLA allows for reconstruction of scenarios from novel viewpoint trajectories, which can be used as ground truth for evaluating pipeline output. Moreover, it enables data collection from sensors placed outside the vehicle.
3- Accurate and Effortless Ground Truth: The simulator provides instant, accurate ground truth for all modalities (RGB, depth, semantics) from any viewpoint, facilitating comprehensive evaluation. This allows us to test various scenarios without the burden of manually generating ground truth and eliminates inaccuracies in the evaluation process.

Real World Datasets: In this study we have selected Waymo Open Dataset [23], Zenseact Open Dataset [1], and KITTI [7] due to their high-quality LiDAR and vision sensor data with fused data and detailed annotations, collected across various driving conditions. We showcased the advantage of the method by extracting a human from one dataset and injecting it into another scenario in a different dataset to demonstrate the method's capability for aggregating object models, an essential feature for industrial use cases, since datasets are sometimes collected by different parties with different sensor setups. Moreover, it enables the industry to share their datasets and treat the objects in them like assets in simulation, which helps with sharing rare agents or behaviors across the industry to better collaborate and improve system performance.

Experimental Setup: We conduct experiments on a suite of four simulated driving scenarios created in CARLA. These scenarios include: (1) Following: the ego vehicle following a lead car on a straight road with a pedestrian on the sidewalk, (2) Intersection: the ego car approaching an intersection where cross-traffic and pedestrians are present, (3) Overtaking: a scenario with a stalled vehicle and the ego vehicle changing lanes to overtake, and (4) Oncoming: a two-way road scenario with an oncoming car that potentially swerves. Each scenario is run for a few seconds, yielding on the order of 200 multi-view frames (1000 images across the 5 cameras). The multi-view images and LiDAR point clouds from each scenario are used to train four separate 3DGS reconstructions. We evaluate several 3D reconstruction methods on these scenarios for comparison: DeformGS [6] as a baseline image-based dynamic NeRF, PVG [3] which uses explicit 3D bounding boxes to guide Gaussian placement, StreetGS [27] which incorporates LiDAR for street scenes, and OmniRe [4] - which uses LiDAR and a scene graph. We use the authors' released implementations for these methods and apply a consistent set of parameters and training iterations for fairness. All reconstructions are supervised with the same set of input views and LiDAR, when applicable. After training, for each method and scenario, we render novel views not seen during training. We define a systematic set of camera poses for novel view testing, which includes view-points along the original trajectory to test interpolation and increasingly distant viewpoints that deviate laterally and vertically from the path to test extrapolation. We categorize novel views into three bins: near (0.5 m from training path), medium (1.6 m), and far (3.2 m).[1]

Metrics Used for Evaluation: To rigorously assess the fidelity of scene reconstruction by each method, we designed a layered evaluation strategy that combines state-of-the-art computer vision metrics with AD perception-oriented measures. This breakdown enables us to better analyze the capabilities and limitations of the methods in reconstructing different objects and aspects of a scene for both original and novel viewpoints. Peak Signal-to-Noise Ratio (PSNR) [4] and Structural Similarity Index Measure (SSIM) [4] are used in computer vision to quantitatively evaluate rendering quality compared to the ground truth. In our study, these two metrics are computed for both the full image and specifically for non-sky regions and dynamic objects (vehicles and pedestrians). However, PSNR and SSIM alone may not fully capture the reconstruction fidelity from the perspective of AD perception. To address this, two state-of-the-art robust object detection models, YOLOv10 [24] and DETR [2], are used to assess how detectable the modified agents are in the scene. The models are real-time and are pre-trained on the COCO dataset, which avoids data leakage from our datasets (CARLA and real datasets). For each rendered novel view, we evaluate detection precision and recall for full images, vehicles, and pedestrians. Lastly, semantic segmentation metrics are calculated to evaluate the quality of pixel-level semantic understanding, which is a crucial capability for safe path planning and collision avoidance. To this end, mean Intersection-over-Union (mIoU) [26] (i.e., an indicator of average overlap) and Dice coefficient [13] (i.e., an indicator of object boundaries) scores are used, which complement the object detection evaluations.

Pass/Fail Criteria for Reconstruction Fidelity: As highlighted in Sect. 2, Fidelity assurance processes for reconstruction techniques are not yet established in existing standards [9–11]. However, the required guidelines for these new approaches can be extrapolated by looking at existing processes with similar purposes. For instance, according to ISO/TR 4804 [12], the model used in a simulation environment can be accurate only to a certain degree, and the pass/fail criteria for accuracy highly depend on the test goal. For example, when testing the behavior of a function, a lightweight and abstract simulation environment could be sufficient, whereas for other tests, even a high-fidelity simulation environment may not suffice, and real-world testing is required. By adapting the process described in Part 8 of ISO 26262 [10], which addresses conventional tools and methods used in the development of safety-related software, the pass/fail criteria can be defined based on the following factors:
1- Reconstruction Error Impact: As reconstructed data can be used for both training and testing of AD software, it is crucial to distinguish between these use cases. Using reconstructed data for training could contaminate the software deployed in the vehicle, making the impact significantly higher than when the same data is used for verification and validation.

---

[1] Our implemented framework, video demo, and hardware and time resources are available at GitHub repository.

2- Reconstruction Error Detection: Errors in synthetic data can often be detected, especially when the software is tested in simulation environments using various available methods. However, detecting the effects of errors in training data is more difficult, and even when such effects are identified, it is hard to trace them back to the training data.
3- Safety Requirements: Safety requirements of the perception system are further broken down into data quality requirements. For instance, in Automated Emergency Braking (AEB), false positives (i.e., brake requests when there is no imminent collision) are assigned a higher Automotive Safety Integrity Level (ASIL) than false negatives (i.e., no brake request when a collision is imminent). This is because, in the case of a false negative, the driver can still react and apply the brakes, resulting in a lower ASIL. However, if the system brakes unnecessarily, the driver may not have enough time to respond, potentially causing a rear-end collision. Therefore, in this context, precision should be prioritized, while lower recall is more acceptable. The criticality of these requirement changes in AD, where both false positives and false negatives must be minimized.

Hence, as shown in Fig. 1, the fidelity of reconstructions should be evaluated for each specific scenario and use case. As reported in Fig. 3, in this study, we assume that if the precision or recall remain above 90%, the rendered images are treated as acceptable for the considered use case.

4 Results and Discussion

Quality of The Spatial Novel View: As shown in Table 1, OmniRe consistently demonstrates the best performance for both original and novel viewpoint reconstruction with respect to PSNR and SSIM for the full image. When looking at specific object classes, StreetGS and DeformGS show the weakest performance for human reconstruction in both original and novel viewpoints, while StreetGS achieves relatively strong performance for vehicles, ranking second after OmniRe. A potential reason for this might be StreetGS's dependency on LiDAR information, which is normally insufficient for humans. The non-rigid nature of the human body requires a high point cloud density, which is generally not available, whereas vehicles, due to their rigid structure, require less. This pattern confirms that different scene elements may require tailored reconstruction strategies. Moreover, OmniRe presents the strongest performance for both pedestrians and vehicles. The results also indicate that OmniRe and StreetGS, which use LiDAR data, achieve overall better performance compared to DeformGS and PVG, which do not use it.

The performance evaluation of the methods is further visualized in Fig. 2.a for different lateral distances from the training trajectory, which shows that all methods experience a clear and consistent drop in PSNR and SSIM. OmniRe consistently maintains higher performance, while the other methods rank next in overall performance, with PVG, StreetGS, and DeformGS following in order. This leads to the conclusion that spatial generalization is fundamentally limited in current 3DGS pipelines, particularly for scenes with dynamic objects and

Table 1. Comparison of different methods on scene reconstruction and average of spatial novel view for 10 cameras over three different lateral distances. The heatmap highlights relative rankings, from green to red indicating best to worst. StreetGS and DeformGS failed to reconstruct human (as they cannot model deformable objects [4]); hence, the corresponding metrics are marked as not applicable (N/A).

Methods	Box	LiDAR	Scene Reconstruction						Spatial Novel View					
			Full Image		Human		Vehicle		Full Image		Human		Vehicle	
			PSNR↑	SSIM↑	PSNR↑	SSIM↑	PSNR↑	SSIM↑	PSNR↑	SSIM↑	PSNR↑	SSIM↑	PSNR↑	SSIM↑
DeformGS(Yang et al., 2023c)	✓		27.82	0.915	N/A	N/A	25.18	0.860	22.08	0.733	N/A	N/A	20.92	0.732
PVG(Chen et al., 2023)	✓		33.32	0.942	19.41	0.621	24.23	0.812	20.04	0.649	15.33	0.645	20.24	0.743
StreetGS(Yan et al., 2024)	✓	✓	31.84	0.937	N/A	N/A	26.18	0.808	20.66	0.656	N/A	N/A	21.43	0.789
OmniRe(Chen et al., 2024)	✓	✓	34.07	0.951	23.90	0.787	28.30	0.876	23.75	0.768	15.16	0.619	21.52	0.806

limited viewpoints on objects, which are critical for safe autonomous driving. Furthermore, depending on the required fidelity limit, the reconstruction validity is bounded to a certain lateral distance from the training trajectory.

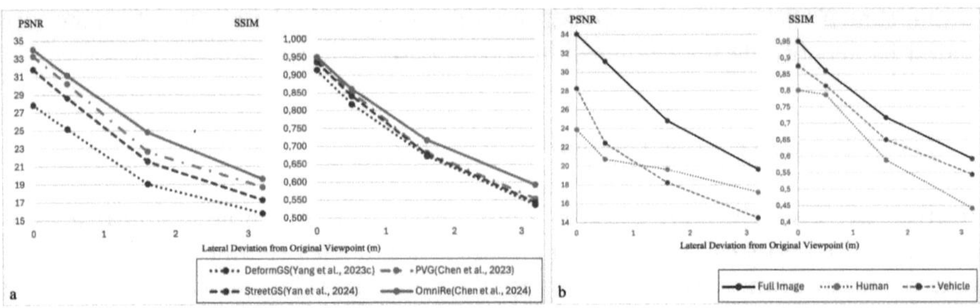

Fig. 2. (a) Presents evaluation of four methods based on distance for PSNR and SSIM metrics. As it is presented the performance of all methods drop by increasing spatial distance in novel viewpoint. Moreover, OmniRe consistently has better performance compared with other three methods. (b) Presents evaluation of OmniRe across spatial distances and object categories. The performance drops as the novel viewpoint goes further, and it is the same trend for the two main dynamic objects (human and vehicle).

OmniRe, as the best performer, is selected for further investigation of how the performance for humans and vehicles differs with respect to the full image, as shown in Fig. 2.b. For these two dynamic objects, the performance is understandably lower, while the drop in performance with increasing lateral distance follows the same slope as for the full image. This highlights that human and vehicle reconstruction requires denser data collection and, if possible, complementary generative techniques to maintain high-fidelity representations in novel views. By looking more closely at Fig. 2.b, it can be seen that PSNR is slightly higher for humans than for vehicles, while for SSIM, the opposite is true. This suggests that while pixel intensities are more accurate (higher PSNR) for humans compared to vehicles, the structural details are less accurate (lower SSIM). This

Table 2. Comparison of different methods on average precision and recall by Yolo and DETR.

Methods	Full Image		Human		Vehicle	
	Precision↑	Recall↑	Precision↑	Recall↑	Precision↑	Recall↑
DeformGS	0.893	0.866	N/A	N/A	0.931	0.893
PVG	0.924	0.654	0.858	0.566	0.978	0.726
StreetGS	0.931	0.931	N/A	N/A	0.970	0.968
OmniRe	0.943	0.915	0.919	0.833	0.963	0.982

is a result of the rigidity of vehicles versus the dynamic and softer structure of pedestrians. This highlights the importance of considering both metrics together, as they capture complementary aspects of reconstruction quality.

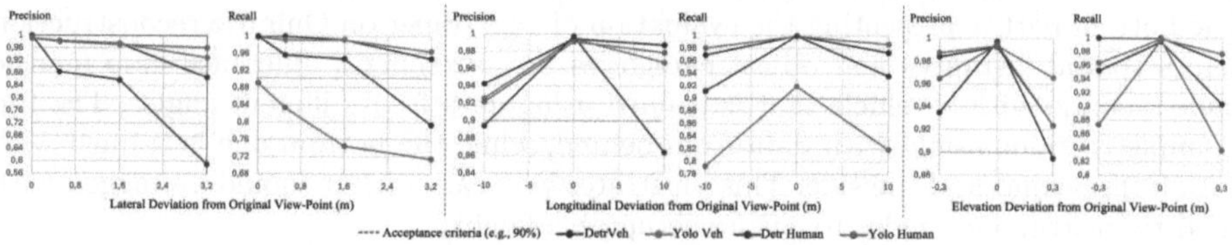

Fig. 3. Presents the evaluation of OmniRe across multiple spatial distances (lateral, longitudinal, and elevation) and object categories, based on precision and recall for DETR and YOLO object detection. The ranges for each direction are selected based on the maximum required deviations from the original viewpoint: 3.2 m for lateral (e.g., lane changes), 10 m for longitudinal, and 30cm for elevation (e.g., due to road elevation or sensor placement on different platforms). As shown, the performance drop in the lateral direction is significantly greater than in the longitudinal direction, even for smaller deviations.

Object Detection Metrics for Rendered Images: According to Table 2 reports precision and recall across the four reconstruction methods, providing a measure of the fidelity of each reconstructed object class from the perspective of object-detection algorithms (i.e., YOLOv10 and DETR). OmniRe on average outperforms the other methods in both human and vehicle object categories for both YOLO and DETR. Comparing the detection metrics highlights the higher fidelity for vehicle classes due to the rigidity of vehicles and their well-defined geometry in reconstruction. On the contrary, the detection performance for humans is significantly lower for PVG. StreetGS and DeformGS fail to reconstruct the human at all, as they cannot model deformable objects [4], which is a significant drawback for AD applications. Moreover, while OmniRe's performance for humans is better than that of the other methods, it is still lower than for vehicles, which is also reflected in the computer vision metrics. Figure 3 presents OmniRe's performance for vehicle and human reconstruction under deviations in different directions, highlighting that the performance drop in lateral deviations is much greater than in longitudinal ones.

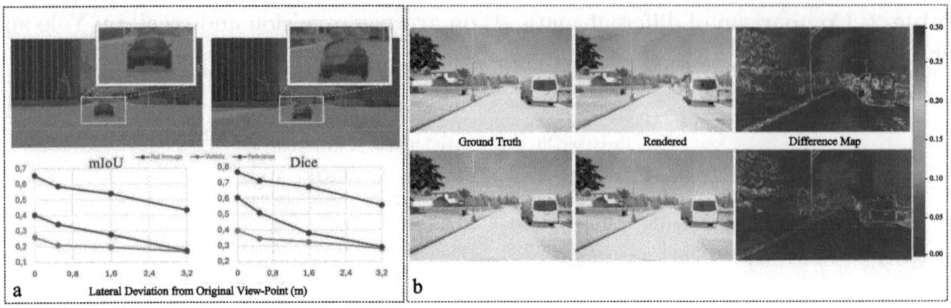

Fig. 4. (a) Presents the Semantic Segmentation KPIs for multiple reconstructions with different lateral deviations. The top-left figure shows novel view with 0.5 m spatial difference with relatively good segmentation quality where the vehicle is correctly identified and segmented with clear boundaries; the right side is presenting 3.2 m spatial difference in novel view showing significant degradation in segmentation performance. The bottom part is presenting the evaluation of SegFormer on OmniRe reconstruction across spatial distances and object categories. (b) Presents a difference map for the same scenario with a vehicle that is trained using data from different angles. The top example is trained only with data from behind, while the bottom one is trained with data from behind and the side. This illustrates why each object should be tagged with valid reconstruction angles to ensure acceptable fidelity.

Semantic Segmentation Performance: As presented in Fig. 4.a, both mIoU and Dice coefficient degrade for OmniRe reconstructions with increasing spatial distance from the original training trajectory. mIoU and Dice remain relatively high for the full image, which includes both static and dynamic objects, while they show far steeper declines for vehicles and humans as the two dynamic objects. These sharp drops in Dice coefficients confirm the structural breakdown in pixel-level understanding. These results confirm that while OmniRe maintained overall consistency with increasing distance, its ability to represent accurate semantics for dynamic objects drops significantly at greater distances.

Viewpoint Coverage and Reconstruction Validity: Difference maps presented in Fig. 4.b show the reconstruction quality for the same object with different training data coverage from various viewpoints. The top example illustrates a vehicle reconstructed using only rear-view training data, while the bottom example shows the same vehicle reconstructed from a separate scenario with more comprehensive viewpoint coverage. This highlights the importance of assigning explicit visibility tags to each object, indicating which angles have sufficient data coverage and are therefore valid for novel view synthesis to satisfy fidelity requirements. This is crucial because, in AD and ADAS, unlike other domains, it is not feasible to collect perfect 360° data for every object. Instead, reconstruction must be done using data with different levels of visibility while still maintaining fidelity. By automatically identifying objects with robust coverage, the pipeline can ensure that only well-supported sides of an object are used in reconstruction.

Scene Editing and Edge Case Generation: So far, we have established the capabilities and limitations of each reconstruction method by analyzing spatial generalization, especially for humans and vehicles, as the two most important object classes in our application. As the final step, the current section demonstrates how our scene editing pipeline can be applied to real-world datasets to ensure visual consistency. Crafting safety-related scenarios from data recorded in the field with an acceptable level of fidelity requires the combination of multiple agents and modifications to their trajectories. These agents include both rigid (e.g., vehicles) and non-rigid (e.g., pedestrians) objects. The bottom of Fig. 1 demonstrates rigid body scene editing applied to real-world datasets. The figure illustrates the lateral shifting and change in speed of a vehicle, effectively transforming one normal scenario into a safety-related cut-in scenario. Figure 1 further showcases human editing that results in a more critical scenario due to changes in pedestrian behavior. In the top example, the pedestrian's trajectory is shifted from walking normally on a zebra crossing to a "jayrunner" scenario near the ego vehicle, which is more critical. As rare scenarios and objects might not be captured by a single developer or organization, they can be extracted and shared with other parties to test their software across companies or be used in simulation environments as replacements for manually modeled assets. Hence, we extended our pipeline to demonstrate cross-dataset agent injection, which is showcased in Fig. 1.

5 Conclusion and Future Work

In this study, we proposed and demonstrated a pipeline for editing and augmenting real-world scenes through agent-level manipulation, including cross-dataset agent injection (addressing *RQ2*). This capability enables the creation of rare or critical safety-related scenarios, which are essential for testing and validating AD software. We conducted a comprehensive evaluation of state-of-the-art scene reconstruction methods, focusing on their applicability in AD and ADAS as safety-critical domains (addressing *RQ1*). Reconstruction fidelity for dynamic objects, especially vehicles and pedestrians, plays a vital role due to their high exposure and risk levels in traffic environments. Our results show that reconstruction fidelity is strongly influenced by viewpoint coverage during data collection for each object type. OmniRe consistently achieved the highest performance across all metrics and object classes, demonstrating strong robustness in novel view synthesis. Further analysis revealed a significant drop in fidelity for pedestrians at lateral distances, underscoring the need for complementary generative techniques. We also highlighted the importance of viewpoint-aware reconstruction. Since full 360° data collection is often infeasible in real-world AD/ADAS operations, our findings emphasize the value of tagging reconstructed objects with valid visibility angles. Such metadata can inform future frameworks for automating high-fidelity scenario generation. Future work could focus on combining 3DGS with diffusion-based methods for enhancing reconstruction quality for unobserved or insufficiently sampled regions, as well as integrating pedestrian motion-generation methods to improve scenario diversity.

Acknowledgments. This work has been partially supported by Sweden's Innovation Agency (Vinnova, diarienummer: 2021-02585), and by the WASP Program funded by the Knut and Alice Wallenberg Foundation.

Disclaimer. The views and opinions expressed are those of the authors and do not necessarily reflect the official policy or position of Volvo Cars or Zenseact.

References

1. Alibeigi, M., et al.: Zenseact open dataset: a large-scale and diverse multimodal dataset for autonomous driving. In: Proceedings of the IEEE/CVF International Conference on Computer Vision (2023)
2. Carion, N., Massa, F., Synnaeve, G., Usunier, N., Kirillov, A., Zagoruyko, S.: End-to-end object detection with transformers. In: Vedaldi, A., Bischof, H., Brox, T., Frahm, J.-M. (eds.) ECCV 2020. LNCS, vol. 12346, pp. 213–229. Springer, Cham (2020). https://doi.org/10.1007/978-3-030-58452-8_13
3. Chen, Y., Gu, C., Jiang, J., Zhu, X., Zhang, L.: Periodic vibration gaussian: dynamic urban scene reconstruction and real-time rendering. arXiv preprint arXiv:2311.18561 (2023)
4. Chen, Z., et al.: Omnire: OMNI urban scene reconstruction. arXiv preprint arXiv:2408.16760 (2024)
5. Dosovitskiy, A., Ros, G., Codevilla, F., Lopez, A., Koltun, V.: Carla: an open urban driving simulator. In: Conference on Robot Learning, pp. 1–16. PMLR (2017)
6. Duisterhof, B.P., et al.: DeforMGS: scene flow in highly deformable scenes for deformable object manipulation. arXiv preprint arXiv:2312.00583 (2023)
7. Geiger, A., Lenz, P., Stiller, C., Urtasun, R.: Vision meets robotics: the Kitti dataset. Int. J. Robot. Res. **32**(11), 1231–1237 (2013)
8. Hess, G., Lindström, C., Fatemi, M., Petersson, C., Svensson, L.: SplatAD: real-time lidar and camera rendering with 3D gaussian splatting for autonomous driving. In: Proceedings of the Computer Vision and Pattern Recognition Conference, pp. 11982–11992 (2025)
9. ISO 21448, Road vehicles — Safety of the intended functionality. Standard, International Organization for Standardization (2022)
10. ISO 26262:2018 (all parts), Road vehicles — Functional safety. Standard, International Organization for Standardization (2018)
11. ISO/PAS 8800:2024, Road vehicles — Safety and artificial intelligence. Publicly available specification, International Organization for Standardization, Geneva, Switzerland (Dec 2024), published December 2024. ISO/PAS 8800:2024 (Edition 1, 172 pages)
12. ISO/TR 4804:2020, Road vehicles – Safety and cybersecurity for automated driving systems – Design, verification and validation methods. Standard, International Organization for Standardization (2020)
13. Jadon, S.: A survey of loss functions for semantic segmentation. In: 2020 IEEE conference on computational intelligence in bioinformatics and computational biology (CIBCB), pp. 1–7. IEEE (2020)
14. Jiang, W., Yi, K.M., Samei, G., Tuzel, O., Ranjan, A.: Neuman: neural human radiance field from a single video. In: Avidan, S., Brostow, G., Cissé, M., Farinella, G.M., Hassner, T. (eds.) ECCV 2022. LNCS, vol. 13692, pp. 402–418. Springer, Cham (2022). https://doi.org/10.1007/978-3-031-19824-3_24

15. Kerbl, B., Kopanas, G., Leimkühler, T., Drettakis, G.: 3D gaussian splatting for real-time radiance field rendering. ACM Trans. Graph. **42**(4), 1–139 (2023)
16. Liu, M., et al.: A survey on autonomous driving datasets: Statistics, annotation quality, and a future outlook. IEEE Trans. Intell. Veh. (2024)
17. Loper, M., Mahmood, N., Romero, J., Pons-Moll, G., Black, M.J.: SMPL: a skinned multi-person linear model. In: Seminal Graphics Papers: Pushing the Boundaries, vol. 2, pp. 851–866 (2023)
18. Mildenhall, B., Srinivasan, P.P., Tancik, M., Barron, J.T., Ramamoorthi, R., Ng, R.: NERF: representing scenes as neural radiance fields for view synthesis. Commun. ACM **65**(1), 99–106 (2021)
19. Nouri, A., Cabrero-Daniel, B., Törner, F., Berger, C.: The devsafeops dilemma: a systematic literature review on rapidity in safe autonomous driving development and operation. arXiv preprint arXiv:2506.21693 (2025)
20. Nouri, A., Cabrero-Daniel, B., Torner, F., Sivencrona, H., Berger, C.: Welcome your new AI teammate: on safety analysis by leashing large language models. In: Proceedings of the IEEE/ACM 3rd International Conference on AI Engineering - Software Engineering for AI. CAIN '24, New York, NY, USA, pp. 172–177. Association for Computing Machinery (2024). https://doi.org/10.1145/3644815.3644953
21. Rong, G., et al.: LGSVL simulator: a high fidelity simulator for autonomous driving. In: 2020 IEEE 23rd International Conference on Intelligent Transportation Systems (ITSC), pp. 1–6. IEEE (2020)
22. Song, Z., et al.: Synthetic datasets for autonomous driving: a survey. IEEE Trans. Intell. Veh. **9**(1), 1847–1864 (2023)
23. Sun, P., et al.: Scalability in perception for autonomous driving: WAYMO open dataset. In: Proceedings of the IEEE/CVF Conference on Computer Vision and Pattern Recognition, pp. 2446–2454 (2020)
24. Wang, A., Chen, H., Liu, L., Chen, K., Lin, Z., Han, J., et al.: Yolov10: Real-time end-to-end object detection. Adv. Neural. Inf. Process. Syst. **37**, 107984–108011 (2024)
25. Wang, L., et al.: Distillnerf: perceiving 3D scenes from single-glance images by distilling neural fields and foundation model features. Adv. Neural. Inf. Process. Syst. **37**, 62334–62361 (2024)
26. Xie, E., Wang, W., Yu, Z., Anandkumar, A., Alvarez, J.M., Luo, P.: Segformer: simple and efficient design for semantic segmentation with transformers (2021). https://arxiv.org/abs/2105.15203
27. Yan, Y., et al.: Street gaussians: modeling dynamic urban scenes with gaussian splatting. In: Leonardis, A., Ricci, E., Roth, S., Russakovsky, O., Sattler, T., Varol, G. (eds.) ECCV 2024. LNCS, vol. 15131, pp. 156–173. Springer, Cham (2024). https://doi.org/10.1007/978-3-031-73464-9_10

Industrial Safety Engineering in the AI-Era

Adoption of Safety Standards in the Japanese Automotive Industry: A Consortium Study

Yutaka Matsuno[1(✉)], Shinobu Ochiai[2], and Fumiaki Kono[2]

[1] Nihon University, Tokyo, Japan
matsuno.yutaka@nihon-u.ac.jp
[2] Japan Automotive Software Platform and Architecture (JASPAR), Tokyo, Japan
shinobu_ochiai@jp.honda, kohnof@hhq.suzuki.co.jp

Abstract. As the number of automotive safety standards increases, including ISO 26262, ISO 21448 (Safety of the Intended Functionality, SOTIF), and UL 4600, practitioners need evidence on how these standards are being adopted in industry. This practical experience report presents a questionnaire-based study conducted within JASPAR, a Japanese automotive consortium, through its Functional Safety Working Group, with 30 responses from 28 of 40 companies (70% company coverage). We compared adoption status across the three standards and analyzed adoption drivers and role allocation between original equipment manufacturers (OEMs) and suppliers for ISO 26262 and SOTIF using non-parametric tests with Holm correction. Adoption stages differ markedly: ISO 26262 is institutionalized (93%), SOTIF is emerging (10% active, 60% preparing), and UL 4600 is pre-adoption in this sample (0% active). Four findings remained significant after multiple-comparison correction: (1) contract requirements and (2) pressure to conform to industry practice are both perceived as stronger drivers of ISO 26262 adoption than of SOTIF; (3) suppliers perceive stronger ISO 26262 contract pressure than OEMs; and (4) despite similar V-model allocation patterns, respondents perceive SOTIF role boundaries as less clear. We derive practical recommendations for OEMs, suppliers, and industry consortia.

Keywords: ISO 26262 · ISO 21448 (SOTIF) · UL 4600 · safety standards adoption · institutional theory

1 Introduction

Assuring the safety of advanced driver assistance systems (ADAS) and automated driving functions has become more important than ever as vehicles increasingly rely on perception, decision-making, and control software. Three major standards address different aspects of automotive safety. ISO 26262 [8] provides a well-established framework for functional safety, focusing on risks arising from system failures or malfunctions. ISO 21448 (SOTIF) [9] addresses

© The Author(s), under exclusive license to Springer Nature Switzerland AG 2027
E. Troubitsyna et al. (Eds.): SAFECOMP 2026, LNCS 16868, pp. 179–194, 2027.
https://doi.org/10.1007/978-3-032-34867-8_12

safety issues caused by functional insufficiencies or unforeseen environmental conditions, even when a function performs as intended. UL 4600 [23] (first published in 2020) provides safety-case guidance for autonomous systems, emphasizing evidential completeness and governance.

The difficulty of assuring such software-intensive systems is illustrated by the 2009–2010 Toyota unintended-acceleration recalls and the investigation that followed. Although NHTSA identified pedal entrapment and sticking accelerator pedals as the known causes, its investigation of possible electronic causes, conducted with technical support from NASA, made visible the assurance burden associated with software-intensive automotive systems [16]. For the Japanese automotive industry, this case highlighted the need for more systematic safety engineering and accountability as vehicles became increasingly software-intensive. ISO 26262 was first published in 2011 and has since become a de facto requirement in the global automotive supply chain.

While ISO 26262 is widely adopted, SOTIF (published in 2022) and UL 4600 are comparatively new. Adoption patterns, interactions with established processes, and the organizational challenges of operating multiple standards concurrently remain underexplored, especially in Japan. We examine the adoption of these three standards in collaboration with JASPAR (Japan Automotive Software Platform and Architecture)[1], whose more than 200 member companies include nearly all major Japanese vehicle manufacturers and leading suppliers. This consortium setting provided cross-organizational access that is rarely feasible through conventional academic channels. Although the data come from Japan, the tiered original equipment manufacturer (OEM)–supplier structure and contract-driven compliance that shape adoption have clear parallels in other regions, notably Germany [19] and South Korea [12]. Germany represents a core European automotive production network with tiered OEM–supplier relationships, while South Korea provides an example of OEM-centered parts-supply networks, illustrated by Hyundai Motor's quasi-vertical integration of internal and external supplier networks. These parallels suggest that the contract-pressure and boundary-clarity findings are most transferable to OEM-centered, tiered automotive supply chains; the UL 4600 adoption pattern and JASPAR's coordinating role are more Japan-specific.

Because UL 4600 emphasizes safety-case governance rather than component-oriented V-model activities, it cannot be directly compared with ISO 26262 and SOTIF on all analytical dimensions. Furthermore, no respondent reported active UL 4600 use, precluding meaningful analysis of drivers or role allocation. Accordingly, UL 4600 appears only in the adoption overview (Sect. 4.1) and the discussion of adoption stages (Sect. 5.1); all other analyses focus on ISO 26262 and SOTIF.

This work arose from a practical need within JASPAR to assess the readiness of its member companies for multi-standard operation. In particular, questions had emerged about how to allocate SOTIF responsibilities across the OEM–supplier interface; the analysis reported here allowed us to examine whether

[1] https://www.jaspar.jp/en/about_us.

this concern was broadly shared across companies and how large the perceived boundary-clarity gap was.

This paper makes three contributions. First, it quantifies an asymmetry in contract pressure between OEMs and suppliers and shows how this asymmetry shapes adoption of safety standards. Second, it grounds the interpretation of multi-standard adoption in institutional theory. Third, it translates the findings into priorities for consortium-level action, including empirical support for JASPAR's ongoing work on SOTIF responsibility-assignment templates. These contributions connect the empirical results to safety-case practice, multi-standard governance, and the organizational dimension of safety engineering.

Section 2 reviews related work and introduces the institutional-theory lens used to interpret adoption stages. Section 3 describes the research questions, questionnaire design, and statistical methods. Section 4 presents the results, including adoption status, benefit–cost perceptions, driver profiles, and OEM–supplier role allocation. Section 5 discusses the findings and identifies practical priorities for industry consortia. Section 6 concludes with directions for future work.

2 Related Work

Several empirical and technical studies have examined safety-standard practice (Table 1); to the best of our knowledge, none directly compares the adoption of ISO 26262, SOTIF, and UL 4600 empirically.

Table 1. Positioning among related empirical and technical studies. • = addressed.

Study	Scope	Method	Standards	Cross-standard	OEM– supplier
Born et al. [1]	Auto (DE)	Exp. report	ISO 26262		
Nair et al. [15]	Cross-domain	Quest.	Multiple		
De la Vara et al. [24]	Cross-domain	Quest.	Safety std		
Huber et al. [7]	Automotive	Quest.	26262+security		Partial[a]
Nouri et al. [18]	Auto (SE)	Exp. report	26262+SOTIF		
Wang et al. [25]	AV/ADAS	Review	SOTIF		
Diemert et al. [4]	Cross-domain	Interv. + quest.	Assurance case		
Soleimani & Sari [21]	AV/ADAS	Framework	3 standards	•	
This study	**Auto (JP)**	**Quest.**	**3 standards**	•	•

[a]OEM/supplier roles collected but not comparatively analyzed. Quest. = questionnaire-based study.

Born et al. [1] reported early industrial experience with the ISO 26262 draft and noted supplier hesitance to commit before OEMs demonstrated acceptance, an observation consistent with the contract-pressure asymmetry we quantify here. Cross-domain questionnaire-based studies by Nair et al. [15] and De la Vara et al. [24] identified operational burden as a systemic issue, but neither compared adoption drivers across standards nor analyzed OEM–supplier role differences. Huber et al. [7] studied automotive safety and security practitioners using a questionnaire and reported coordination challenges, though without comparing adoption patterns across standards.

For SOTIF, Wang et al. [25] reviewed research and practice, focusing on the technical state of the art rather than organizational adoption. Soleimani and Sari [21] proposed a unified framework integrating ISO 26262, SOTIF, and UL 4600, addressing technical integration at the artifact level. Yet how organizations actually adopt and coordinate these standards remains an open question. Nouri et al. [18] reported friction between safety-process requirements and rapid development cycles, a concern related to the operational burden described by our respondents.

Assurance cases, which are central to UL 4600 and increasingly used for ISO 26262, have also been studied empirically [2,3,22]. Diemert et al. [4] found that barriers include additional workload and inadequate guidance, findings that complement our results. Recent work on automated-driving safety argumentation [6,13,14] addresses specific methodological aspects rather than cross-standard adoption patterns.

To characterize differing adoption stages of international standards, we draw on institutional theory [5], which distinguishes three mechanisms that drive organizational adoption: *coercive* pressure (adoption mandated through contracts or regulations), *mimetic* pressure (adoption driven by imitating industry peers), and *normative* pressure (adoption driven by professional recognition of technical necessity). In Sect. 5, we use these mechanisms to explain why ISO 26262, SOTIF, and UL 4600 are at different stages of adoption.

3 Research Design and Methodology

3.1 Research Questions

Drawing on prior literature on safety-standard practice and adoption [1,7,15,18,24] and practical concerns raised in JASPAR Functional Safety Working Group meetings, we formulated three research questions:

- **RQ1 (Benefits and Costs)**: Do perceptions of safety improvement, operational burden, and cost-effectiveness differ between ISO 26262 and SOTIF?
- **RQ2 (Adoption Drivers)**:
  - **RQ2a**: Do the relative weights of adoption drivers differ across standards?
  - **RQ2b**: Do OEMs and suppliers differ in their driver profiles?
- **RQ3 (Role Allocation)**:
  - **RQ3a**: How are V-model activities allocated between OEMs and suppliers for each standard?
  - **RQ3b**: Are role boundaries perceived as less clear for SOTIF than for ISO 26262?

3.2 Questionnaire Design and Data Collection

Data were collected through a web questionnaire administered to members of the JASPAR Functional Safety Working Group. Quantitative items used 5-point scales; anchors ranged from 1 = Strongly disagree to 5 = Strongly agree for driver and benefit/cost items, from 1 = OEM only to 5 = Supplier only for role allocation, and from 1 = Much less clear to 5 = Much clearer for boundary clarity. Open-ended items elicited qualitative comments. The questionnaire covered respondent characteristics, adoption status, standard-specific items (ISO 26262, SOTIF, UL 4600), assurance-case practice, and multi-standard challenges. Branching logic directed respondents only to questions for standards they were using or preparing to adopt.

The questionnaire was distributed in December 2025 to the 40 companies participating in the JASPAR Functional Safety Working Group, and we received 30 responses from 28 distinct companies (70% company coverage). Two companies contributed two respondents each. Because the questionnaire was anonymous, individual responses cannot be linked to specific companies. Participation was voluntary, informed consent was obtained on the entry page, and no personally identifiable information was collected. To reduce social-desirability bias, respondents were instructed to answer on the basis of their own practical experience rather than provide official company positions. Table 2 summarizes respondent characteristics. For the OEM–supplier comparisons in subsequent analyses, the 10 OEM respondents form the OEM group, and the remaining 20 respondents (Tier 1, Tier 2, and Tier 1 & 2) form the Supplier group.

To support reproducibility, the companion repository[2] provides questionnaire materials, analysis outputs, and analysis scripts.

3.3 Statistical Methods

Given the small sample size and the ordinal nature of the data, we used non-parametric tests throughout (all two-tailed). Specifically, the Wilcoxon signed-rank test was used for paired comparisons (e.g., ISO 26262 vs. SOTIF ratings from the same respondent), the Mann–Whitney U test for independent-group comparisons (OEM vs. supplier), the Friedman test for comparing multiple drivers within a standard, and the one-sample Wilcoxon test for testing whether a rating differs from the neutral midpoint (3).

Effect sizes are reported as $r = |Z|/\sqrt{N}$ [20], where $r \geq 0.5$ indicates a large effect[3]. To guard against false positives from multiple comparisons, we applied Holm correction within each research-question family. When a planned test could not be computed (e.g., too few differing responses), it was still counted, making the correction more conservative. Results that are significant only before correction are treated as exploratory. We excluded "Not applicable" responses

[2] https://github.com/yutakaJssst/SafeComp2026SurveyQuestionnaire.

[3] For the Wilcoxon test, N is the number of respondents whose paired ratings differ ("non-zero differences"); for Mann–Whitney, $N = n_1 + n_2$ with tie-corrected variance.

Table 2. Respondent characteristics ($n = 30$)

Category	Item	n	%
Company type	OEM	10	33%
	Tier 1 supplier	17	57%
	Tier 2 or below	2	7%
	Tier 1 & Tier 2	1	3%
Role	Safety assessor	11	37%
	Engineer	8	27%
	Quality assurance	4	13%
	Manager/Other	7	23%
Automotive SPICE status	Operating	15	50%
	Not operating	9	30%
	Preparing	5	17%
	No response	1	3%
ISO 26262 experience	5+ years	26	87%

item-wise and used pairwise-complete data. Means and standard deviations are reported as descriptive summaries; Norman [17] provides justification for interpreting means of ordinal scales in this manner.

An AI assistant (Claude, Anthropic) assisted with initial scripting and preliminary coding of free-text comments; all results were independently verified by the authors using reproducible scripts. AI-generated initial codes for the 28 free-text comments were reviewed, revised, and consolidated into themes through author discussion on the original Japanese-language responses.

4 Results

Because branching logic skipped questions for standards a respondent was not using or preparing to adopt, sample sizes vary: ISO 26262 items typically have $n = 25$–28, whereas SOTIF items have $n = 12$–20.

4.1 Adoption Overview

ISO 26262 is nearly universally adopted, with 28 of 30 respondents (93%) reporting active use. In contrast, SOTIF is still in an early phase of adoption: only 3 respondents (10%) reported active use, whereas 18 (60%) reported that they were preparing to adopt it. No respondent reported active UL 4600 use. Only 2 respondents (7%) reported preparing for UL 4600 adoption, and 16 (53%) reported no plans to adopt it. UL 4600 is therefore included only in this adoption overview; it is excluded from the driver and role-allocation analyses that follow. Figure 1 illustrates these adoption patterns.

Among ISO 26262 adopters, Part 4 (system-level product development) and Parts 7–9 (production, supporting processes, and Automotive Safety Integrity

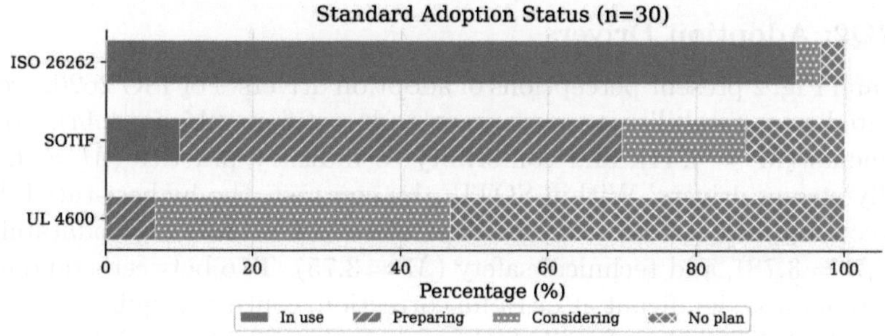

Fig. 1. Standard adoption status ($n = 30$). ISO 26262 shows near-universal adoption (93%), while SOTIF is in early adoption (10% active, 60% preparing). No respondent reported active UL 4600 use.

Level (ASIL)-oriented analysis) showed the highest adoption at 93%, followed by Part 2 (management of functional safety) at 89%. Part 3 (concept phase) was notably lower at 46%, likely because 67% of respondents are suppliers (Table 2) and concept-phase activities are typically performed by OEMs in the Japanese supply chain.

4.2 RQ1: Perceived Benefits and Costs

Table 3 presents perceptions of benefits and costs. The key pattern is an effort–benefit imbalance: both standards were rated as high burden ($M \geq 4.17$), but SOTIF's perceived safety improvement fell below the scale midpoint ($M = 2.92$), whereas ISO 26262 was rated moderately positively ($M = 3.73$). Respondents thus see SOTIF as nearly as burdensome as ISO 26262 but less beneficial, which may help explain why SOTIF is harder to promote internally (see also Sect. 5). The difference in perceived safety improvement was suggestive ($p = 0.039$) but did not survive Holm correction ($p_{\text{Holm}} = 0.117$). As a supplementary check, we compared companies operating Automotive SPICE ($n = 15$) with those not yet operating it ($n = 14$; Preparing or Not operating) to see whether process maturity influenced these perceptions, but found no significant differences (all $p > 0.2$).

Table 3. Perceived benefits and costs (5-point scale; Holm correction within RQ1). † denotes uncorrected $p < 0.05$ only.

Item	ISO 26262		SOTIF		n_{pair}	p
	M (n)	SD	M (n)	SD		
Safety improvement	3.73 (26)	0.96	2.92 (13)	0.95	12	.039†
Operational burden	4.23 (26)	0.82	4.17 (12)	0.39	12	.844
Cost-effectiveness	3.20 (25)	1.00	2.92 (12)	1.00	11	.438

4.3 RQ2: Adoption Drivers

Table 4 and Fig. 2 present perceptions of adoption drivers. For ISO 26262, respondents rated accountability to end users and certifiers ($M = 4.44$), contract requirements ($M = 4.44$), and conformity to industry practice ($M = 4.25$) as especially strong drivers. Within SOTIF, by contrast, the highest-rated drivers were accountability to end users and certifiers ($M = 4.37$), accountability to clients ($M = 3.78$), and technical safety ($M = 3.75$). Two between-standard differences remained significant after Holm correction: contract requirements scored higher for ISO 26262 than for SOTIF ($M = 4.44$ vs. 3.47; $p = 0.002$, $r = 0.89$, a large effect), and so did conformity to industry practice ($p = 0.004$). Technical safety showed a similar tendency ($p = 0.049$, $r = 0.63$) but did not survive correction.

Table 4. Adoption drivers by standard (5-point scale). ** = significant after Holm correction. Effect size r is shown only when at least 10 respondents rated the two standards differently; otherwise omitted.

Driver	ISO 26262		SOTIF		p	r
	M (n)	SD	M (n)	SD		
Technical safety	4.18 (28)	0.90	3.75 (20)	1.12	.049	.63
Accountability to clients	4.08 (26)	1.26	3.78 (18)	1.26	.109	—
Accountability to end users/certifiers	4.44 (27)	0.70	4.37 (19)	0.60	.109	—
Contract requirements	4.44 (25)	1.16	3.47 (17)	1.28	.002**	.89
Conformity to industry practice	4.25 (28)	0.80	3.58 (19)	0.84	.004**	—

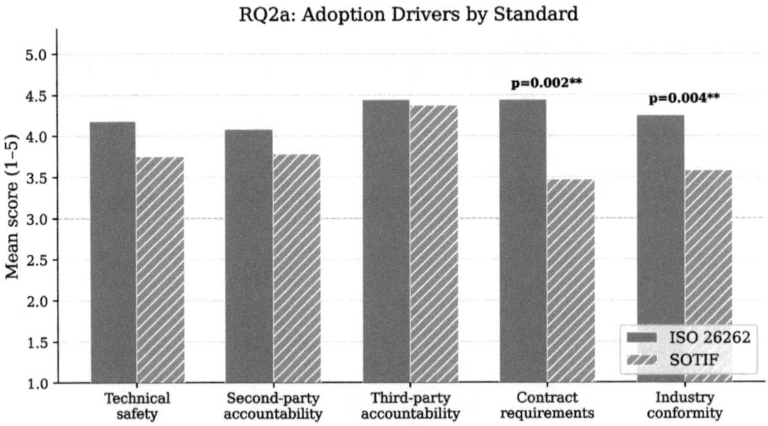

Fig. 2. Adoption drivers by standard. Contract requirements ($p = 0.002$) and conformity to industry practice ($p = 0.004$) are significantly stronger for ISO 26262 than SOTIF (Holm-corrected). The dashed line indicates the neutral midpoint (3).

A Friedman test, which assesses whether the five drivers are rated equally within a standard, indicated significant differences among drivers for SOTIF

($\chi^2 = 13.18$, $p = 0.010$, $n = 17$), confirming that certain drivers (notably accountability and technical safety) stand out more than others. For ISO 26262, the difference did not reach significance ($\chi^2 = 7.61$, $p = 0.107$, $n = 25$), consistent with all drivers being rated similarly high.

Table 5 shows OEM–supplier differences for contract requirements, the only driver that reached significance after correction (full results for all five items are in the companion repository). For ISO 26262, suppliers reported significantly stronger contract pressure ($M = 4.89$, $n = 18$) than OEMs ($M = 3.29$, $n = 7$; $p = 0.002$, $r = 0.63$). This asymmetry reflects the supply-chain structure: OEMs face expectations from end customers and certification bodies, whereas suppliers face OEM-imposed contractual requirements. For SOTIF, a similar tendency was observed (OEM $M = 2.50$ vs. Supplier $M = 4.00$; $p = 0.038$, $r = 0.52$) but did not survive Holm correction, suggesting that a comparable asymmetry may be emerging.

Table 5. OEM vs. Supplier differences in contract requirements (Mann–Whitney U; Holm-corrected over all 5 drivers per standard; only contract requirements shown as the sole significant driver). Supplier = Tier 1 + Tier 2 + Tier 1 & 2.

Standard	OEM M (n)	Supplier M (n)	U	p	r	Holm
ISO 26262	3.29 (7)	4.89 (18)	22.0	.002	.63	Sig. **
SOTIF	2.50 (6)	4.00 (11)	12.5	.038	.52	n.s.

4.4 RQ3: Role Allocation and Boundary Clarity

Table 6 presents the perceived allocation of V-model activities. On a scale where 1 = OEM only and 5 = Supplier only, the pattern is consistent for both standards: upstream activities (requirements definition and validation) are OEM-led ($M \approx 1.8$), midstream activities (detailed design) lean toward suppliers ($M \approx 3.5$), and downstream activities (implementation and unit testing) are supplier-led ($M \approx 4.2$). Integration testing shows mixed responsibility ($M \approx 3.0$).

Most respondents assigned identical allocations for both standards, choosing the same scale point for ISO 26262 and SOTIF on each activity. This left only 3–5 respondents per activity whose paired ratings differed, too few for a meaningful statistical test. The high proportion of identical responses is itself informative, however, and may reflect organizational inertia: once roles are established for ISO 26262, organizations apply them to SOTIF without renegotiation.

Despite similar role allocations, boundary clarity differs markedly. Respondents were asked to compare SOTIF boundary clarity with ISO 26262 on a scale from 1 = Much less clear through 3 = Same to 5 = Much clearer. Of the 15 respondents, 73% (11/15) rated SOTIF boundaries as "less clear" or "much less clear," and none rated them clearer (Fig. 3). The mean rating was $M = 1.93$ ($SD = 0.80$), significantly below the neutral midpoint ($p = 0.001$, $r = 0.88$); this result remained significant after Holm correction ($p_{\text{Holm}} = 0.006$).

Table 6. V-model activity allocation (1 = OEM only, 5 = Supplier only). Paired tests are omitted because only 3–5 respondents per activity rated the two standards differently.

Activity	ISO 26262		SOTIF	
	M (*n*)	*SD*	*M* (*n*)	*SD*
Safety goals/upper req.	1.73 (26)	0.83	1.77 (13)	0.83
Requirements validation	1.85 (26)	0.67	1.83 (12)	1.11
Detailed design	3.69 (26)	0.79	3.25 (12)	1.22
Implementation/unit test	4.23 (26)	0.65	4.17 (12)	0.72
Integration/system test	3.08 (26)	0.93	2.92 (12)	1.24

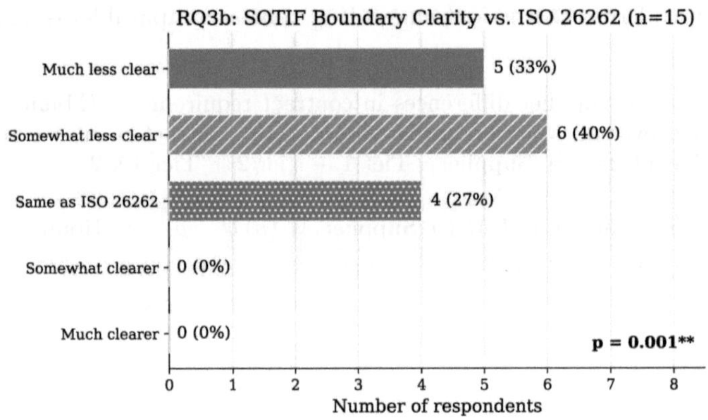

Fig. 3. SOTIF boundary clarity compared to ISO 26262 ($n = 15$). 73% of respondents rated SOTIF boundaries as less clear, with none rating them clearer ($p = 0.001$, $r = 0.88$).

This pattern is notable because if organizations assign the same roles as for ISO 26262, one might expect comparable boundary clarity. The gap likely arises because SOTIF requires system-level scenario analysis whose scope depends on the OEM's operational design domain (ODD) definition. In contrast, ISO 26262 requirements can be decomposed per component, making responsibility boundaries easier to define. When OEM specifications are incomplete, it becomes unclear which party is accountable for a gap in scenario coverage.

4.5 Supplementary Findings

Table 7 shows the challenges in operating multiple standards; the most frequently cited were requirement overlap (47%), scope understanding (40%), and document integration (37%). Regarding integration approaches, 30% used ISO 26262 as a foundation, 20% operated standards independently, and 17% had no clear policy.

Table 7. Challenges in multi-standard operation ($n = 30$, multiple selection)

Challenge	%
Requirement overlap	47%
Scope understanding	40%
Document integration	37%
Terminology confusion	33%
Cross-department coordination	33%
Priority judgment	27%
Customer requirement differences	27%
No challenges	13%

Assurance Case Practice. Only 11 of 30 respondents (37%) reported active use of assurance cases for ISO 26262; 10 (33%) had no plans. Plain documents were the most common format ($n = 14$), followed by Goal Structuring Notation (GSN) or Claims-Arguments-Evidence (CAE) ($n = 6$). Comments revealed mixed views: some saw long-term value but warned of "enormous effort," whereas others felt assurance cases are "still created mainly for accountability rather than genuine safety benefit."

Qualitative Themes. The free-text material was limited but useful for interpreting the quantitative results. Across the open-ended fields, we obtained 28 comments: ISO 26262 challenges (10), SOTIF challenges (4), multi-standard operation (7), assurance-case practice (4), and JASPAR activities (3). We therefore use these comments as contextual evidence rather than as a separate qualitative study.

Three themes were most relevant to the statistical findings. First, respondents repeatedly described *workload and organizational burden*, including work-product creation, resource shortages, and training difficulties; one respondent noted that "the workload doubles with each new standard" (all quotations translated from the original Japanese responses). Second, several comments pointed to *OEM–supplier coordination friction*, especially delayed or unclear upper-level requirements: "OEMs tell us to comply, but they are slow to issue upper-level safety requirements." Third, SOTIF comments reflected uncertainty about value and mandate. Respondents described SOTIF as documenting practices already performed, or noted that it is difficult to promote internally because it is "not clearly required by regulations." These comments are consistent with the quantitative findings that SOTIF has weaker contract pressure and less clear role boundaries than ISO 26262.

5 Discussion

The main practical message of this study is that the emerging challenge lies less in reallocating V-model activities than in negotiating clear SOTIF responsibilities across the OEM–supplier interface. Although the direction of the four significant findings (Sect. 4) may align with practitioner intuition, the effect magnitudes are large, particularly for contract pressure ($r = 0.89$, Table 4) and boundary ambiguity ($r = 0.88$). These magnitudes have not previously been quantified and are relevant for calibrating consortium-level interventions.

5.1 Interpretation

An important caveat applies: because ISO 26262 and SOTIF differ markedly in adoption maturity (93% vs. 10% active use), the observed differences may partly reflect maturity effects rather than inherent properties of the standards themselves. With this limitation in mind, the data suggest a three-stage picture.

ISO 26262 can be characterized as *institutionalized*: adoption is high (93%), and all drivers are rated highly ($M \geq 4.08$), although no single driver stood out as significantly stronger than the others. High operational burden ($M = 4.23$) combined with only moderate perceived safety improvement ($M = 3.73$) may indicate that compliance is seen as burdensome relative to its benefits.

SOTIF can be characterized as *emerging*: 70% of respondents are either using it or preparing to adopt it, but weaker contract pressure and less clear role boundaries make implementation more difficult. Qualitative comments (Sect. 4.5) also identify the lack of regulatory mandate as a practical obstacle.

UL 4600 remains in a *pre-adoption* stage in this sample (0 active users; 53% with no plans), likely reflecting both the Functional Safety Working Group sample composition and the absence of coercive pressure from regulatory references in Japan. Only 37% reported assurance-case use even for ISO 26262, with plain-document approaches being the most common.

These stages align with the institutional mechanisms introduced in Sect. 2. ISO 26262 is more strongly associated with *coercive* ($p = 0.002$) and *mimetic* pressure ($p = 0.004$) than SOTIF, for which *normative* drivers (accountability, technical safety) are more prominent. UL 4600 shows limited evidence of any pressure type in this sample.

Regarding RQ1, the effort–benefit imbalance noted in Sect. 4.2 is consistent with comments such as "SOTIF is mainly documenting what we already do." This comment and the boundary-clarity finding are complementary rather than contradictory: before SOTIF, the underlying activities (e.g., scenario analysis, sensor-performance evaluation) were likely performed without a dedicated standard, under ISO 26262 processes or OEM-specific requirements, without explicit responsibility allocation. SOTIF's formalization makes the previously hidden ambiguity visible, which explains why practitioners feel they are "just documenting" existing practice while simultaneously reporting unclear boundaries. With only 10% reporting active SOTIF use, many respondents may not yet have faced the challenges of situation-dependent safety reasoning for ADAS/AD functions that cannot be decomposed as in ISO 26262. This perception may change as adoption progresses. The effort–benefit imbalance nonetheless makes SOTIF harder to justify internally without the coercive pressure that drove ISO 26262.

Turning to the supply-chain perspective, qualitative responses echo the friction identified in RQ2b: suppliers are asked to comply before upper-level guidance has been fully specified, which may contribute to SOTIF boundary ambiguity.

5.2 Implications for Practice

For OEMs, timely provision of upper-level safety requirements and early clarification of SOTIF-specific role boundaries should precede demands for supplier compliance. For suppliers, building on established ISO 26262 processes when preparing for SOTIF and proactively seeking clarification of role boundaries may help reduce perceived overhead.

For industry consortia such as JASPAR, the findings point to five practical priorities:

1. *Artifact-mapping tables* that cross-reference ISO 26262, SOTIF, and UL 4600 work products (addressing requirement overlap, 47%).
2. *Terminology-harmonization guides* (terminology confusion, 33%).
3. *RACI (Responsible, Accountable, Consulted, Informed) templates for SOTIF* that define responsibility assignments at the OEM–supplier interface.
4. *Minimal safety-case templates* with reusable GSN/CAE patterns.
5. *Role-based training programs.*

Of these, responsibility-assignment templates are the most urgent given the large boundary-clarity effect ($r = 0.88$); a JASPAR working group is already developing such templates, and our results reinforce the urgency of this effort.

The JASPAR Functional Safety Working Group includes a dedicated UL 4600 study group; the pre-adoption findings reported here will serve as a baseline for its efforts to identify adoption barriers.

5.3 Threats to Validity

Internal Validity. Social-desirability bias was mitigated through anonymity and individual-level response instructions. Two companies contributed two respondents each (4 of 30), introducing minor non-independence. Small paired-sample sizes (SOTIF $n_{\text{pair}} = 11$–20) limit precision and may inflate effect sizes.

Construct Validity. Cross-standard comparisons are confounded with adoption maturity (see Sect. 5.1); longitudinal studies are needed to disentangle these factors. The questionnaire did not ask respondents to estimate the planned year of SOTIF adoption, so the timing gap relative to ISO 26262 may also confound the cross-standard comparison. The study scope is limited to functional and operational safety standards and does not cover adjacent standards such as ISO/SAE 21434 (cybersecurity engineering) [11] or ISO/PAS 8800 (safety and artificial intelligence) [10]; interactions with cybersecurity and AI/ML safety standards are important directions for future work.

External Validity. Twelve of the 40 companies did not respond and may differ systematically. Non-response bias could skew the results toward more favorable adoption perceptions. The 40 companies in the JASPAR Functional Safety Working Group include nearly all major Japanese vehicle manufacturers and leading Tier 1 suppliers active in functional-safety practice, so the 28 responding companies provide near-census coverage of the safety-engineering community within this consortium setting. However, the sample does not cover smaller suppliers, non-JASPAR firms, or regions without comparable OEM-centered supply chains. In particular, the UL 4600 findings should not be generalized to all Japanese companies.

Statistical Validity. Although $n = 30$ limits statistical power, the sample covers 70% of the target population. We address power limitations by reporting effect sizes, using non-parametric tests, applying Holm correction, and distinguishing robust findings from exploratory tendencies.

6 Conclusion

Using questionnaire responses from the JASPAR Functional Safety Working Group ($n = 30$, 70% company coverage), we examined how Japanese automotive safety practitioners perceive the adoption of ISO 26262, SOTIF, and UL 4600. Contract requirements and pressure to conform to industry practice appear to motivate ISO 26262 adoption more strongly than SOTIF adoption, and suppliers report stronger contract pressure than OEMs. SOTIF role boundaries are perceived as less clear than those of ISO 26262 despite similar allocation patterns. Adoption follows a three-stage pattern: ISO 26262 is institutionalized (93%), SOTIF is emerging (70% in use or preparing), and UL 4600 remains at a pre-adoption stage in this sample. High operational burden combined with uncertain benefit was a recurring concern.

Based on these findings, we identify five practical priorities for industry consortia, with SOTIF responsibility-assignment templates as the highest priority. Future work should evaluate such templates in OEM–supplier projects, track SOTIF/UL 4600 diffusion longitudinally, and replicate this study in other regions to assess transferability.

Acknowledgments. We thank the shepherd and the anonymous reviewers for their constructive comments, which helped improve the clarity and presentation of this paper. We also thank the JASPAR Functional Safety Working Group members for their participation.

Disclosure of Interests. The authors have no competing interests to declare that are relevant to the content of this article.

References

1. Born, M., Favaro, J., Kath, O.: Application of ISO DIS 26262 in practice. In: Proceedings of the 1st Workshop on Critical Automotive Applications: Robustness & Safety (CARS), pp. 3–6. ACM (2010)
2. Cheng, J., Goodrum, M., Metoyer, R., Cleland-Huang, J.: How do practitioners perceive assurance cases in safety-critical software systems? In: Proceedings of the 11th International Workshop on Cooperative and Human Aspects of Software Engineering (CHASE), pp. 57–60 (2018)
3. Dardar, R., Gallina, B., Johnsen, A., Lundqvist, K., Nyberg, M.: Industrial experiences of building a safety case in compliance with ISO 26262. In: 23rd IEEE International Symposium on Software Reliability Engineering Workshops, ISSRE Workshops, Dallas, TX, USA, November 27-30, 2012, pp. 349–354 (2012)
4. Diemert, S., Shortt, C., Weber, J.H.: How do practitioners gain confidence in assurance cases? Inf. Softw. Technol. **185**, 107767 (2025)
5. DiMaggio, P.J., Powell, W.W.: The iron cage revisited: Institutional isomorphism and collective rationality in organizational fields. Am. Sociol. Rev. **48**(2), 147–160 (1983)
6. Fenn, J., Hawkins, R., Nicholson, M.: A new approach to creating clear operational safety arguments. In: Ceccarelli, A., Trapp, M., Bondavalli, A., Schoitsch, E., Gallina, B., Bitsch, F. (eds.) SAFECOMP 2024 Workshops. LNCS, vol. 14989, pp. 227–238. Springer, Cham (2024). https://doi.org/10.1007/978-3-031-68738-9_17
7. Huber, M., Brunner, M., Sauerwein, C., Cârlan, C., Breu, R.: Roadblocks on the highway to secure cars: an exploratory survey on the current safety and security practice of the automotive industry. In: Gallina, B., Skavhaug, A., Bitsch, F. (eds.) SAFECOMP 2018. LNCS, vol. 11093, pp. 157–171. Springer, Cham (2018). https://doi.org/10.1007/978-3-319-99130-6_11
8. ISO: ISO 26262:2018 Road vehicles – Functional safety (Parts 1–12) (2018)
9. ISO: ISO 21448:2022 Road vehicles – Safety of the intended functionality (2022)
10. ISO: ISO/PAS 8800:2024 Road vehicles – Safety and artificial intelligence (2024)
11. ISO/SAE: ISO/SAE 21434:2021 Road vehicles – Cybersecurity engineering (2021)
12. Kim, C., Jeong, J.H., Jo, H.J.: Detecting dynamic changes in Hyundai Motor's parts supply system as an industry latecomer: the quasi-vertical integration of internal and external networks. J. Asian Sociol. **50**(1), 55–89 (2021). https://doi.org/10.21588/dns.2021.50.1.004
13. Kodama, H., Matsuno, Y., Takai, T., Ota, H., Okada, M., Tsuchiya, T.: A case study of continuous assurance argument for level 4 automated driving. In: Ceccarelli, A., Trapp, M., Bondavalli, A., Bitsch, F. (eds.) SAFECOMP 2024. LNCS, vol. 14988, pp. 150–165. Springer, Cham (2024). https://doi.org/10.1007/978-3-031-68606-1_10
14. Matsuno, Y., Hayashi, M., Tsuchiya, T.: Consensus building in level 4 automated driving field trials through assurance cases. In: Gallina, B., Törngren, M., Bitsch, F. (eds.) SAFECOMP 2025. LNCS, vol. 15954, pp. 33–47. Springer, Cham (2025). https://doi.org/10.1007/978-3-032-01241-8_3
15. Nair, S., de la Vara, J.L., Sabetzadeh, M., Falessi, D.: Evidence management for compliance of critical systems with safety standards: a survey on the state of practice. Inf. Softw. Technol. **60**, 1–15 (2015)
16. National Highway Traffic Safety Administration: Technical assessment of Toyota electronic throttle control (ETC) systems. Technical report, U.S. Department of Transportation (2011). https://www.nhtsa.gov/sites/nhtsa.gov/files/nhtsa-ua_report.pdf. Accessed 24 May 2026

17. Norman, G.: Likert scales, levels of measurement and the "laws" of statistics. Adv. Health Sci. Educ. **15**(5), 625–632 (2010)
18. Nouri, A., Berger, C., Törner, F.: An industrial experience report about challenges from continuous monitoring, improvement, and deployment for autonomous driving features. In: Proceedings of the 48th Euromicro Conference on Software Engineering and Advanced Applications (SEAA), pp. 358–365 (2022)
19. Pavlínek, P.: Restructuring and internationalization of the European automotive industry. J. Econ. Geogr. **20**(2), 509–541 (2020). https://doi.org/10.1093/jeg/lby070
20. Rosenthal, R.: Meta-Analytic Procedures for Social Research, 2nd edn. SAGE Publications, Newbury Park, CA (1991)
21. Soleimani, M., Sari, A.A.: A unified safety framework for automated vehicle development: integrating ISO 26262, SOTIF, and UL 4600. Transp. Res. Interdisc. Perspect. **36**, 101831 (2026)
22. Törner, F., Öhman, P.: Automotive safety case: a qualitative case study of drivers, usages, and issues. In: Proceedings of the 11th IEEE High Assurance Systems Engineering Symposium (HASE), pp. 313–322 (2008)
23. UL: UL 4600: Standard for Safety for the Evaluation of Autonomous Products (2023)
24. de la Vara, J.L., Borg, M., Wnuk, K., Moonen, L.: An industrial survey of safety evidence change impact analysis practice. IEEE Trans. Softw. Eng. **42**(12), 1095–1117 (2016)
25. Wang, H., Shao, W., Sun, C., Yang, K., Cao, D., Li, J.: A survey on an emerging safety challenge for autonomous vehicles: safety of the intended functionality. Engineering **33**, 17–34 (2024)

SmartStandards: Efficient Extraction and Operationalization of Requirements from Industry Safety Standards Using Generative AI

Daniel Hillen(✉), Pascal Gerber, Marc Lorenz, and Jan Reich

Fraunhofer IESE, Kaiserslautern, Germany
{daniel.hillen,pascal.gerber,marc.lorenz,jan.reich}@iese.fraunhofer.de

Abstract. Developing safety-critical systems require compliance with multiple standards which impose numerous requirements on development and assurance processes. However, understanding, filtering, and operationalizing relevant requirements demands substantial effort and expertise. The shortage of experienced engineers and the growing number of applicable standards and regulations call for new methods to manage this complexity and improve efficiency.

This study investigates how generative AI can support engineers in efficiently extracting concrete requirements from a set of standards. We developed a generative AI-based tool that automatically analyzes standards and proposes a list of requirements based on predefined relevance criteria. The tool is embedded into a human-in-the-loop workflow in which domain experts verify and validate the extracted requirements to ensure high quality and maintain traceability to the original standards.

We evaluated our approach to three safety standard documents for automated driving systems in the context of verification and validation. AI-generated requirements were first reviewed by a novice safety engineer, and we measured both the tool's processing time and the engineer's verification time. We compared this AI-supported workflow to fully manual analysis in an industry project by engineers with three to eight years of experience where 130 out of 314 clauses are declared as relevant. We observed an 84% increase in efficiency, while 7% of the entries were incorrectly specified after novice verification. Subsequent expert review resolved the remaining errors and resulted in a net efficiency improvement of 60%. The evaluation shows that a collaborative human–AI approach can improve the efficiency of standards analysis significantly and reduce the effort required from expert engineers, while maintaining a high level of quality suitable for safety-critical development.

Keywords: ChatGPT · Automated Driving · Safety

1 Introduction

Safety-critical systems must comply with multiple standards to assure safe operation across their lifecycle. These standards define requirements for develop-

© The Author(s), under exclusive license to Springer Nature Switzerland AG 2027
E. Troubitsyna et al. (Eds.): SAFECOMP 2026, LNCS 16868, pp. 195–209, 2027.
https://doi.org/10.1007/978-3-032-34867-8_13

ment, validation, and management processes. However, they are often written in a generic and technology-agnostic way. Interpreting these generic clauses for a concrete system and deriving actionable, project-specific requirements requires substantial effort and expertise.

In current practice, engineers manually scan standards, extract relevant clauses, and derive concrete requirements from them, often maintaining traceability in spreadsheets or requirements management tools. Different stakeholders focus on different subsets of requirements: for example, test engineers are primarily interested in verification and validation (V&V) clauses, whereas architects focus on requirements related to system and software architecture. This manual process is time-consuming and error-prone, and it demands experts who are well-trained in the particular standards. With the growing complexity of safety-critical systems, the increasing number of applicable standards and regulations, and a global shortage of experienced safety engineers, this approach does not scale. Traditional tool support has been limited. Requirements management tools mainly support storage and traceability, while search and filtering capabilities are typically restricted to keyword-based queries. As a result, the core activities of understanding, filtering, and operationalizing standard clauses remain largely manual.

Recent advances in generative AI, and in particular large language models (LLMs), open up new opportunities for supporting complex text processing tasks. LLMs have shown promising results in tasks such as code generation, summarization, and the structured extraction of information from unstructured text. In other domains, tasks where complex documents are analyzed by natural language processes (NLP) show significant efficiency improvements [1,3]. On the other hand, in safety engineering, first studies have demonstrated that LLMs can significantly improve the efficiency of hazard analysis and risk assessment [10,15]. Similar results were observed for other safety engineering activities [5,9,16]. All studies mentioned concern in the trustworthiness of the result and recommend human oversight. Another study investigated the benefit of agent-based Retrieval-Augmented Generation (RAG) to improve the quality of the derivation of safety requirements from standards [2]. However, they focus on improving the quality of the results by improving the LLM components.

In this paper, we investigate whether and to what extent LLMs can improve the efficiency of safety standard analysis without compromising quality. We focus on the following research questions:

- **RQ1:** To what extent can an LLM-based, human-in-the-loop workflow reduce the effort required for standards analysis compared to a fully manual expert-based process?
- **RQ2:** How accurately can an LLM-based tool identify relevant clauses and derive capabilities from safety standards for a given use case (here: verification and validation)?

To address these questions, we present *SmartStandards*, a generative AI-supported tool for standard analysis. SmartStandards takes a standard document (PDF) and a user-provided criteria catalog as inputs, automatically extracts and classifies clauses based on the catalog, and derives a list of engineering capabilities for each relevant clause. The tool embeds these steps in a human-in-the-loop workflow: engineers can review and correct the AI-generated classifications and capabilities via a graphical user interface, ensuring traceability and quality.

We evaluate SmartStandards on three safety standard documents for automated driving systems, with a focus on V&V. In the AI-supported workflow, a novice safety engineer conducts the analysis using SmartStandards. We compare this to a fully manual analysis performed by four engineers with three to eight years of experience, measuring both effort and quality. Our results show that the AI-supported workflow can substantially reduce effort while maintaining a high level of quality after human review.

The contributions of this paper are threefold: (i) we introduce SmartStandards, a generic, criteria-driven tool for LLM-assisted analysis of safety standards; (ii) we propose a human-in-the-loop workflow that integrates LLM-based automation with expert review to ensure trustworthy results in safety-critical contexts; and (iii) we report an empirical evaluation on three automotive safety standards, comparing AI-supported analysis with a manual industry baseline in terms of effort and quality and quantifying the reduction in expert effort.

The remainder of this paper is structured as follows. Section 2 provides an overview of related work on LLMs and safety engineering. Section 3 presents SmartStandards and the underlying analysis process in detail. Section 4 describes our evaluation setup and reports the results. Finally, Sect. 5 discusses limitations and concludes the paper.

2 Related Work

Understanding and analyzing complex text documents and retrieving requirements is a labor intensive task in various domains that requires significant expertise. In finance, studies show increased efficiency gain analyzing financial reports for their compliance to the respective laws [3]. Complex regulations need to be analyzed by the LLM and then applied to the reports. The study identified high efficiency gains and high accuracy. In the legal domain, Ariai et al. [1] survey NLP and LLM techniques for tasks such as document summarization, text classification, and question answering over regulatory and statutory texts and report substantial workload reductions. They also emphasize that challenges such as hallucination, lack of interpretability, and data scarcity are particularly critical when model outputs have consequential implications [4]. Building on this, recent legal-informatics approaches apply LLMs to extract and operationalize normative statements from legislation. Dal Pont et al. and Galli et al. [6,8] combine traditional NLP, LLMs, and knowledge graphs to identify deontic obligations in EU digital regulations (GDPR, DSA, AI Act) and represent them in searchable obligation networks. Ioannidis et al. [11] present a GPT-4-based compliance platform that automatically constructs obligation registers from heterogeneous regulatory material with mandatory human review; and Singhal and Breaux [17] translate data-breach notification laws into an executable Python requirements model capturing obligations, conditions, exceptions, and rule relationships, validated via unit tests. These works show that LLMs can extract and structure obligations from complex normative texts, but they focus on legal regulations rather than technical safety standards and do not address role-specific clause selection or capability derivation which are central to our work.

In order to improve the quality of the LLM output, Masoudifard et al. [14] published a review where they investigated how different generative AI techniques can improve the quality of the results for safety compliance tasks. In their review, they found that prompt engineering techniques like chain of thoughts and tree of thoughts can improve the results of the LLM. They recommend combining these prompt engineering techniques with an advanced Graph-RAG framework to optimize output quality and to ensure that the LLM stays within the context of the standards. This combination showed significant improvements over using standard RAG. For safety analysis, various studies evaluated the benefits of LLMs to reduce the effort and to support engineers. These studies include the Hazard analysis and risk assessment (HARA) [15], scenario generation [5], failure mode and effect analysis (FMEA) [9] and System-Theoretic Process Analysis (STPA) [16]. All of these studies come to the conclusion, that the results by the LLM alone are not trustworthy enough so that human review and oversight is mandatory. Therefore, other researches studied human-in-the-loop approaches [10,15]. They observed a high potential in LLM-human collaboration for improving the efficiency and maintaining the trustworthiness of the result.

Balu et al. [2] investigate an agent-based Retrieval-Augmented Generation (RAG) architecture for deriving safety requirements in an automated driving case study. Their system retrieves context from a corpus of automotive safety standards and system documentation and shows that an agent-based RAG setup can obtain more relevant and better-used context than a default RAG pipeline by using several RAG-specific metrics. In contrast to our work, which evaluates a human-in-the-loop workflow for standards analysis and quantifies efficiency gains at the task level, Balu et al. focus on improving retrieval and grounding quality around the LLM using RAG-specific metrics. Their pipeline is complementary to our approach and could potentially enhance the generative AI component of SmartStandards. Our contributions lie in a human-in-the-loop workflow for standards analysis, role-specific clause extraction, capability derivation, and a quantitative assessment of efficiency gains.

Overall, existing work demonstrates that LLMs can support safety-relevant engineering tasks and that human-in-the-loop workflows are essential to maintain trustworthiness. However, to the best of our knowledge, there is no quantitative study of efficiency gains for the analysis of safety standards themselves, including role-specific views and capability derivation. With RQ1 and RQ2, we specifically address these gaps by evaluating both the quality and the efficiency of an LLM-based, human-in-the-loop standards analysis workflow.

3 SmartStandards GenAI Standard Analysis Tool

In order to explore the potential of efficient LLM-supported standard analysis, we developed SmartStandards. SmartStandards is a tool with a graphical user interface that enables collaborative standard analysis between human engineers and large language models. An overview of the SmartStandards is shown in Fig. 1. SmartStandards takes two main inputs: (i) the standard to be analyzed (typically a PDF file), and (ii) a criteria catalog that defines what constitutes a relevant clause for a given use case such as V&V. The output of the tool is a list of clauses with relevancy classifications, assigned criteria, and derived capabilities, presented in a graphical user interface so that an expert can review and refine the LLM output. In the following subsections, we describe the different aspects of the tool in more detail.

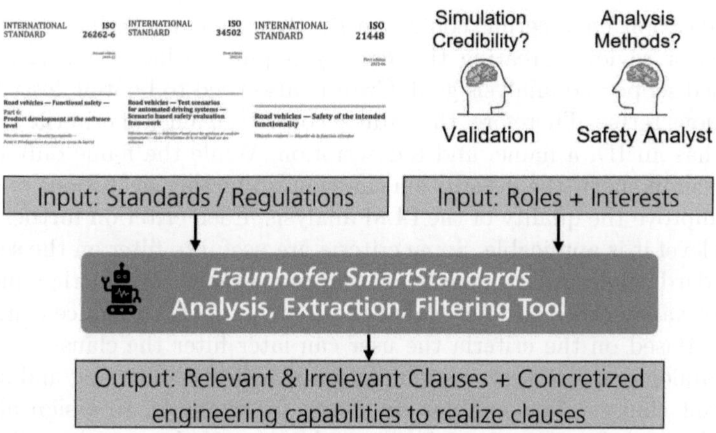

Fig. 1. General process of our tool.

3.1 Input

Standards are usually available in PDF format. In a manual analysis, engineers often copy relevant text from the PDF into separate documents or tools to facilitate later lookup and traceability of requirements. SmartStandards therefore needs to (i) reliably read PDF standards and (ii) precisely cite the relevant parts. What is considered relevant depends on various aspects such as system properties, the engineering role, and the specific analysis objective.

Our tool is designed to analyze various standards for different use cases and engineering roles such as verification, validation, or safety engineer. Therefore, the second input to SmartStandards is the criteria catalog. This catalog is provided by the user of the tool as an excel table, which makes it easy to tailor the analysis to the needs of a particular engineer or project. A relevance criterion

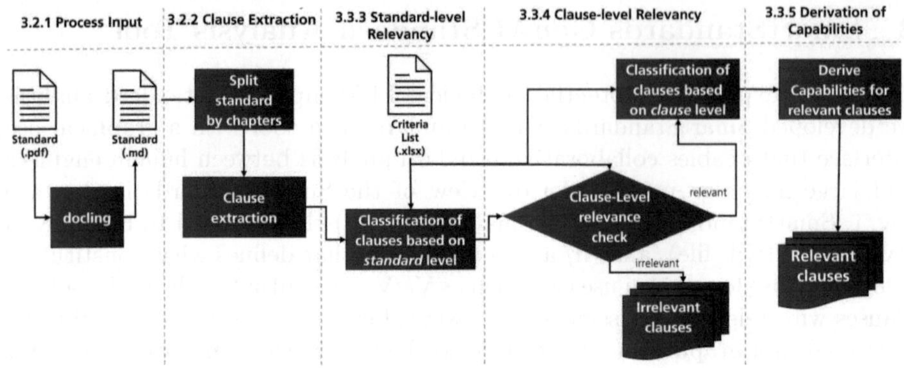

Fig. 2. Steps during the analysis of a standard.

for a verification engineering could specify that a clause defines requirements for testing or review. Creating this catalog requires a lot of experience and a well defined scope and analysis goal. Criteria also need to be well described with unambiguous terms. Therefore, the catalog consists of a list of criteria where each criterion has an ID, a name, and a description. While the name only describes the criterion in short, the description can contain further definitions or explanations to improve the quality of the LLM analysis. Each criterion further specifies on which level it is applicable. Some criteria are useful to filter on the scope level of a standard while others classify on the clause level. We further included a general relevancy check to filter clauses on a higher level to reduce computation resources. Based on the criteria the user can later filter the clauses so that the criteria catalog can also serve as foundation for effective filtering and sorting of the relevant clauses. This enables engineers, for example, to assign all clauses with architecture focus to the software architect and those with verification to the testing team.

3.2 Analysis

The analysis steps are illustrated in Fig. 2 and described in the following.

Standard Pre-processing. The first step pre-processes the standards provided as PDF files. In preliminary experiments, we observed that the LLM did not reliably handle raw PDF inputs due to layout artifacts such as multi-column text and headers or footers. Therefore, we convert the PDF standards into markdown format using Docling [7], an external document processing library that performs layout analysis and OCR-based text extraction. Our LLM handles markdown documents significantly more reliably.

However, this conversion is not error-free. Typical issues include misinterpreted headings, broken lists, and incorrectly converted tables. For best results, an engineer manually checks the generated markdown and corrects obvious errors before the automated analysis is started. In our evaluation, we did not correct the markdown, however, this effort can reduce the efficiency for other standards.

Clause Extraction. We assume that a given standard may be analyzed multiple times for different use cases and stakeholders. Therefore, we extract individual clauses and store them in a database for reuse in future analyses. Each clause record contains the clause text, its identifier, and the surrounding chapter or section heading. In later processing steps, the entire chapter can be provided to the LLM as additional context to increase the accuracy of classification and capability derivation.

Standard-level Relevancy Check. Next, standard-level criteria are determined from the scope, foreword, and introduction of the standard. These high-level parts of the document are analyzed by the LLM to identify which standard-level criteria apply (e.g., *"applies to automated driving systems"* or *"addresses V&V processes"*). Each clause within the standard is then automatically associated with the identified standard-level criteria. This is particularly useful when multiple standards with different scopes are analyzed, as it enables subsequent filtering of clauses based on standard-level scope.

Clause-level Relevancy Check. In the next step, clauses are checked for relevance on a general level. A lightweight LLM prompt applies a small set of coarse-grained criteria to filter out obviously irrelevant clauses (e.g., purely informative text or references to other standards). Example criteria at this stage include "mandatory clause" or "clause related to architecture".

The main purpose of this step is to reduce the number of clauses that require more expensive, fine-grained analysis, thereby saving tokens and computation time. Clauses that pass this initial check are then classified according to the full set of clause-level criteria defined in the criteria catalog.

Derivation of Capabilities. For all clauses that are deemed relevant, SmartStandards derives a list of capabilities. In our context, a capability is a concise description of an action or competency that an organization or system must exhibit to satisfy the clause (e.g., *"establish a process for scenario-based testing"*).

The LLM generates, for each clause, a bullet-point list of capabilities that expresses the clause in more operational terms, supports the derivation of concrete tasks and requirements in downstream engineering activities, and provides a structured basis for comparing clauses across standards, for example to identify overlapping capabilities.

As a running example, Table 1 illustrates how SmartStandards processes Clause 4.7.3.5.2 of ISO 34502. This clause requires sufficient coverage of test scenarios after testing. From the criteria catalog the following criteria are assigned:

- **CRIT003 - Operational safety**: Requirements for a product to achieve operational safety. Tasks involve assessing operational risks, implementing controls, conducting safety audits, stress testing, and maintaining safety over the product lifecycle. Level: Standard
- **CRIT006 - Scenario-based V&V**: Requirements for scenario-based V&V. Tasks include creating scenarios using knowledge-based and data-driven

Table 1. Example mapping from a standard clause to criteria and derived capabilities in SmartStandards. Bold capabilities are analog to the manually derived capabilities.

Element	ISO 34502 clause 4.7.3.5.2 (test scenario coverage)
Criteria	CRIT003 – operational safety; CRIT011 – V&V related to ODD; CRIT006 – scenario-based V&V.
Capabilities	**1. Define the relevant scenario space for the automated driving system (ADS) on limited access highways.** 2. Identify all critical and relevant scenarios within the defined scenario space. 3. Develop a comprehensive set of test scenarios that cover the identified critical and relevant scenarios. 4. Ensure that each test scenario is designed to evaluate specific safety aspects of the ADS. 5. Conduct a gap analysis to identify any missing scenarios that could impact safety. 6. Document the rationale for the selection of test scenarios and their coverage of the scenario space. 7. Execute the test scenarios in a controlled environment to validate the safety performance of the ADS. **8. Monitor and record the results of each test scenario to ensure they meet the predefined safety criteria. 9. Review and analyze the test results to confirm that the test scenario coverage is sufficient.** 10. Iterate the testing process if necessary to address any gaps or deficiencies in the test scenario coverage. 11. Prepare a final report summarizing the test scenario coverage and the results of the safety evaluation.

approaches, executing tests, analyzing outcomes, refining scenarios, and documenting findings. Level: Clause

- **CRIT011 - V&V related to ODD**: Requirements for monitoring and handling ODD exits. Tasks include analyzing exit strategies, conducting transition tests, assessing process and method reliability, and documenting lessons learned. Level: Clause

CRIT003 is assigned at the standard level, and CRIT011 and CRIT006 at the clause level. SmartStandards provided a more detailed list of capabilities covering all capabilities identified by the engineers. This example demonstrates the correspondence between standard text, criteria catalog, and expected and derived capabilities.

3.3 Manual Review

After the automated analysis, the relevant clauses are presented in the graphical user interface. For each clause, the interface shows the original text, assigned criteria, and derived capabilities. Engineers can manually adjust the assigned criteria, refine or add capabilities, or mark an entire clause as irrelevant. Conversely, they can also inspect clauses initially marked as irrelevant and reclassify them as relevant. This manual review step is essential to ensure that the result of the tool is trustworthy and that no important clauses are missed. The results can be exported as an Excel sheet, which then serves as a traceable artifact for further safety engineering activities.

3.4 Generative AI

Throughout the analysis, we use different prompts tailored to each process step to ensure reliable output. As shown in Fig. 2, the core analysis steps are LLM-driven, while document pre-processing, including PDF conversion with Docling, is performed using dedicated document parsing and text extraction tools.

The subsequent process steps are executed using a locally hosted large language model deployed via vLLM [13], a high-throughput inference and serving framework for transformer-based models. We selected the Qwen/Qwen2.5-72B-Instruct-GPTQ-Int4 model, a 72B-parameter instruction-tuned model with 4-bit GPTQ quantization, as it provided the best trade-off between reasoning performance and computational resource requirements on our internal hardware at the time of the experiments. We employ standard prompting patterns using system and human messages and provide explicit instructions to restrict the model's reasoning to the supplied context.

For context, we provide not only the single clause text but also a broader context by including the surrounding chapter whenever possible. To increase robustness, we rely on an OpenAI-compatible client that supports structured output binding, which enforces schema-validated responses (e.g., fixed JSON structures for criteria and capabilities). This approach also enables us to switch to more powerful models in the future with minimal changes to the implementation.

4 Evaluation

To evaluate our approach, we analyzed three safety standard documents (ISO 26262-4, ISO 26262-6, ISO 34502) using SmartStandards in combination with a novice safety engineer and compared the results to a fully manual analysis by experienced engineers in the context of a finalized industry project. The goals of the evaluation were (i) to quantify the effort reduction enabled by SmartStandards (RQ1) and (ii) to assess the quality of the generated analysis in terms of correctness and completeness (RQ2). The baseline, tool output and full criteria catalog cannot be released due to contractual non-disclosure with our industrial partner.

In the AI-supported setting, a novice safety engineer with approximately one year of experience operated the tool, performed the pre-processing steps, initiated the automated analysis, and conducted the manual review of the generated results. We measured the time required for these activities. In the manual baseline, four engineers with three to eight years of experience performed a traditional, manual analysis of the same standards together. For both settings, we measured the total person-time required to produce an analysis artifact consisting of identified relevant clauses, assigned criteria, and derived capabilities.

4.1 Evaluation Criteria

Before starting the evaluation, we defined criteria to measure both quality and efficiency so that we could compare the manual and the (semi-)automated analysis. The standard analysis must identify all relevant clauses, since missing clauses can lead to expensive correction measures or, in the worst case, to harm to people or property. Therefore, we measured not only efficiency (RQ1) in terms of effort but also several quality aspects (RQ2).

The following quality attributes were assessed:

- **E1: Citation correctness:** Is the clause text and identifier (e.g., clause number) correctly cited from the standard?
- **E2: Clause relevance:** Is the clause correctly identified as relevant or irrelevant for the chosen use case (both before and after human review)?
- **E3: Criteria assignment:** Are the correct criteria from the catalog assigned to the clause?
- **E4: Capabilities generation:** Are the derived capabilities sound and semantically comparable to those produced in a manual analysis?

4.2 Clause Relevancy Criteria Catalog

As relevancy criteria we used 18 criteria that define the relevancy of clauses for V&V. An excerpt of criteria within this catalog is shown in the example in Sect. 3.2. The table is directly provided as an Excel input to the tool, and the list is then stored in a database so that the same criteria catalog can be reused for other standards.

Each criteria catalog is stored with a version ID so that improvements to the catalog can be tracked and compared to previous analysis results. For our evaluation, we used the same criteria catalog for all three standards and for both the manual and AI-supported analyses, ensuring comparability between conditions.

For E1 (citation correctness), all clauses were correctly extracted and cited from the PDF documents. This shows that the combination of Docling and manual markdown correction was sufficient to ensure reliable citation of clause text and identifiers.

For E2 (clause relevance), approximately half of the clauses were correctly classified as relevant or irrelevant by the LLM without human intervention. After the review and correction by the novice engineer, the correctness of relevance classification increased to 58% which indicates a moderate improvement. The 7% incorrectly excluded clauses are critical because they can cause that important requirements are not fulfilled. 35% of the incorrectly included clauses can lead to an overall increase in effort and thus need to be reduced. An improvement of the automatic classification can involve refinement of the relevance criteria catalog or a better LLM component.

For E3 (criteria assignment), the assigned criteria were fully correct or partially correct in 75% of the cases after the automated analysis. The manual

Table 2. Overview of the evaluation results after applying SmartStandards by a novice safety engineer. * ISO 34502 was only analyzed by the LLM without human review

Number of	ISO 26262-4	ISO 26262-6	ISO 34502*	SUM	%
Total clauses	94	139	81	314	
Correctly included clauses	46	30	31	107	34%
Incorrectly included clauses	17	66	26	109	35%
Correctly excluded clauses	21	35	19	75	24%
Incorrectly excluded clauses	10	8	5	23	7%
Relevant clauses	56	38	36	130	
Clauses with correctly assigned criteria	23	22	0	45	35%
Clauses with incorrectly assigned criteria	17	8	7	32	25%
Clauses with partially correct assigned criteria	16	8	29	53	40%
Included clauses	63	96	57	217	
Correctly cited clauses	63	95	57	216	99.5%
Incorrectly cited clauses	0	1	0	1	0.5%

review improved the rate of fully correct assignments by approximately 7%. Incorrect assignments typically involved missing one relevant criterion, rather than assigning entirely wrong criteria.

For E4 (capabilities generation), we manually inspected the generated capabilities and compared them to capabilities derived in the manual analyses. Due to the open-ended nature of this task, the capabilities cannot be compared one-to-one. Instead, we classified them as *sound and semantically comparable* (capturing essentially the same meaning as the manually derived capabilities) or *not fitting*. We found that 92% of the generated capabilities were sound and semantically comparable, while 8% needed revision. Overall, given the complexity of the task, these results are promising for a first iteration of the approach.

4.3 Evaluation Results

The measured time indicates an effort reduction of approximately 84% when using SmartStandards with a novice engineer, compared to the purely manual analysis by experienced engineers. This includes the time for pre-processing, running the tool, and manually reviewing the results. However, the AI-supported analysis still contained some incorrect or incomplete entries after the novice review. To estimate the net effort required to reach similar quality as in the fully manual analysis, we additionally considered an expert review step that corrects remaining issues shown in Table 2. We approximated the additional effort for this expert review by adding the effort that one expert needed to analysis one standard manually.

Taking the novice effort and the estimated expert review effort together, the net effort required to produce an artifact of comparable quality is around 63% of the effort of the fully manual analysis. This corresponds to a significant reduction in expert effort while still achieving a high-quality standards analysis.

Assuming that a manual analysis of each standard, including writing all clauses into a table, categorizing them, describing potential capabilities, and conducting at least one review, requires five person-days, our approach reduces the effort by approximately two person-days per standard. In the current landscape of standards for autonomous systems, 10–20 standards are potentially relevant for certain activities. Therefore, one to two person month could be saved.

5 Discussion and Conclusion

5.1 Discussion

Our current implementation of SmartStandards is primarily tailored to the structure of automotive ISO standards. This has several implications for the generality and limitations of our results. The tool assumes a relatively regular structure of standards with well-defined clauses and chapters, as is typical for automotive ISO documents. While this is mainly relevant for text extraction and clause identification, different document structures may also influence the correctness of criteria assignments and relevancy checks. We expect that other technical standards will require only minor adjustments, but regulations and legal texts often use different structures and terminology and may require more extensive adaptation. Given the increasing number of regulations relevant to safety-critical systems, extending the tool to better support such documents is a promising direction. Effort required for tuning the framework was not measured during our evaluation and we expect non-negligible effort required for extending SmartStandards for other standards and regulatory texts. On the other hand, this tuning effort is mainly required for the initial setup and digitization of the standards. In our use case, we assume that standards are analyzed for different roles and different relevancy criteria so that we only need the tuning effort once.

The non-deterministic nature of LLMs can cause variability in the results. We applied various prompt engineering techniques and used structured output binding to increase consistency, but a certain level of variability remains. This limitation is reflected in our evaluation results, where the LLM alone correctly classified only about half of the clauses with respect to relevance (E2), and some criteria assignments (E3) still required human correction. Future enhancements to the prompting strategy and the use of more advanced models could further improve robustness and reduce the remaining manual effort.

We employed Qwen2.5-72B as a locally hosted large language model due to cost considerations and the need to satisfy privacy, confidentiality, and data sovereignty requirements. While more capable proprietary models, such as recent GPT variants, or emerging reasoning-oriented models (e.g., Qwen 3), are likely to yield improved results with reduced prompt engineering effort, their use was outside the scope of this study. Importantly, our architecture is based on an OpenAI-compatible client interface and schema-validated outputs, enabling seamless substitution of the underlying LLM. As a result, the proposed approach can directly benefit from future advances in LLM capabilities without requiring architectural changes. The limited context length of the model required us to split the input into smaller chunks. This chunking can affect output quality because clauses may be analyzed without full access to their broader document context. A model with a larger context window, capable of processing entire chapters or even full standards at once, could improve the quality of relevancy checks and capability derivation. Finally, the PDF input format posed challenges. Our model could not reliably process raw PDFs, so we converted documents to markdown using Docling. This conversion sometimes introduced errors that had to be manually corrected, particularly for tables and complex layouts. While the safety standards we used contained few tables with essential requirements, other standards may rely heavily on tabular requirements, which could lead to worse results if the conversion is not reliable. Either manually curated markdown versions of standards or models that can robustly read PDF documents would further improve the overall quality and reduce pre-processing effort.

Our evaluation is further limited to three automotive safety standard documents and the V&V use case. We did not assess SmartStandards for other roles (e.g., safety management, cybersecurity, supplier management) or other downstream activities such as safety case structuring or test planning. Moreover, only a single novice engineer operated the tool, and the manual baseline stems from a separate industry project rather than a controlled A/B comparison. As a result, the reported efficiency gains should be interpreted as indicative for this setting, not as universally representative across organizations, standards, or use cases. From a safety perspective, the observed 7% of incorrectly excluded clauses are particularly critical, since they may lead to missing requirements with potential safety impact. Incorrectly included clauses primarily increase effort, but do not directly reduce safety margins. Our current mitigation is to require human review and to preserve traceability to the original standard text and identifiers. However, we did not investigate human factors such as automation bias or reviewer workload, which could affect how reliably engineers detect AI errors. Future work should therefore define and evaluate explicit review procedures when using SmartStandards in safety-critical projects.

Overall, our evaluation shows that despite these limitations, SmartStandards can substantially reduce effort (RQ1), while archiving a similar quality then the manual analysis especially when combined with human review (RQ2). The observed errors and limitations highlight where future improvements should focus: better handling of document structure, more capable LLMs, and enhanced context management. Furthermore, using generative AI tools in safety critical systems imposes compliance challenges itself. In the automotive domain, ISO 26262 demands tool qualification which is currently not feasible for generative AI tools. Therefore, a human engineer has to review the results and approve them which creates additional challenges to ensure that the review guarantees the required quality. While AI-based tools promise significant benefits to support safety engineering and development tasks, current standardization activities for ISO/IEC AWI TS 22440-1 [12] discuss the usage of (generative) AI in safety engineering processes.

5.2 Conclusion

In this study, we examined whether generative AI can support engineers in analyzing safety standards and (semi-)automatically extracting relevant requirements for different stakeholders. To this end, we proposed SmartStandards, a tool that analyzes standards, filters their clauses for relevancy based on a user-provided criteria catalog, and derives capabilities that operationalize the intent of the clauses. The tool integrates a graphical user interface that enables human engineers to review and correct the AI-generated results, thereby ensuring trustworthy and reliable output.

In our evaluation, we compared the effort required to analyze three different automotive safety standards using SmartStandards (operated by a novice engineer) with the effort of a traditional manual analysis performed by four experienced engineers. The results indicate a substantial reduction in effort for standards analysis while maintaining a high level of quality in terms of citation correctness, clause relevance, criteria assignment, and capabilities generation. A novice safety engineer with only one year of experience was able to create a high-quality first iteration artifact that, after limited expert review, was comparable to the artifact produced by the group of experienced engineers.

We therefore conclude that generative AI, when embedded in a human-in-the-loop workflow, can significantly reduce the required effort for standards analysis and decrease the reliance on scarce expert time. In answer to RQ1, SmartStandards achieves high citation correctness and largely sound capability derivation, with remaining errors mitigated by human review. In answer to RQ2, our evaluation indicates that the AI-supported workflow can reduce the net effort for standards analysis to around two thirds of the effort of a fully manual expert analysis. Future work will focus on extending SmartStandards to additional domains (e.g., other industries and regulatory documents), integrating more advanced LLMs, and further automating consistency checks to reduce the remaining manual review effort.

Acknowledgments. This research was funded by the Leistungszentrum Simulations- und Software-based Innovation in Kaiserslautern, Germany. The authors acknowledge its support with gratitude. (https://www.leistungszentrum-simulation-software.de/)

References

1. Ariai, F., Mackenzie, J., Demartini, G.: Natural language processing for the legal domain: a survey of tasks, datasets, models, and challenges. ACM Comput. Surv. **58**(6), 1–37 (2025). https://doi.org/10.1145/3777009
2. Balu, B.V., et al.: Towards automated safety requirements derivation using agent-based rag. In: Proceedings of the AAAI Symposium Series, vol. 5, no. 1, pp. 299–307 (2025). https://doi.org/10.1609/aaaiss.v5i1.35605
3. Berger, A., et al.: Towards automated regulatory compliance verification in financial auditing with large language models. In: 2023 IEEE International Conference on Big Data (BigData), pp. 4626–4635 (2023)

4. Bonnin, H., Trevey, K., Geneau, T., Xiao, J.: On the challenges of responsible usage of LLM-based tools in safety engineering. In: Proceeding of ERTS 2026 on Embedded Real Time Systems. Proceeding of ERTS 2026 on Embedded Real Time Systems, Toulouse, France (2026). https://doi.org/10.82331/ERTS.2026.26
5. Cyffka, E., et al.: Can large language models assist with Sotif scenario generation? In: IEEE International Conference on Recent Advances in Systems Science and Engineering (RASSE), pp. 1–8 (2025)
6. Dal Pont, T., Galli, F., Sartor, G., Contissa, G.: "lost in EU regulation? don't worry, AI found the obligation": Extracting and representing legal obligations in the GDPR the DSA and the AI act. In: Proceedings of the Twentieth International Conference on Artificial Intelligence and Law. ICAIL '25, New York, NY, USA, pp. 129–138. Association for Computing Machinery (2026). https://doi.org/10.1145/3769126.3769260
7. Deep Search Team: Docling technical report. Technical report (2024). https://doi.org/10.48550/arXiv.2408.09869
8. Galli, F., Dal Pont, T.R., Sartor, G., Contissa, G.: Approaching the AI act... with AI: LLMs and knowledge graphs to extract and analyse obligations. Comput. Law Secur. Rev. **60**, 106230 (2026). https://doi.org/10.1016/j.clsr.2025.106230
9. Hassani, I.E., Masrour, T., Kourouma, N., Tavčar, J.: AI-driven FMEA: integration of large language models for faster and more accurate risk analysis. Des. Sci. 11 (2025)
10. Hillen, D., Helten, C., Reich, J.: Towards LLM-augmented situation space analysis for the hazard and risk assessment of automotive systems. In: INFORMATIK 2024, pp. 709–714. Gesellschaft für Informatik e.V., Bonn (2024). https://doi.org/10.18420/inf2024_58
11. Ioannidis, J., Harper, J.M., Quah, M.S., Hunter, D.: Gracenote.ai: Legal generative AI for regulatory compliance. In: LegalAIIA@ICAIL (2023)
12. ISO: ISO/IEC CD TS 22440-1 1Artificial intelligence — Functional safety and AI systems Part 1: Requirements (2026)
13. Kwon, W., et al.: vLLM: Easy, Fast, and Cheap LLM Serving with PagedAttention (2023). https://arxiv.org/abs/2309.06180
14. Masoudifard, A., Sorond, M.M., Madadi, M., Sabokrou, M., Habibi, E.: Leveraging graph-rag and prompt engineering to enhance LLM-based automated requirement traceability and compliance checks (2024)
15. Nouri, A., Cabrero-Daniel, B., Törner, F., Sivencrona, H., Berger, C.: Engineering safety requirements for autonomous driving with large language models. In: IEEE 32nd International Requirements Engineering Conference (RE), pp. 218–228 (2024)
16. Qi, Y., Zhao, X., Khastgir, S., Huang, X.: Safety analysis in the era of large language models: a case study of STPA using ChatGPT. Mach. Learn. Appl. **19**, 100622 (2025). https://doi.org/10.1016/j.mlwa.2025.100622
17. Singhal, A., Breaux, T.: Legal requirements translation from law (2025). https://arxiv.org/abs/2507.02846

Engineering Safety Systems

Chaining Unsafe Control Actions in STPA

Nicholas Petrunti[1](✉), Spencer Deevy[2], Vera Pantelic[1], Mark Lawford[1], Richard F. Paige[1], and Alan Wassyng[1]

[1] McMaster Centre for Software Certification, Department of Computing and Software, McMaster University, Hamilton, Canada
{petruntn,pantelv,lawford,paigeri,wassyng}@mcmaster.ca

[2] NAV CANADA, Ottawa, Canada
spencer.deevy@navcanada.ca

Abstract. The increasing complexity of safety-critical systems demands rigorous hazard analysis, especially for capturing unsafe interactions between subsystems. System-Theoretic Process Analysis (STPA) is an effective way to identify unsafe system interactions, associated hazards, and loss scenarios, and then derive safety requirements. Becoming proficient with STPA requires significant understanding of control actions, which is a challenging concept to novices. We leverage repetitions in the derivation of unsafe control actions and loss scenarios of STPA to propose modifications that help more analysts get the most out of STPA. These modifications result in a more systematic approach to conducting STPA. They also help lay the groundwork for the risk assessment that many companies must perform. We demonstrate the application of the modified STPA in an industrial application – a remote and virtual tower system for air traffic services.

1 Introduction

The *remote and virtual tower (RVT)* is an advanced technology that has emerged to replace traditional Air Traffic Service (ATS) facilities at airports. In RVTs, the operator's traditional out-of-the-window view of an airport and its surroundings is replaced by screens that present an augmented view, synthesized from real-time sensor data. RVTs simplify service in remote areas, improve operation and maintenance efficiency, and can increase safety of existing services [4,10]. These advanced systems will be more difficult to motivate if requisite system hazard analysis significantly increases the time and resources required to safely deploy them. Thus, we must find ways to improve efficiency of the analysis while maintaining or improving the quality of the results.

NAV CANADA, the main civil air navigation service provider in Canada, is deploying RVT technology across Canada under their Digital Aerodrome Air Traffic Services (DAATS) program. System-Theoretic Process Analysis (STPA)

Partially supported by the Natural Sciences and Engineering Research Council of Canada and Mitacs.

© The Author(s), under exclusive license to Springer Nature Switzerland AG 2027
E. Troubitsyna et al. (Eds.): SAFECOMP 2026, LNCS 16868, pp. 213–227, 2027.
https://doi.org/10.1007/978-3-032-34867-8_14

[6], has proven to be an effective hazard analysis technique for complex safety-critical systems. While STPA has been used in aviation, such as in [1,2,7], RVT is a novel field of application for STPA. The goal of our research is to investigate how STPA can be used to complement traditional hazard analysis techniques currently used for ATS. While applying STPA to NAV CANADA's RVT system, we discovered that even experienced safety analysts were frustrated with the ad hoc nature of some of the steps. They also wanted more support for risk analysis. This motivated us to see if we could help these analysts and still remain true to STPA principles. We discovered that we can use inherent repetition in STPA steps corresponding to unsafe control action and loss scenario identification to propose enhancements to STPA. The main contributions of the work are as follows:

- To the best of our knowledge, this is the first publication on applying STPA to a real industrial example in the RVT domain.
- We propose modifications to STPA that can increase the efficiency of its unsafe control action and loss scenario identification steps. These modifications make the process more systematic and easier to do well, enabling more analysts to achieve useful results with STPA.
- The proposed modifications also establish foundations for an easier integration of STPA results into a company's risk assessment framework.

2 Background

2.1 System-Theoretic Process Analysis

STPA was initially proposed as a top-down hazard analysis technique with the main goal of identifying emergent properties in complex systems [6]. The technique has been successfully used in many applications, such as in [11,12], which apply it to rail transport and nuclear power plants, respectively. The four steps in STPA are described in some detail in [8]. Briefly these steps are: i) identify *losses* and *system-level hazards*; ii) model a control structure showing how functional *controllers* interact with *controlled processes* through *control actions* and *feedback*; iii) identify and analyze *unsafe control actions* (UCAs) that will lead to a hazard; and iv) identify *loss scenarios*, which are situations that may lead to UCAs and by proxy, system-level hazards.

The last stage is considered the least structured of the four stages [3,8,14,15], and methods of completing it vary.

2.2 Remote and Virtual Towers

The RVT concept involves replacing conventional ATS facilities such as air traffic control towers with sensor suites affixed to structure(s) at the aerodrome. This may be accomplished with a dedicated sensor mast and/or retrofitting existing structures to improve aerodrome coverage. Video and audio streams are then sent to an external facility to be rendered on panoramic displays to replicate the visual experience at the aerodrome. This concept allows consolidating ATS operators into centralized "remote centres", where multiple remote aerodromes are virtualized and each provides services that would normally require a dedicated conventional ATS facility. RVTs open the door to new features such as augmentation of human visual capabilities including on-screen aircraft and vehicle tagging. Enhanced vision in low visibility conditions and elimination of blind spots become possible with advanced camera technology.

RVT technology is still in early stages of adoption, and it has only been in use or trialed in the recent few years in several countries that have different operational environments from that of Canada [5]. NAV CANADA has been actively working in furthering the RVT concept with the DAATS program, but recognizes that the novelty of the concept and technology introduces new risks. These risks are complex and there is limited historical data to help understand them. As a means of providing additional safety assurance and continual improvement of NAV CANADA's safety management system, STPA was chosen to augment safety assessment processes.

3 Application of STPA for RVTs

Over the course of approximately one year, an STPA was completed on portions of NAV CANADA's RVT system. More precisely, the analysis focused on the novel interactions introduced into an existing system by the RVT concept. The application of the STPA process, including the modified Steps 3 and 4, involved approximately 20 sessions with personnel from NAV CANADA, including safety engineers, systems engineers, human factors specialists, operational system requirements specialists and ATS operators.

The proposed modifications to STPA are to Steps 3 and 4. The STPA application started with the first two steps as described in the Handbook [8].

3.1 Step 1

In the first step, the losses and system-level hazards for the RVT system were identified. These included safety-critical losses, such as loss of life, health, or property, as well as loss of mission, sustainability, and user confidence. These reflect the company's need to provide safe, effective, and efficient service. Three system-level hazards were identified: loss or degradation of visual (*H1*) and auditory (*H2*) information of the airfield, and loss or degradation of ability to provide air navigation service (*H3*). The most relevant hazard for the remainder of this paper is *H1 – Air traffic control decisions are made in the RVT system with incomplete, degraded, or misleading visual information about the airfield environment.*

3.2 Step 2

A detailed control structure was created to map all functional interactions between ATS Operators, the aerodrome, aircraft, and other air navigation tools. Figure 1a shows a representative control structure for an RVT system that has been obfuscated for simplicity of explanation. The main novelty of RVTs lies in the interaction between the *ATS Operator* – the individual who monitors air traffic and provides instructions to aircraft – and the *aerodrome cameras*, which ensure sufficient coverage of the airfield. This is done through the *RTOS (Remote Tower Optical System) Interface*, which includes functionality to control the aerodrome cameras as well as the screens, which display the camera visuals of the airfield.

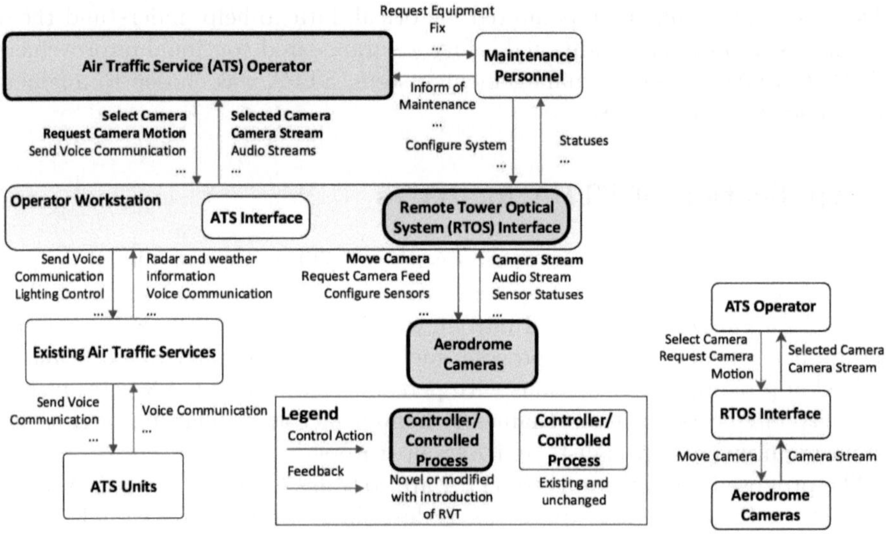

(a) The entire ATS system

(b) Novel portions of the ATS system

Fig. 1. STPA control structures for an ATS system.

A portion of this control structure serves as the example in this paper, and is illustrated in Fig. 1b. In this system, the ATS Operator interacts with the RTOS Interface to select one of multiple pan-tilt-zoom (PTZ) cameras to control (*Select Camera*). Once selected, the operator requests the panning, tilting, or zooming of the camera (*Request Camera Motion*). The RTOS interface then issues corresponding commands to the aerodrome cameras (*Move Camera*), where an electromechanical actuator physically moves the camera as requested. The information about which camera, if any, is selected (*Selected Camera*), and the feed from the camera (*Camera Stream*), are displayed on the RTOS interface.

3.3 Step 3

In Step 3, Unsafe Control Actions (UCAs) are identified. STPA provides a set of guide words to help with identifying UCAs for each control action in the control structure: *"Not providing"*, *"Providing [in an unsafe way]"*, *"Providing too early or late"*, and *"Applying for too long or stopping too soon"* [8]. Table 1 is an extract of a UCA table for the control structure in Fig. 1b. For brevity, it contains some of the identified UCAs for only the first two categories of the guide words, *"Not providing"* and *"Providing in an unsafe way"*. Each UCA has an ID indicating its associated control action and guide word for ease of searching.

Table 1. Traditional UCA table for the control actions in Fig. 1b

Control Action	Not Providing	Providing in an Unsafe Way	...
CA1: Select Camera	*UCA_SlctCam_NP1*: ATS Operator does not Select Camera before attempting to control it [H1].	*UCA_SlctCam_Pv1*: ATS Operator does Select Camera, but different from the camera they intend to control [H1].	...
CA2: Request Camera Motion	*UCA_CamMtn_NP1*: ATS Operator does not Request Camera Motion when they want to move a camera [H1].	*UCA_CamMtn_Pv1*: ATS Operator requests Camera Motion when no camera is selected [H1]. *UCA_CamMtn_Pv2*: ATS Operator Requests Camera Motion when the wrong camera is selected [H1].	...
CA3: Move Camera	*UCA_MoveCam_NP1*: RTOS Interface does not issue Move Camera when requested by ATS Operator [H1].	*UCA_MoveCam_Pv1*: RTOS Interface issues Move Camera, but a different camera from that intended by ATS Operator is moved [H1]. *UCA_MoveCam_Pv2*: RTOS Interface issues Move Camera, but moves it differently from intended by ATS Operator [H1].	...

3.4 Step 4

Step 4 is where *loss scenarios* are identified – these are specific scenarios involving causal factors that have the potential to lead to a UCA or a system-level hazard. As mentioned in Sect. 2.1, there are many different approaches in use. However, all approaches involve identifying which causal factors in the system can lead to a given UCA, and how a given control action can lead to a hazard without any failure occurring [8].

The STPA Handbook [8] introduces "*a control model to assist in loss scenario generation*", which includes a simple control loop and 15 mechanisms for causal factors inside or adjacent to the loop that can lead to a hazard. Figure 3a is a simplified version of that diagram. A common method of identifying loss scenarios uses this model as guidance.

For *UCA_CamMtn_NP1* from Table 1 (i.e. not providing a request to control the camera when required), we list some example loss scenarios and their associated mechanisms from Fig. 3a:

- **LS1:** RTOS Interface provides incorrect feedback on the selected camera, leading to the ATS Operator being unable to issue control actions to the desired camera [*Provided Feedback*].
- **LS2:** ATS Operator misinterprets the current position of the camera and as a result does not reposition to a more advantageous view, leading to a degraded visual of the airspace [*Process Model*].
- **LS3:** ATS Operator issues the correct command to move the camera, but that message is not received by the camera due to a loss of connection between the RTOS Interface and the camera [*Received Control Action*].

4 UCA Chains

In this section, we propose modifications to STPA's Steps 3 and 4; we refer to this technique as *"UCA Chains"*. We motivate the modifications, demonstrate how this technique can be used in Step 3 of STPA, and examine how this makes Step 4 more systematic. We introduce another modification to STPA – identifying unsafe feedback – to make UCA chains more useful.

4.1 Motivation

The primary motivation for UCA Chains was to create a more systematic process for identifying loss scenarios in STPA's Step 4. While the STPA Handbook [8] does provide some guidance on identifying loss scenarios, this step lacks rigor when compared with the previous three steps [3,15]. This limitation was addressed by one of the authors of the STPA Handbook, John Thomas, in a 2024 lecture, stating, "*One of the comments we've gotten from folks is they love the first three steps in STPA, especially Step 3, where you have a lot of rigor... can you make STPA Step 4 look a bit more like this?*" [15]. In the same lecture, he then proposes another viable strategy to loss scenario identification with a formal structure similar to how UCAs are identified.

We propose another strategy to introduce rigor to loss scenario identification. The main inspiration comes from Step 3 of STPA. First, we notice that many UCAs associated with controllers at the top of the control structure funnel to one of a few UCAs of lower authority controllers. For example, in Fig. 1b, UCAs in either selecting the camera or requesting camera motion can lead to UCAs in moving the camera. As a result, significant repetition occurs when considering how these UCAs lead to hazards. An experienced analyst may recognize that, since the camera not moving when needed can lead to a hazard, any other

UCA that causes this situation can therefore also lead to a hazard. Since UCAs directly feed into loss scenarios, all of the described patterns also apply in Step 4. Once again, instead of independently identifying loss scenarios from scratch for each UCA, we can reuse them between UCAs. For example, loss scenario *LS3*, as defined in Sect. 3.4, involves the causal factor of a loss of communication between the RTOS and aerodrome. This causal factor would also contribute to a loss scenario for *UCA_SlctCam_NP1* (ATS Operator does not Select Camera before attempting to control it), and many others, but only needs to be identified once.

To leverage the UCA/scenario repetition, we propose *UCA Chains.* The approach involves identifying which UCAs have the potential to cause others. Following this, the causal factors of the causal UCA can therefore also cause the affected UCA by proxy. We now describe modified STPA Steps 3 and 4 taking into account UCA Chains.

4.2 Introduction to UCA Chains

In Step 3 of STPA, [8] instructs the user to link each UCA to hazards by determining if a UCA can lead to a hazard. Instead, we require that a UCA is linked to hazards and/or *other UCAs* it can lead to.

Our approach modifies Table 1 to Table 2. (The table makes use of another concept, *unsafe feedback*, introduced in Sect. 4.3). Table 2 indicates, for each UCA of CA1 and CA2 (higher controller), if they cause a UCA of CA3 (lower controller). For example, *UCA_SlctCam_Pv1* involves the ATS Operator selecting the wrong camera. This can lead to *UCA_MoveCam_NP1*, where the intended camera is not moved, and *UCA_MoveCam_Pv1*, where the wrong camera is moved. Both of these UCAs can then lead to an inadequate camera stream, an unsafe feedback which is a direct cause of H1, where the chain terminates.

4.3 Unsafe Feedback

To unlock the full potential of UCA chains, we examine feedback in Step 3. We introduce a new concept, called *unsafe feedback (UFB)*, which is identified using the same process as UCAs, via the guide words. We can apply the same guide words to identify similar deficiencies in feedback paths, since feedback is designed into the system and under the designers' control. We can then use UFBs within the UCA chains to understand how different UCAs may be caused, and how they may cause other UCAs. By including UFBs at this stage, the connection between UCAs is made more clear in situations where one control action is dependent on feedback from another controller. Analyzing unsafe feedback at this stage is not strictly necessary, but doing so reduces the amount of loss scenarios needed to be identified per UCA.

We include UFBs for the Selected Camera and Camera Stream in Table 2. Including feedback at this step makes its contribution to the system and potential hazards explicit. For example, in Table 2, UFB of Selected Camera can lead to

UCAs of Select Camera and Request Camera Motion, and UCAs of Move Camera can lead to UFBs of the Camera Stream feedback.

4.4 Modified STPA Step 3 with UCA Chains

Table 2. Modified Table 1 Using UCA Chains and Unsafe Feedback

Control Action/ Feedback	Not Providing	Providing in an Unsafe Way
FB1: Selected Camera	*UFB_SlctCam_NP1*: RTOS Interface does not display Selected Camera when requested by ATS Operator [*UCA_SlctCam_NP1*, *UCA_CamMtn_Pv1*, *UCA_CamMtn_Pv2*].	*UFB_SlctCam_Pv1*: RTOS Interface displays incorrect Selected Camera at any time [*UCA_SlctCam_NP1*, *UCA_SlctCam_Pv1*, *UCA_CamMtn_Pv1*, *UCA_CamMtn_Pv2*].
CA1: Select Camera	*UCA_SlctCam_NP1*: ATS Operator does not Select Camera before attempting to control it [*UCA_MoveCam_NP1*, *UCA_MoveCam_Pv1*].	*UCA_SlctCam_Pv1*: ATS Operator does Select Camera, but different from the camera they intend to control [*UCA_MoveCam_NP1*, *UCA_MoveCam_Pv1*].
CA2: Request Camera Motion	*UCA_CamMtn_NP1*: ATS Operator does not Request Camera Motion when they want to move a camera [*UCA_MoveCam_NP1*].	*UCA_CamMtn_Pv1*: ATS Operator requests Camera Motion when no camera is selected [*UCA_MoveCam_NP1*]. *UCA_CamMtn_Pv2*: ATS Operator Requests Camera Motion when the wrong camera is selected [*UCA_MoveCam_Pv2*].
CA3: Move Camera	*UCA_MoveCam_NP1*: RTOS Interface does not issue Move Camera when requested by ATS Operator [*UFB_CamFeed_Pv1*].	*UCA_MoveCam_Pv1*: RTOS Interface issues Move Camera, but a different camera from that intended by ATS Operator is moved [*UFB_CamFeed_Pv1*]. *UCA_MoveCam_Pv2*: RTOS Interface issues Move Camera, but moves it differently from intended by ATS Operator [*UFB_CamFeed_Pv1*].
FB2: Camera Stream	*UFB_CamStrm_NP1*: RTOS Interface does not provide Camera Stream while ATS Operator is dependent on it [H1].	*UFB_CamStrm_Pv1*: RTOS Interface provides degraded or inadequate Camera Stream while ATS Operator is dependent on it [H1].

Table 2 is iteratively populated starting from the bottom. First, we identify which control actions or feedback can directly lead to a hazard without necessarily causing other UCAs or UFBs. We call these 'terminating UCAs' and 'terminating

UFBs', respectively. Then, the four UCA guide words provided in the Handbook [8] should be used to identify the terminating UCAs/UFB. In the next iteration, we identify which UCAs/UFBs may lead to the previously identified terminating UCAs/UFBs. This is isolated to control actions and feedback that act as input to the terminating controllers, but may use any of the guidewords. The process continues iteratively, following along the control structure until all control actions and feedback have been considered. To prevent loops, if a given UCA/UFB has already been identified previously within the same chain, it need not be added again.

Following this process removes implicitness in the connection between UCAs and hazards. Consider *UCA_SlctCam_NP1* in Table 1. It is not explicit how failing to select a camera can lead to degradation of visual information. One must walk through the chain of events from the camera not being selected, to the request to control the camera not being executed, and finally to the camera being in the wrong position being a degradation of visual information. This is challenging for large systems with many interactions; UCA chains make this sequencing explicit.

4.5 Modified STPA Step 4 with UCA Chains

The modifications to Step 3 expedite identification of loss scenarios in Step 4. If UCA A causes UCA B, then any causal factor that leads to UCA A may also lead to UCA B.

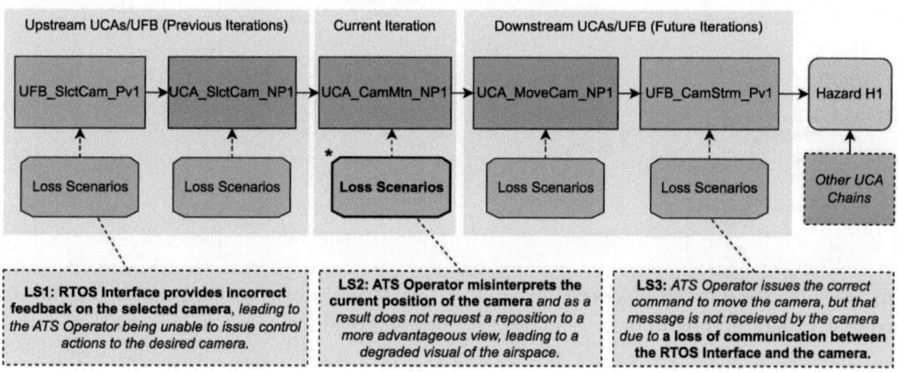

Fig. 2. An iteration of identifying loss scenarios with UCA chains. In this iteration, the loss scenarios for *UCA_CamMtn_NP1* are identified. All causal factors that lead to previous UCAs, labelled as Upstream UCAs/UFB, are assumed to be applicable and need not be reconsidered. Downstream UCAs/UFB can also be omitted as they are considered in future iterations. Therefore, only the loss scenarios labelled with a * are considered in this iteration. In the loss scenario examples, causal factors are bolded.

Figure 2 illustrates a UCA chain from left to right, such that the resulting hazard is the right-most part of the chain. The process of identifying loss scenarios with UCA chains flows from left to right, starting with the left-most UCA or UFB. In the first iteration, loss scenarios that may directly lead to the first UCA/UFB are identified. The next iteration moves along this chain to the next UCA/UFB, until the entire chain has been considered. A causal factor that may lead to one UCA/UFB may also lead to others later in the chain. Thus, all of the causal factors of prior and subsequent UCAs/UFBs do not need to be considered for a given iteration, and instead get inherited.

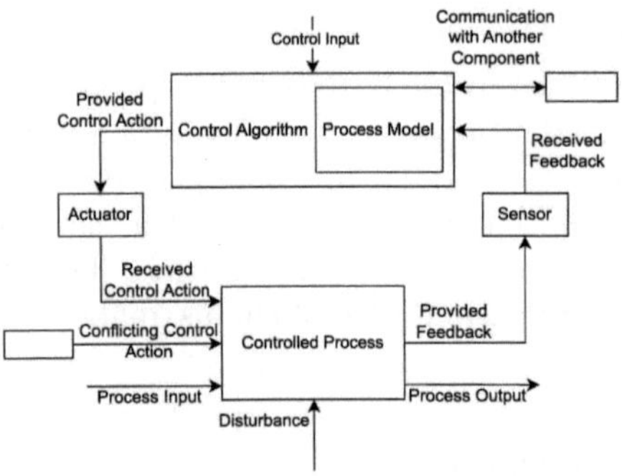

(a) The entire control model (loop) for a UCA (original STPA).

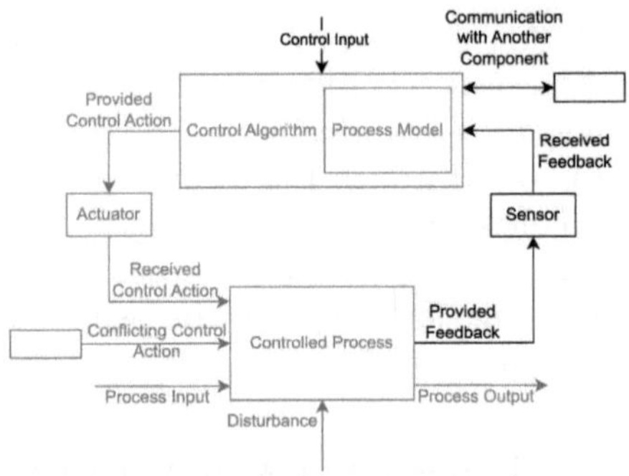

(b) Mechanisms implicitly covered by prior events highlighted.

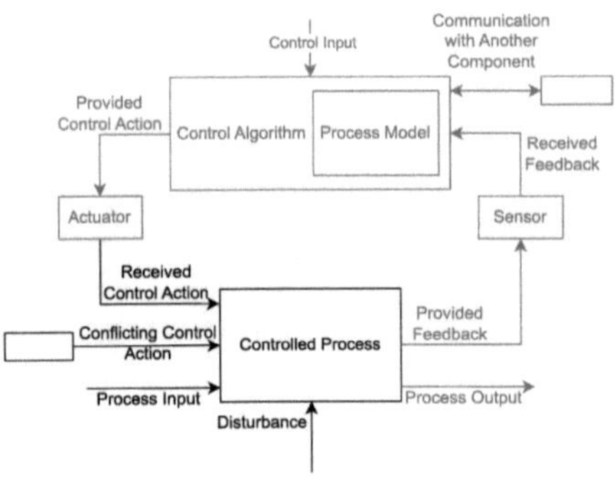

(c) Mechanisms implicitly covered by subsequent events highlighted.

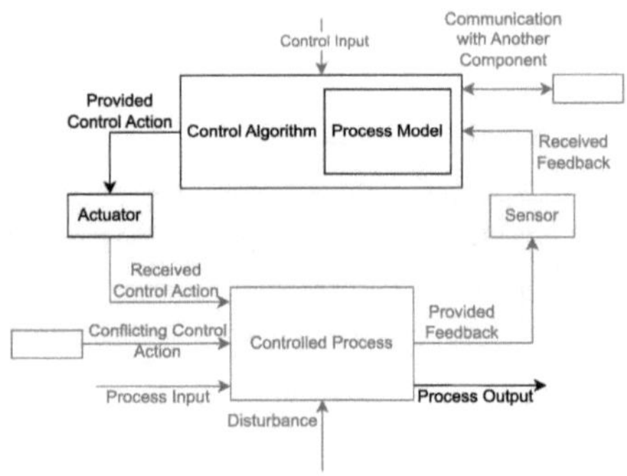

(d) Remaining mechanisms highlighted, to be considered in the current iteration.

Fig. 3. Revised "*Control model to assist in loss scenario generation*", as described in [8], with examples of mechanisms that may be considered in prior (3b), subsequent (3c), and current (3d) iterations of loss scenario identification.

In each iteration, the process isolates the loss scenario identification to the causal factors that *directly* lead to a UCA/UFB without causing another UCA/UFB first. Revisiting the vanilla STPA control model (loop) for a UCA from Fig. 3, we can assess which mechanisms need to be considered for a given iteration of this process. When assessing different UCAs along a chain, the mechanisms associated with different parts of the system will change. For example, what is identified as the 'Provided Control Action' for UCAs of one control action may be the 'Control Input' or 'Received Control Action' for UCAs of another control action. Since we consider only the direct loss scenarios, many of these mechanisms can be eliminated from consideration. Figures 3b and 3c highlight (in black) the mechanisms that are covered implicitly by creating UCA chains, as they are upstream and downstream causal factors, respectively. This leaves the five mechanisms shown in Fig. 3d to be considered in each iteration ('*Process Output*' is relevant for a terminal UCA only); the remaining ten mechanisms are covered by previously identified causal factors for upstream UCAs/UFBs.

In practice, it might be unclear how to classify a causal factor into one of the 15 mechanisms. For example, there may be ambiguity as to whether the cable between the RTOS interface and the aerodrome is classified as an 'actuator' or a 'received control action'. It is not important to which mechanism the cable is classified; only that it is classified consistently to avoid being missed.

We can now revisit the loss scenarios identified in Sect. 3.4. LS1 pertains to the Selected Camera feedback; a direct cause of *UFB_SlctCam_Pv1*. LS2 involves an error of the ATS Operator; a direct cause of *UCA_CamMtn_NP1*. LS3 leads directly to the camera stream being inadequate; a direct cause of *UFB_CamStrm_Pv1*. In the iteration of the loss scenario identification process analyzing *UCA_CamMtn_Pv1*, as illustrated in Fig. 2, only LS2 would need to be identified. The causal factors of all three loss scenarios could lead to *UCA_CamMtn_Pv1* and any others in the chain, but the efficiency gain comes from using the UCA chain to identify each of them only once and having other UCAs inherit it. Contrast this with vanilla STPA, where each of the UCAs in this chain would be treated independently, and therefore the process of identifying these loss scenarios would be repeated for each. An experienced STPA analyst may know to reuse previously identified causal factors for other applicable UCAs, but there is no explicit guiding process.

Often, a given UCA will be part of multiple chains, where the same set of direct loss scenarios may be used within each chain. It is important to document which loss scenarios are direct causes of each UCA. The list of loss scenarios for a given UCA includes each causal factor of every UCA in all related chains.

Figure 3 illustrates that our approach does not miss any causal factor/loss scenario that vanilla STPA identifies. However, it does provide more guidance and eliminates repetition inherent in the original STPA.

5 Observation and Insights of Initial Application

We now discuss benefits observed from our initial application, and feedback from our industrial partner. Since STPA was adopted to augment safety assessment processes specifically for the introduction of RVTs, only the novel parts of the ATS system and its interface were assessed using STPA (the portions in Fig. 1b). Safety of the original system had already been assured using other methods.

Effectiveness and Efficiency. In Step 3 of vanilla STPA, analysts identify all UCAs that may lead to hazards until no more can be identified. This sometimes creates difficulties as there can be a large separation between a UCA and the hazard it causes. UCA chains help to make explicit the sequence of events that lead from a UCA to the hazard. By constructing chains, an analyst can take advantage of the inherent repetition in the UCA identification process and apply reuse. In our initial application, many UCAs were found by identifying how they could lead to another UCA that was part of a chain. Analysts found this more effective than identifying the entire chain of events leading to a hazard for each UCA. Reducing the repetition also promoted a more efficient analysis.

These benefits extended to Step 4, where the subset of mechanisms to consider as potential causal factors for loss scenarios was much smaller than in vanilla STPA, and limited to those that can directly cause a UCA. While an experienced analyst may have been able to get the same results with vanilla STPA, we find our process more targeted and less reliant on analyst expertise.

Actionability. STPA results can be used for developing requirements and strategies for elimination or mitigation of hazards. When using UCA chains, all identified causal factors are direct causes for the UCAs or UFBs they are associated with. In our initial application, this meant that they all provided insight into what causes them and where and how they may be mitigated.

The UCA chain structure also makes functional common causes for hazards explicit. We found this to be an important risk management tool for prioritization. While complete risk management is not within the remit of STPA, prioritization of mitigation strategies is a reality in industrial applications, including RVT. Vanilla STPA does result in information about which causal factors can lead to hazards, but because each is identified independently, there is no notion of prioritization. For example, our analysis made it explicit that unsafe interactions between ATS Operators and the RVT interface appeared in many chains as a common cause of downstream effects, indicating the importance of proper design of RVT's interface, sufficient training, and operational procedures.

Traceability. Traceability between UCAs and between loss scenarios is the main enabler of the previous benefits we discussed: effectiveness, efficiency, and actionability. Furthermore, as traceability enables change impact analysis efforts, we speculate that this will also be the case for modified STPA. We hope to confirm this hypothesis in future work.

6 Related Work

Since STPA has been standardized as a four-step process [7], many modifications/additions to the technique have been proposed to improve the analysis, including the aforementioned modification to Step 4 by the original authors [15]. While this work does propose a viable alternative to vanilla loss scenario identification, it does not bring the improvements in efficiency and traceability that chaining UCAs does. A future research opportunity is to compare this method with UCA chains and examine how these methods may complement each other. Others modifications include context tables [16], integration with fault tree analysis [13], and incorporating finite state machines [9]. However, to the best of our knowledge, none of the existing work utilizes repetition in Steps 3 and 4.

While STPA has been used in aviation [7], including application in the traditional air traffic services environment [1], as well as in remotely controlled vehicles and ships [2,17], RVT is a novel application field for the technique.

7 Discussion

While UCA chains may structurally resemble event chains, they are conceptually different and designed to serve a different purpose. The chaining of UCAs is a way to formalize system interactions. It is not considering the subsystems in isolation, but the interactions of control actions as a whole. We are still doing the usual STPA consideration of individual UCAs and their potential impacts on the entire system. Our process uses the same mechanisms as vanilla STPA to identify loss scenarios. It is not solely an event-based analysis and is still performed within the STPA framework.

In theory, chaining UCAs works for all applications of STPA. We found it to be beneficial in a complex system where the links between UCAs and the hazards they cause are sometimes difficult to discover. Because these modifications are exclusively 'additive', an analyst may choose to not use them in portions of the system and still fall back to the traditional process.

Our approach does not have the track record of being successfully used and refined that traditional STPA has, so it is possible that an edge-case comes up in a specific system that UCA chains cannot handle. However, as long as the requirement that all UCA chains terminate in a hazard is met, the UCA chains can be 'broken' into vanilla STPA UCAs which can lead to the terminating hazard. As with any hazard analysis, the quality of the results is still contingent on the skill of the analyst performing the analysis, but our modified process is more guided than traditional STPA, so it is generally easier to not miss things.

8 Conclusions

This paper introduces UCA Chains, leveraging the repetition in the identification of UCAs and loss scenarios in Steps 3 and 4 of STPA, respectively. This is done through identifying unsafe feedback in Step 3 alongside UCAs, and creating chains by identifying which UCAs/UFBs can happen as a result of others. These chains remove the complexity in identifying UCAs with unclear or long paths to hazards, and guide the loss scenario identification stage by putting focus on only the causal factors that directly lead to a UCA and inheriting the upstream and downstream effects from others. Our intent is to provide more guidance rather than change the fundamentals of STPA. This modification was demonstrated in an industry application on remote & virtual towers, where improvements in effectiveness, efficiency, actionability, and traceability were observed.

Acknowledgments. The authors would like to thank Kai Deng, Vaibhavi Gupta, the DAATS program team, and more broadly, NAV CANADA, for their support.

References

1. Fleming, C.H., Spencer, M., Thomas, J., Leveson, N., Wilkinson, C.: Safety assurance in nextgen and complex transportation systems. Saf. Sci. **55**, 173–187 (2013)
2. Hans, O., Radlak, K., Adamy, J.: Human-factor focused application of STPA to remotely driven vehicles. In: 2024 IEEE World Forum on Public Safety Technology (WFPST), pp. 120–125. IEEE (2024)
3. Hobbs, K.L., et al.: Systems theoretic process analysis of a run time assured neural network control system. In: AIAA SCITECH 2023 Forum. AIAA (01 2023)
4. Josefsson, B., Polishchuk, T., Polishchuk, V., Schmidt, C.: Scheduling air traffic controllers at the remote tower center. In: 2017 IEEE/AIAA 36th Digital Avionics Systems Conference (DASC), pp. 1–10. IEEE (2017)
5. Kearney, P., Li, W.C.: Multiple remote tower for single European sky: The evolution from initial operational concept to regulatory approved implementation. Transp. Res. Part A: Policy Practice **116**, 15–30 (2018)
6. Leveson, N.G.: Engineering a safer world: systems thinking applied to safety. The MIT Press, Cambridge, MA, USA (01 2012)
7. Leveson, N.G.: Safety analysis in early concept development and requirements generation. In: INCOSE International Symposium, vol. 28, pp. 441–455. Wiley Online Library (2018)
8. Leveson, Nancy G., Thomas, John P.: STPA Handbook. Manual, Massachusetts Institute of Technology, Cambridge, MA, USA (03 2018)
9. Marliere, T.A., de Azevedo Castro Cesar, C., Hirata, C.M.: Extending the STPA to model the control structure with finite state machine. J. Saf. Sci. Resilience **6**(4), 100214 (2025)
10. Martins, M., Silva, J., Alves, C.: Remote air traffic control towers: an application in Portugal. J. Airline Airport Manage. **14**(1), 1–17 (2024)
11. Oginni, D., Camelia, F., Chatzimichailidou, M., Ferris, T.L.J.: Applying System-Theoretic Process Analysis (STPA)-based methodology supported by systems engineering models to a UK rail project. Saf. Sci. **167**, 106275 (2023)
12. Shin, J., Choi, J.G., Lee, J.W., Lee, C.K., Song, J.G., Son, J.Y.: Application of STPA-SafeSec for a cyber-attack impact analysis of NPPs with a condensate water system test-bed. Nucl. Eng. Technol. **53**(10), 3319–3326 (2021)
13. Shorthill, T., Bao, H., Zhang, H., Ban, H.: A redundancy-guided approach for the hazard analysis of digital instrumentation and control systems in advanced nuclear power plants. Nucl. Technol. **208**(5), 892–911 (2022)
14. Thomas, J.: Developing Systematic Procedures to Identify Causal Factors in Systems-Theoretic Process Analysis. In: INFORMATIK 2014: Big Data – Komplexität meistern. Lecture Notes in Informatics (LNI), vol. 232, pp. 641–642. Gesellschaft für Informatik (GI), Bonn, Germany (2014), presented at Software Engineering 2014 (SE 2014)
15. Thomas, J.: STPA Step 4: Building Scenarios — A Formal Scenario Approach (2024)
16. Thomas, I.V., J.P.: Extending and automating a systems-theoretic hazard analysis for requirements generation and analysis, Massachusetts Institute of Technology (2013). Ph.D. thesis

17. Yamada, T., Sato, M., Kuranobu, R., Watanabe, R., Itoh, H., Shiokari, M., Yuzui, T.: Evaluation of effectiveness of the STAMP/STPA in risk analysis of autonomous ship systems. In: Journal of Physics: Conference Series, vol. 2311, p. 012021. IOP Publishing (2022)

SubSeaCure: Grid-Based Risk Mapping of Europe's Subsea Data Cable Network

Jonas Franken(✉), Martin Hagemeier, Timon Dörnfeld, Julian Löffler, Paula Meissner, and Christian Reuter

Science and Technology for Peace and Security (PEASEC), Technical University of Darmstadt, Pankratiusstraße 2, 64289 Darmstadt, Germany
franken@peasec.tu-darmstadt.de

Abstract. Subsea data cables carry the vast majority of intercontinental data traffic, yet existing regional risk assessments have relied on simulated inputs or narrow-area studies. SubSeaCure provides a grid-based methodology that uses observational data to quantify and visualize external risks to subsea data cables across Europe's adjacent seas. Six risk factors (fishing, anchorage, dredging, seismic activity, volcanic activity, and seabed ruggedness) were mapped to a raster grid and normalized to a common scale. Using a weighted-sum model derived from the Analytic Hierarchy Process (AHP), the factors were aggregated into a combined risk score dominated by fishing and anchorage as the most prevalent causes of cable incidents. The resulting risk map revealed pronounced spatial heterogeneity. The most extensive high-risk belt spans the North Sea and English Channel, with particularly elevated risk around Denmark, Germany, and the Netherlands. Further high-risk concentrations were also observed in the northeastern Mediterranean. Offshore, deeper Atlantic areas, and parts of the Black Sea exhibited comparatively low risk levels. These results provide contextual input for public and private investment considerations, route engineering, targeted protection and burial, surveillance prioritization, or plausibility checks during incident assessment. The SubSeaCure methodology is scalable and provides a template for finer-grained risk mapping of seabed infrastructure networks in European seas and beyond.

Keywords: Subsea data cables · Risk assessment · Geospatial risk modeling · Maritime critical infrastructure protection

1 Introduction

Forming a network that connects all permanently inhabited continents, fiber-optic subsea data cables (SDCs) transmit more than 99% of intercontinental data traffic [47]. With incidents of suspected malicious activity on the rise and digitized societies' heavy reliance on this critical infrastructure (CI), its security has come under increasing scrutiny [24]. Recent incidents in the Baltic Sea [35,42] and South Pacific [5,43] highlight the vulnerability of SDCs. Furthermore, the

© The Author(s), under exclusive license to Springer Nature Switzerland AG 2027
E. Troubitsyna et al. (Eds.): SAFECOMP 2026, LNCS 16868, pp. 228–242, 2027.
https://doi.org/10.1007/978-3-032-34867-8_15

European Commission recently agreed on the "EU Action Plan on Cable Security" [9], which covers prevention, detection, response, and deterrence as the four steps of a resilience cycle approach. Key proposed prevention measures include coordinated risk assessments by the European Union and its member states, underscoring the need for a robust assessment framework to safeguard assets that span roughly 1.5 million km and form the world's most extensive fixed infrastructure network [3,16]. More recent EU work has further operationalized this agenda: the Submarine Cable Infrastructures Informal Expert Group prepared an EU-level mapping and consolidated risk assessment of SDC infrastructures, including threats, vulnerabilities, dependencies, and initial stress-test guidance [11]. Building on this, the Expert Group's 2026 report proposed a Cable Security Toolbox of various mitigation measures and identified spatial and sectoral priorities for "Cable Projects of European Interest" (CPEIs), including new SDC routes and additional capabilities such as a renewal of the maintenance fleet and modularized repair equipment stocks [12]. These developments underline the growing need for scalable, spatially explicit risk assessment methods that can support EU-wide SDC resilience planning.

Our work introduces SubSeaCure, a risk assessment approach that quantifies and visualizes external risks using a weighted sum model. It includes six key threats – fishing, anchorage, dredging, seismic activity, volcanoes, and terrain ruggedness – which are used to calculate individual and combined risk scores, integrated into an existing map [18]. Ultimately, we aim to enhance maritime domain awareness, making high-risk areas easily visible, and support informed decision-making and proactive risk mitigation by uncovering previously hidden vulnerabilities in SDC infrastructure. Our work is guided by the research question: **How can risks to the SDC network in European seas be quantified and visualized?**

To address this question, the paper proceeds with an overview of current research on threats to SDCs and risk assessment methodologies in Sect. 2, followed by the implementation of selected risk factors resulting in a combined risk score in Sect. 3. Section 4 presents the spatial distribution of risks, both visually and quantitatively, which is discussed in Sect. 5, highlighting the novel contributions and limitations of our approach. Section 6 concludes the paper.

2 Related Work

As extensive, spatially distributed material infrastructures, SDCs are vulnerable to various external factors of harm. While the consequences of an SDC incident are often inconceivable to end users [25] due to varying degrees of redundancy [17,46], understanding the associated risks from their spatial location is essential – both for SDC route design and for assessing the plausibility of incidents, such as cases of suspected intentional sabotage. Of about 150-200 yearly incidents, the most frequent causes are human activities, specifically fishing and anchoring. Natural hazards, such as geological events and abrasion, also pose notable risks, while component failures account for a smaller proportion of reported faults [4].

So far, Muhlbauer & Murray [27], Ariany et al. [2], and Zhao et al. [48] have developed approaches related to a spatial risk assessment of SDCs.

Muhlbauer & Murray [27] provide a comprehensive overview of risk in the context of onshore and offshore pipelines. Given the similar situation of offshore pipelines and SDCs, these insights are also relevant for understanding risk for the latter - even though offshore pipelines are typically shorter and follow different installation patterns and route design choices [27]. To manage risk, defined as the product of event likelihood and consequence, Muhlbauer & Murray propose a five-step process. The first step is to acquire a risk model that uses algorithms and rules to measure risk levels. Data collection includes construction details, environmental conditions, and past failures. Because risk varies, the pipeline is segmented into sections with constant risk characteristics for targeted analysis. The risk assessment model is then applied to each segment to produce comparable risk scores, and finally, risk management decisions are made based on these assessments. Hence, Muhlbauer & Murray view risk assessments as a measurement process for quantifying risks, while emphasizing that these assessments are inherently uncertain and time-varying [28].

Ariany et al. conducted a risk assessment of a single subsea power cable (SPC) in the South Sumatera Province, Indonesia [2]. The study focused on three scenarios: the impact of dropped anchors, the risk posed by sinking vessels, and the potential damage from dragged anchors. For our work, we aim to include more factors and a spatial approach, instead of a single infrastructure focus.

They quantified the likelihood and consequences of these events, considering factors such as ship traffic and anchor weights, and categorized various scenarios into risk matrices. The findings indicate that although risks are generally low to medium, mitigation measures such as a minimum burial depth of two meters would effectively reduce these risks.

To plan cost-effective and reliable routes for SDCs, Zhao et al. [48] propose a methodology that assesses costs and risks. The six risks they analyzed include fishing activity, shipping activity, earthquakes, volcanoes, submarine slopes, and marine organisms. Each risk was categorized into five distinct categories. Each category consisted of a risk damage level and its occurrence probability, which enabled the creation of a risk matrix for each risk. The study utilized the Analytic Hierarchy Process (AHP) by Saaty [37] to derive weights for each risk factor. Although the methodology relied on simulated data, its framework highlighted the integration of diverse risk factors and adjustable weights to visualize risks.

For our analysis, we omit marine organisms, as we could not identify any record of fish bites resulting in faults of SDCs since the introduction of more sophisticated insulation and armor in the 1990s, which became standard thereafter [26].

Commercial solutions for SDC risk assessment are offered by companies such as *Global Marine* [22] and *OceanIQ* [31]. These companies offer a range of services, including cable fault analysis, route engineering, and GIS-based risk mapping, demonstrating the commercial viability of this field. *OceanIQ*, for example, provides cable fault data and generates cable fault rate heat maps using their

GeoCable®GIS [31] of extensive historical fault data to identify high-risk areas and aid in maintenance planning [30].

Previous research and commercial solutions have either focused on cable- or risk-factor-specific analyses using real-world data or on small-area analysis using simulated data. Both approaches limit their practical applicability when scaled to larger regions. We address the gap by developing a risk assessment methodology based on real-world data from multiple sources and applying it on a continent-wide scale. We chose Europe for two reasons. First, because political actors within Europe recently requested a technical mapping of SDCs and risk factors [8,9,11]. Second, Europe, with access to a large ocean body (Atlantic) but also multiple marginal seas (Baltic, Mediterranean, Black, and North Sea), offers a great variation of water bodies, for which the EU Commission outlined a prioritization scheme of maritime areas for future SDC public (co-)funding decisions [12]. Decision-making may be informed and supported by our work, as it identifies high-risk areas that may have been neglected by the SDC industry or could help support the prioritization order of CPEIs [6]. These decision-making processes within the Connecting Europe Facility (CEF) will be allocating up to 180 million EUR of public funds until 2027 [10], adding to the immense sum of 1 billion EUR in EU backbone investments since 2021 [7].

3 Methods

This section outlines the design and integration of a risk assessment for European SDCs into a browser-based Leaflet map [18,19] and the subsequent analyses. The underlying risk model calculates a weighted sum of individual risk factors mapped to a spatial grid.

To establish this spatial framework, a bounding box is defined spanning longitudes -35° to 45° and latitudes 25° to 75° to encompass the major European maritime regions. At a spatial resolution of 0.1° × 0.1° (approximately 11 × 11 km per grid cell at the equator), the resulting matrix provides 400,000 grid cells for granular spatial analysis.

The datasets selected for this assessment focus on the leading causes of cable faults. Anthropogenic drivers are dominated by fishing (40.8%) and anchorage (15.8%), with dredging accounting for a smaller fraction (0.8%) [4]. To model these human activities, fishing and anchorage records (2012–2024) were retrieved via the Global Fishing Watch API [20,21], and dredging data were sourced from the EMODnet Human Activities portal [13]. Natural hazard factors, which account for 4.8% of recorded incidents [4], include seismic activity and volcanic eruptions. Seismic data (2000–2024) were sourced from the USGS [45], utilizing a minimum magnitude threshold of 3.0 to isolate events capable of triggering destructive submarine landslides and mass flows [32]. Volcanic activity comprising 1,264 Holocene-epoch eruptions was retrieved from the Smithsonian Institution's Global Volcanism Program [40]. Lastly, because terrain ruggedness directly influences abrasion and suspension damage (4.6%) [4], we calculated the Terrain Ruggedness Index (TRI) [34] using QGIS [33] on bathymetric data from the GEBCO Seabed 2030 Project [39].

To standardize the datasets, all data points were mapped to the predefined grid and converted into localized intensity values (e.g., fishing hours or seismic magnitude). Because the physical nature of each risk factor varies, applying a universal mathematical operator to aggregate these points within a single cell would misrepresent the underlying risk. For instance, summing fishing hours could artificially inflate risk estimates based on sensor sampling density, while averaging dredging volumes would obscure the total amount of physical material removed from the seabed. To ensure representational fidelity, aggregation operators are tailored to the underlying data type: continuous intensities (fishing and seismic activity) are summarized by the arithmetic mean, volumetric quantities (dredging) by the sum, discrete categorical risks (volcanoes) by the maximum severity, and stationary point events (anchorage) by the total event count.

Following spatial aggregation, statistical normalization is applied to each risk factor. To prevent outliers from skewing the distribution, the 1st and 99th percentiles are computed across all grid cells for a given factor. Each cell's aggregated value is then winsorized to these boundaries and rescaled to a 0–1 range using min-max normalization. This enables comparability across different risk factors and the calculation of the combined risk score.

To calculate the combined risk score $x_{i,j}$ for a specific grid cell at row i and column j, we calculate a weighted sum of the six risk factors:

$$x_{i,j} = w_1 \cdot f_{i,j} + w_2 \cdot a_{i,j} + w_3 \cdot d_{i,j} + w_4 \cdot s_{i,j} + w_5 \cdot v_{i,j} + w_6 \cdot t_{i,j}\,, \quad (1)$$

where the variables represent the normalized values for the risk factors within that specific cell (fishing, anchorage, dredging, seismic activity, volcanic activity and terrain ruggedness, respectively), and $w_1, \ldots, w_6$ represent their assigned weights.

To calculate the weights, the Analytic Hierarchy Process (AHP) is used. This process conducts a pairwise comparison of risks to calculate weights that reflect their varying relevance. Saaty proposes a nine-point scale for comparisons [36]. If risks A and B are equally important, a value of 1 is assigned. If risk A is more important, it receives a value between 2 and 9, reflecting increasing importance. Next, the dot product is calculated for each cell, repeating until convergence. Finally, the priority vector is obtained by averaging the normalized values across each row of the pairwise comparison matrix to determine the weights, which are then used to calculate the combined risk score. The weights are shown in Table 1.

We use AHP rather than historical fault shares as the primary weighting method because the available incident statistics aggregate faults over several decades and therefore reflect changing SDC designs, burial practices, maritime activity patterns, and reporting conditions. Directly translating these historical shares into model weights would risk reproducing a temporally skewed picture of past SDC failures rather than supporting a current regional risk assessment. Moreover, historical fault occurrence captures realized incidents, whereas our model aims to estimate spatial exposure across European seas, including areas where few incidents have occurred so far but where relevant risk factors are present. As it provides a transparent and adjustable procedure for weighting

Table 1. Pairwise comparison matrix and the AHP weights

Risk	Fishing	Anchorage	Seismic	Volcanoes	Dredging	Terrain	Weight
Fishing	1	4	6	7	8	9	0.508
Anchorage	0.25	1	4	5	6	7	0.251
Seismic	0.167	0.25	1	2	3	4	0.100
Volcanoes	0.143	0.2	0.5	1	2	3	0.066
Dredging	0.125	0.167	0.333	0.5	1	2	0.044
Terrain	0.111	0.143	0.25	0.333	0.5	1	0.031

the selected, spatially operationalized risk factors, AHP is better suited for the purpose of our work. This also allows the weighting scheme to be adapted to different stakeholder priorities or updated threat assessments without treating historical incident ratios as fixed or universally valid.

4 Results

This section presents the findings of the European risk assessment for SDCs in three parts: Building on the risk scoring described above, we begin with a (1) qualitative analysis of the spatial distribution of individual risk variables to identify high- and low-risk regions, followed by a (2) qualitative overview of the combined risk score across the European seabed. Lastly, we apply (3) quantitative distribution analysis to complement the visual analysis.

The spatial distribution of **fishing** risk across the European region, as partially depicted in Fig. 1 a), reveals significant variations in intensity. Areas of high fishing risk, indicated by darker red shading, are concentrated mostly in coastal areas and in the less regulated high seas beyond the 200-nautical-mile Exclusive Economic Zone limit. Notably, the waters surrounding the Iberian Peninsula, the Mediterranean Sea, and the North Sea exhibit high fishing risk. Conversely, areas with lower fishing risk, indicated by lighter yellow, are prevalent in parts of the Atlantic Ocean farther offshore, as well as in the Black and Baltic Seas. While some SDCs traverse through areas of high fishing risk, it is notable that almost all SDCs originate or terminate in cable landing stations located in high-risk coastal regions.

Anchorage risk shows a clear tendency to be concentrated along the shores, as shown in Fig. 1 b). While substantial areas with anchorage risk are observed away from SDC routes, a consistent pattern emerges at cable landing points, where a high density of anchorage activity is present. Notably, the coastal waters of the Netherlands exhibit a particularly high intensity of anchorage risk. Several major ports stand out with very high levels of anchorage, including Rotterdam, Amsterdam, Antwerp, Algeciras, Valencia, Le Havre, and Piraeus. Furthermore, areas that serve as significant landing points for multiple SDCs, often associated with large urban centers such as Lisbon, Valencia, and Marseille, also exhibit elevated anchorage risk.

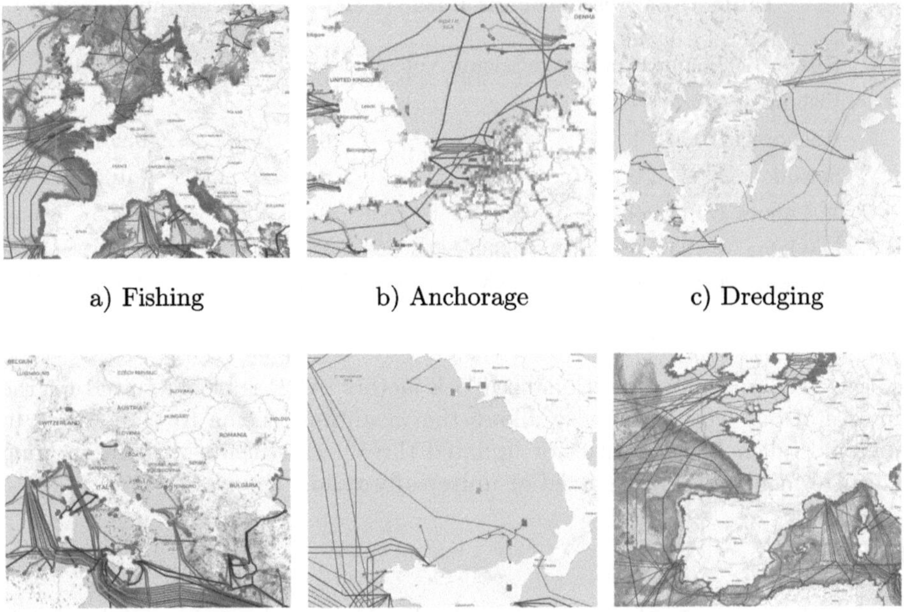

a) Fishing b) Anchorage c) Dredging

d) Seismic activity e) Volcanic activity f) Terrain ruggedness

Fig. 1. Risk map details for the six individual risk scores with cable routes from TeleGeography [44].

Dredging presents a sparse spatial distribution almost exclusively along coastlines. While some dredging occurs near cable landing points, a notable amount also occurs in coastal areas where no SDCs appear to make landfall. Regions with more substantial dredging activity include the vicinities of Valencia, Gibraltar, Bilbao, and Barcelona. A higher density of dredging activity is observed around the coast of Finland (see Fig. 1 c), yet with a lower material volume. In general, the intensity of dredging risk appears to be comparatively low across European waters.

Seismic risk exhibits a distinctive pattern illustrated in Fig. 1 d). Most Atlantic coastal areas exhibit relatively low seismic risk, with notable exceptions in Iceland, near Lisbon and Gibraltar, while the mid-Atlantic ridge is a zone of intense tectonic activity. The Mediterranean is more heterogeneous; Greece and northern Sicily show a high concentration of seismic risk points. The distribution of **volcanoes** is very sparse across Europe, both on land and at sea. A concentration of these cells is observed in Iceland and in Italy. This includes Vesuvius near Naples, Etna in Sicily, and the volcanoes in the Aeolian Islands, each with varying risk (see Fig. 1 e). Submarine volcanoes visible within this layer include the Marsili volcano, located approximately 175 km south of Naples, alongside the adjacent Palinuro volcano.

Terrain ruggedness varies substantially across the European seabed, see Fig. 1 f). Coastlines generally exhibit higher ruggedness than the oceanic seafloor. The Baltic, North, and Adriatic Seas, however, are comparatively less rugged overall. By contrast, the mid-Atlantic Ridge stands out with high ruggedness, and much of the Atlantic seabed also displays considerable variation in topography. Along France's west coast, ruggedness tends to occur several kilometers offshore, whereas around Iberia, rugged zones extend closer to shore.

The distribution of individual and combined risk scores is shown in Fig. 2. Fishing is the most widespread human activity (n = 181,363 cells), yet risk remains low across most of that area, with a median of only 0.0025. Dredging spans far fewer cells (n = 737) and remains low throughout, apart from a few isolated high-intensity sites. Anchorage has a smaller footprint than fishing (n = 4,384), but risk in anchorage cells is markedly higher, with a median of 0.033 that exceeds that of fishing by an order of magnitude. Seismic activity covers a moderate area (n = 7,068) at consistently high risk (median 0.358), as expected in tectonically active regions. Volcanic risk is higher still but covers only 35 cells, too few to support reliable conclusions. Terrain ruggedness covers nearly the whole study area (n = 228,648) and scores predominantly low, because the seabed is mostly flat and rises sharply only on ridges and coastal margins.

The combined risk score shows low risk in the majority of cells, with a median below 0.001 and only a few cells reaching the maximum of 0.794. Because the combined score is a weighted average, each risk factor's influence depends on both its weight and how many cells it covers. Fishing carries by far the largest weight (0.508) and a broad footprint, so its predominantly low scores establish the low baseline across the study area. Terrain covers even more cells but has the smallest weight (0.031), so it is present in nearly every cell yet exerts negligible influence on the result. Localized factors, primarily anchorage (0.251) and seismic activity (0.100), carry substantial weights but narrow spatial footprints, thereby amplifying the risk score only in isolated cells. That the maximum reaches 0.794 rather than 1 indicates that no cell scores highly across all risk factors simultaneously. This distribution is reflected in the combined risk map (Fig. 3), which reveals distinct spatial patterns of risk across European waters. High-risk areas are particularly concentrated around densely trafficked shipping lanes and coastal waters. The North Sea and the English Channel exhibit particularly high-risk zones. In particular, the waters surrounding Denmark, Germany, and the Netherlands have extensive high-risk areas. These regions are characterized by intense fishing and anchoring activity. In contrast, lower-risk areas, such as offshore regions of the Atlantic Ocean west of Ireland's continental shelf or parts of the Black Sea, suggest reduced interaction with human activities or risk of natural events that pose a threat to SDCs. In the Mediterranean Sea, risk is primarily focused on coastal areas, and is especially concentrated around Italy, Greece, and Turkey.

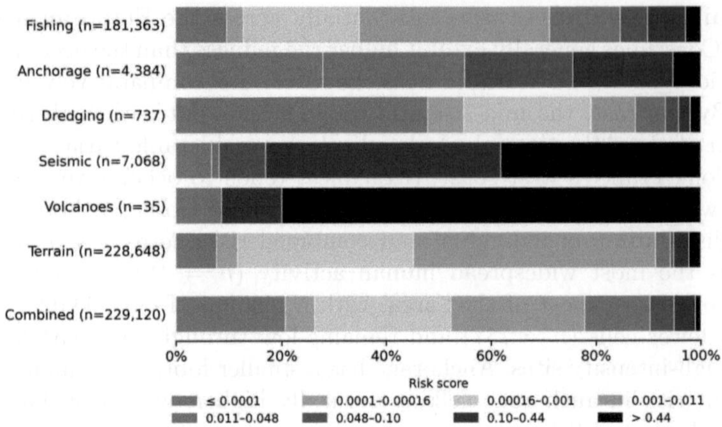

Fig. 2. Distribution of risk scores on European seabed grid cells.

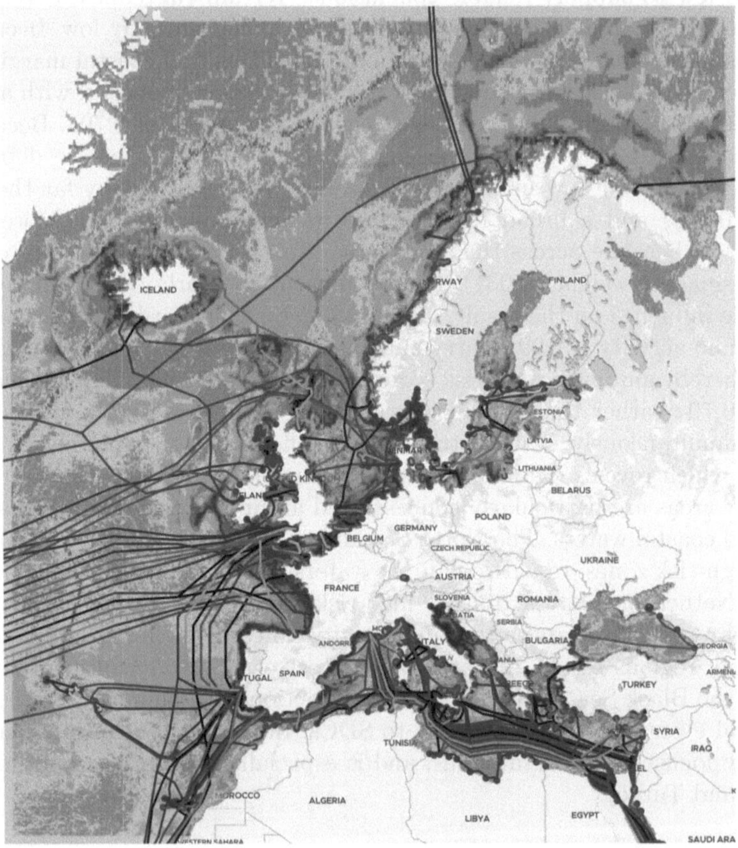

Fig. 3. European combined risk map with cable routes from TeleGeography [44].

5 Discussion

Overall, the analysis showed that European SDCs face location-dependent risks dominated by fishing and anchorage in nearshore areas, with dredging, seismic, volcanic, and terrain-related hazards generally more localized. Ariany et al. [2] focused on a single cable and three risk factors , whereas our cell-based approach covers a larger spatial extent and integrates six distinct risk factors, including anchorage frequency, for a more comprehensive assessment. At the same time, their work highlights a future direction for usable risk visualization: to refocus towards SDCs as referent objects. For this, we suggest a longitudinal risk assessment along the length of an SDC system, which, in some cases, could offer a more directly operational view of risk than cell-based analysis alone. For example, a cable-specific risk score could be introduced by sampling points along existing and planned SDC routes. Compared to Zhao et al. [48], we conducted a risk assessment using real-world data and applied it on a broader scale. While both approaches consider similar risk factors, this work omits marine organisms and adds dredging. Despite these differences, the obtained risk weights were similar.

In general, risk assessments involve assumptions, simplifications, data limitations, and scope restrictions that consequently lead to limitations. One major constraint of the chosen approach is that the data used is static. Although the data is aggregated over several years, it provides only a historical overview rather than serving as a real-time risk alert system. Implementing a dynamic system for live events would require performance optimization, as each layer requires several minutes to compute, making the system unsuitable for time-critical use cases. Automating the calculation of the individual risk scores and thus the combined risk score on a daily or weekly basis could provide a more up-to-date risk assessment. A hybrid approach where, for example, terrain ruggedness remains static (since sampling this data is quite compute-intensive and does not change significantly over time), while earthquake or fishing data is fetched daily (since this is feasible via an API) could be used. This approach could offer a risk assessment that is grounded in historical data and adapts to real-time events.

A further point of consideration is the robustness of the weighting methodology. While the AHP provides a structured baseline, it inherently relies on subjective, rank-based ratios. To ensure these ratios do not misrepresent the physical magnitude of risks, we benchmarked the AHP outputs against a model variant utilizing historical fault distributions [4] as direct weights. The resulting spatial risk distributions were highly consistent, empirically validating the AHP weights. Ultimately, the AHP framework is retained as it mitigates the temporal skew inherent in the historical data, which aggregates records dating back to 1959 [4]. By avoiding reliance on historical fault statistics, the AHP enables the system to adapt to current and emerging regional threats. Furthermore, stakeholders may wish to adjust the weights according to their priorities. For example, industry actors might prioritize fishing accident data, whereas government agencies could focus on areas with a high risk of sabotage through anchor drags. Such elements could also be integrated into a more dynamic system.

Consequently, future improvements could integrate new risk factors, such as traffic density of sub-standard shipping and shadow fleet vessels as a proxy for potential underway anchor drags, or refine existing metrics by isolating high-risk bottom-trawling from general fishing data. Operationally, the framework would benefit from SDC-centered visualizations that generate continuous risk profiles along existing and planned routes. Utilizing region-specific adaptable weights or enhancing spatial resolution could further optimize the model's precision. With the expected increase in polar SDC infrastructures [6,15,38], where traditional latitude-longitude grids approach their limits, alternative cell representations could also be adopted [41]. Ultimately, automating the data processing pipeline to ingest real-time feeds would transform this static assessment into a dynamic monitoring tool [19].

The results of this risk assessment can help stakeholders, including regulators, investors, operators, and planners, identify high-risk regions where regulatory and technical protection and surveillance measures should be prioritized, as proposed in the EU Toolbox [12]. While the International Cable Protection Committee (ICPC) offers best practices that provide general guidance [23], the implemented framework adds spatially explicit risk visualizations, enabling stakeholders to target mitigation measures more strategically. High-risk areas can be prioritized for SDC burial, or the establishment of cable protection zones [23,24], with surveillance efforts focused on these regions – for example, naval crewed and uncrewed patrols and SDC-integrated sensor systems in nearshore and offshore high-risk zones [1]. Cable protection zones, similar to those implemented in New Zealand's Cook Strait [29] or obligatory burial zones and depths as in Singapore [24], can enforce stricter regulations and raise awareness of activities such as fishing and anchorage, which remain major sources of SDC damage [14]. In addition, the map enables ex-post analysis of the plausibility of suspicious incidents, and by adjusting the weights of individual risk factors and the spatial resolution, stakeholders can emphasize the threats and level of detail most relevant to their specific needs, thereby supporting more effective and targeted enhancement of SDC security.

Beyond supporting operational protection measures, the framework can also inform strategic planning and investment decisions. Unlike approaches that primarily rely on historical SDC faults, the presented methodology can assess risks in areas where no SDC infrastructure currently exists, no route surveys have been undertaken, or incident records are sparse. This makes the approach particularly relevant for future route planning, publicly funded SDC projects, and the prioritization of protection measures in emerging connectivity corridors [6,12]. Against the backdrop of substantial EU backbone and connectivity investments [7], the SubSeaCure approach provides a transparent evidence base for prioritizing where future funding, route planning, and protection measures may be most needed. By combining multiple risk factors in a common spatial framework, stakeholders can identify areas where environmental conditions and human activity patterns indicate elevated future risk, even if few incidents have materialized historically. The resulting maps therefore complement commercial fault-rate products and provide a transparent basis for continent-scale comparison and decision-making.

6 Conclusion

This paper has illuminated the complex risk landscape that SDCs face in European seas. The grid-based SubSeaCure approach enabled the visualization and quantification of individual risk factors, whose distributions revealed distinct spatial patterns and substantial variability. Building on this, the calculation of a combined risk score provided a clear overview of the overall risk environment. The results highlight that risks are as widely distributed as the SDC network itself. In general, lower risk scores are predominant across most cells beyond continental shelves, and higher risk scores are mostly found close to coastlines. Additional risk factors, alternative weights, and finer grid resolutions can be implemented in future work to further strengthen this risk assessment approach. Ultimately, this study contributes to a deeper understanding by making the risks surrounding SDCs visible and laying the foundation for prioritizing protective measures for those infrastructures on which intercontinental data traffic critically depends.

Acknowledgments. This research has been funded by the German Federal Ministry of Research, Technology and Space and the Hessian Ministry of Science and Research, Arts and Culture within their joint support of the National Research Center for Applied Cybersecurity ATHENE.

Disclosure of Interests. The authors have no competing interests to declare that are relevant to the content of this article.

References

1. Agarwal, S.: Underwater domain awareness for and by submarine communication cables. In: Chauhan, P., Vachaparambil, J.J. (eds.) Blue Economy and Resilience, pp. 142–157. Maritime Perspectives 2025, National Maritime Foundation, New Delhi, India (2025). https://maritimeindia.org/underwater-domain-awareness-for-and-by-submarine-communication-cables/. Accessed: 30 May 2026
2. Ariany, Z., Santoso, B., Rahardjo, W.D., et al.: Risk assessment of subsea power cables in south sumatera province waters. In: IOP Conference Series: Earth and Environmental Science, vol. 1081, p. 012003. IOP Publishing (2022). https://doi.org/10.1088/1755-1315/1081/1/012003
3. Bueger, C., Liebetrau, T., Franken, J.: Security threats to undersea communications cables and infrastructure – Consequences for the EU. European Parliament, Brussels (2022). https://www.europarl.europa.eu/thinktank/en/document/EXPO_IDA(2022)702557. Accessed 30 May 2026
4. Clare, M.: An update from the icpc. Submarine Cable Protection and the Environment **2** (3 2021). https://www.iscpc.org/publications/submarine-cable-protection-and-the-environment/, accessed: 30 May 2026
5. Clare, M., Yeo, I.A., Watson, S., Wysoczanski, R., Seabrook, S., Mackay, K., et al.: Fast and destructive density currents created by ocean-entering volcanic eruptions. Science **381**(6662), 1085–1092 (2023). https://doi.org/10.1126/science.adi3038

6. Delaunay, M.: Polar Connect, the first submarine cable to cross the Arctic thanks to the European Union support? The Polar J., 1–23 (2026). https://doi.org/10.1080/2154896X.2025.2603867
7. European Commission: Additional €142 million to support submarine networks and connectivity infrastructure under Connecting Europe Facility (CEF) Digital | Shaping Europe's digital future (Dec 2024). https://digital-strategy.ec.europa.eu/en/news/additional-eu142-million-support-submarine-networks-and-connectivity-infrastructure-under. Accessed: 30 May 2026
8. European Commission: Commission recommendation of 26.2.2024 on secure and resilient submarine cable infrastructures (2024). https://digital-strategy.ec.europa.eu/en/library/recommendation-security-and-resilience-submarine-cable-infrastructures. Accessed: 30 May 2026
9. European Commission: Joint communication to the European Parliament and the Council: EU action plan on cable security (2025). https://eur-lex.europa.eu/legal-content/EN/TXT/?uri=celex:52025JC0009. Accessed: 30 May 2026
10. European Commission: Backbone connectivity for digital global gateways - works (CEF-DIG-2026-GATEWAYS-WORKS) (Mar 2026). https://ec.europa.eu/info/funding-tenders/opportunities/portal/screen/opportunities/topic-details/CEF-DIG-2026-GATEWAYS-WORKS. ccessed: 30 May 2026
11. European Commission, Submarine Cables Expert Group: Report on security and resilience of EU submarine cable infrastructures - Mapping, risk assessments, stress tests. Tech. rep., European Commission, Brussels (Oct 2025). https://digital-strategy.ec.europa.eu/en/library/report-security-and-resilience-eu-submarine-cable-infrastructures. Accessed: 30 May 2026
12. European Commission, Submarine Cables Expert Group: Submarine cable security toolbox and cable projects of European interest. Tech. rep., European Commission, Brussels (2026). https://digital-strategy.ec.europa.eu/en/library/submarine-cable-security-toolbox-and-cable-projects-european-interest. Accessed: 30 May 2026
13. European Marine Observation and Data Network: EMODnet human activities - Dredging aggregated dataset (2025), https://emodnet.ec.europa.eu/geonetwork/srv/eng/catalog.search#/metadata/d3e86612-35a7-4c0f-a995-245062fd2792. Accessed: 30 May 2026
14. Franken, J., Backes, T., Reuter, C.: Mapping the maps: Safety-security trade-offs in the charting of subsea data cables. In: Troubitsyna, E., de Andrés, D., Schoitsch, E., Gracia Morán, J., Bitsch, F. (eds.) Computer Safety, Reliability and Security – SAFECOMP 2026 Workshops – SENSEI (2026)
15. Franken, J., Flamm, P., Reuter, C.: Subsea cables to Antarctica: Technological challenges and geopolitical implications of connecting the material internet's last frontier. Telematics Inf. Rep. **22**, 100329 (2026). https://doi.org/10.1016/j.teler.2026.100329, https://www.sciencedirect.com/science/article/pii/S277250302600040X
16. Franken, J., Reinhold, T., Dörnfeld, T., Reuter, C.: Hidden structures of a global infrastructure: Expansion factors of the subsea data cable network. Technol. Forecast. Soc. Chang. **215**, 124068 (2025)
17. Franken, J., Reinhold, T., Reichert, L., Reuter, C.: The digital divide in state vulnerability to submarine communications cable failure. Int. J. Crit. Infrastruct. Prot. **38**, 100522 (2022). https://doi.org/10.1016/j.ijcip.2022.100522

18. Franken, J., Reuter, C.: The subsea data cable security map – Fusing public information for enhanced critical maritime infrastructure security. In: Proceedings of the 4th European workshop on maritime systems, resilience and security 2024 (MARESEC 24). Bremerhaven, Germany (Nov 2024). https://doi.org/10.5281/zenodo.14216270
19. Franken, J., Romer, K., Dörnfeld, T., Meissner, P., Reuter, C.: Leveraging large language models for information extraction of subsea data cable news. In: MARESEC 2025 (2025). https://doi.org/10.5281/zenodo.17119936
20. Global Fishing Watch: Dataset and code: Anchorage (2025). https://globalfishingwatch.org/datasets-and-code-anchorages/. Accessed: 30 May 2026
21. Global Fishing Watch: Dataset and code: Fishing effort (2025). https://globalfishingwatch.org/dataset-and-code-fishing-effort/. Accessed: 30 May 2026
22. Global Marine: Cable fault analysis. https://globalmarine.co.uk/services/cable-fault-analysis/. Accessed: 30 May 2026
23. International Cable Protection Committee: Government best practices for protecting and promoting resilience of submarine telecommunications cables (2024). https://www.iscpc.org/publications/icpc-best-practices/, accessed: 30 May 2026
24. Kavanagh, C., Franken, J., He, W.: Achieving depth: Subsea telecommunications cables as critical infrastructure. Tech. rep., United Nations Institute for Disarmament Research, Geneva, Switzerland (2025). https://unidir.org/publication/achieving-depth-subsea-telecommunications-cables-as-critical-infrastructure/. Accessed: 30 May 2026
25. Liu, S., Bischof, Z.S., Madan, I., Chan, P.K., Bustamante, F.E.: Out of sight, not ou of mind: a user-view on the criticality of the submarine cable network. In: Proceedings of the ACM Internet Measurement Conference, pp. 194–200. IMC '20, ACM, New York, October 2020. https://doi.org/10.1145/3419394.3423633
26. Marra, L.: Sharkbite on the SL submarine lightwave cable system: history, causes and resolution. IEEE J. Oceanic Eng. **14**(3), 230–237 (1989)
27. Muhlbauer, W.K.: Pipeline risk management manual, vol. 3. Elsevier, London (1996)
28. Muhlbauer, W.K., Murray, J.: Pipeline risk management. In: De França Freire, J.L., Rennó Gomes, M.R., Guedes Gomes, M. (eds.) Handbook of Pipeline Engineering, pp. 939–957. Springer International Publishing, Cham (2024). https://doi.org/10.1007/978-3-031-33328-6_23
29. New Zealand Transpower: Cook Strait Cable Protection Zone (CPZ) (2024). https://www.transpower.co.nz/cook-strait-cable-protection-zone-cpz. Accessed: 30 May 2026
30. OceanIQ: Cable fault data. https://oceaniq.co.uk/services/cable-fault-data/. Accessed: 30 May 2026
31. OceanIQ: Geocablegis cable fault rate heatmaps, https://oceaniq.co.uk/geocablegis-cable-fault-rate-heatmaps/, accessed: 30 May 2026
32. Pope, E.L., Talling, P.J., Carter, L.: Which earthquakes trigger damaging submarine mass movements? Insights from a global record of submarine cable breaks. Mar. Geol. **384**, 131–146 (2017). https://doi.org/10.1016/j.margeo.2016.01.009
33. QGIS Development Team: QGIS Geographic Information System. QGIS Association (2025), https://www.qgis.org
34. Riley, S.J., DeGloria, S.D., Elliot, R.: Index that quantifies topographic heterogeneity. Intermountain J. Sci. **5**(1–4), 23–27 (1999)

35. Ringbom, H., Lott, A.: Sabotage of critical offshore infrastructure: A case study of the Balticconnector incident. In: Lott, A. (ed.) Maritime Security Law in Hybrid Warfare, pp. 155–194. Brill/Nijhoff (Jul 2024). https://doi.org/10.1163/9789004707993_008
36. Saaty, T.L.: The Analytical Hierarchical Process: Planning, Priority Setting. Resource Allocation. McGraw-Hill, New York, NY, USA (1980)
37. Saaty, T.L.: Analytic hierarchy process. In: Encyclopedia of operations research and management science, pp. 52–64. Springer (2013). https://doi.org/10.1007/978-1-4419-1153-7_31
38. Saunavaara, J., Salminen, M.: Geography of the global submarine fiber-optic cable network: The case for arctic ocean solutions. Geographical Rev. **113**(1), 1–19 (2023). https://doi.org/10.1080/00167428.2020.1773266
39. Seabed 2030: Seabed 2030: The Nippon Foundation-GEBCO Seabed 2030 Project (2025). https://seabed2030.org/. Accessed: 30 May 2026
40. Smithsonian Institution: Holocene volcano list (2025) https://volcano.si.edu/volcanolist_holocene.cfm. Accessed: 30 May 2026
41. Snyder, J.P.: An equal-area map projection for polyhedral globes. Cartographica **29**(1), 10–21 (1992). https://doi.org/10.3138/27H7-8K88-4882-1752
42. Soldi, G., Gaglione, D., Raponi, S., Forti, N., d'Afflisio, E., Kowalski, P., et al.: Monitoring of critical undersea infrastructures: the Nord Stream and other recent case studies. IEEE Aerosp. Electron. Syst. Mag. **38**(10), 4–24 (2023). https://doi.org/10.1109/MAES.2023.3285075
43. Speidel, U.: The Hunga Tonga Hunga Ha'apai eruption - A postmortem. In: Proceedings of the 17th Asian Internet Engineering Conference, pp. 70–78. ACM, Hiroshima Japan (2022). https://doi.org/10.1145/3570748.3570759
44. TeleGeography: Submarine cable map (2026). https://www.submarinecablemap.com/. Accessed: 30 May 2026
45. United States Geological Survey: Earthquakes API (2025). https://earthquake.usgs.gov/fdsnws/event/1/. Accessed: 30 May 2026
46. Wang, C., Zhang, Y., Dong, Q., Carisimo, E., Durairajan, R., Bustamante, F.E.: Threading the ocean: Mapping digital routes across submarine cables using calypso. In: Proceedings of the ACM SIGCOMM 2025 Conference pp. 1260–1262. SIGCOMM '25, ACM, New York, NY, USA, August 2025. https://doi.org/10.1145/3718958.3750512
47. Ye, H., Wang, S., Li, D.: Impact of international submarine cable on internet routing. In: Proceedings of the IEEE INFOCOM 2023 - IEEE Conference on Computer Communications. IEEE, New York City, NY, USA (May 2023).https://doi.org/10.1109/INFOCOM53939.2023.10229024
48. Zhao, Z., Wang, J., Gao, G., Wang, H., Wang, D.: Multi-objective optimization for submarine cable route planning based on the ant colony optimization algorithm. Photonics **10**(8), 896 (2023). https://doi.org/10.3390/photonics10080896

Practice of Hazard Analysis and Risk Assessment

Five Measures to Counter Obscure Hazardous Failure Conditions in Foundation Model Systems: A Practical Guide for Safety Engineers

Florian Geissler(✉), Josef Jiru, Johanna Schmidhuber, Poulami Sinhamahapatra, Florian Butsch, and Reinhard Stolle

Fraunhofer Institute for Cognitive Systems (IKS), Munich, Germany
florian.geissler@iks.fraunhofer.de

Abstract. Machine learning models for natural language processing have achieved remarkable performance across many domains, yet their adoption in safety engineering remains limited. A key barrier are hazardous failure conditions - such as hallucinations - which arise from an insufficient functionality of a foundation model system and must be controlled in safety-critical contexts. This article examines such behavior through a safety lens and traces root causes for obscure hazardous failure conditions in generic foundation model setups. From this analysis, we distill five broadly applicable concepts that provide a structured way to counter these challenges (narrow domain, setup diligence, uncertainty estimation and interpretability, iterative approaches, hybrid approaches). Together, they offer a principled starting point for deriving concrete design and assessment practices for safety engineers working with language-model-based systems. We illustrate our approach through three representative use cases that demonstrate how these five countermeasures can be applied in real-world safety engineering scenarios.

Keywords: Safety engineering · Hazardous failure condition · Foundation model systems · Retrieval-augmented generation

1 Introduction

Recent years have seen spectacular progress in Artificial Intelligence (AI), leading to widespread application across professional and private domains. This surge is driven largely by Generative AI, such as Large Language Models (LLMs), which now perform complex cognitive tasks at or above human benchmarks. However, generations of these models are not inherently grounded in a reference truth, hence they remain notoriously prone to *obscure Hazardous Failure Conditions (HFC)* such as factual misrepresentations (e.g., hallucinations). The prevalence of these errors raises a critical question [15]: Under what conditions, if at all, is it appropriate to apply generative AI in safety-relevant domains? In sectors such as

© The Author(s), under exclusive license to Springer Nature Switzerland AG 2027
E. Troubitsyna et al. (Eds.): SAFECOMP 2026, LNCS 16868, pp. 245–259, 2027.
https://doi.org/10.1007/978-3-032-34867-8_16

autonomous mobility, digital health, or industrial robotics, even marginal error rates bear the risk of considerable harm to human life and the environment. A naïve approach to this dilemma would be to ban the use of AI for safety-relevant problems. Yet, ignoring the innovative potential of such a powerful technology appears reckless. Failing to leverage the analytical and generative capabilities of generative AI might itself be a missed opportunity to improve safety.

One of the core powers of a broader class of Foundation Models (FM) - operating directly on natural language (NL) tokens - is also a primary weakness. Their strength is their ability to perform complex tasks without translation into a formal reasoning system. Their weakness is the lack of formal semantics, as FMs have no explicit model of the meaning of inputs, outputs, or the function mapping one to the other. This creates profound challenges for safety engineers: Without formal semantics, there is no direct way to formally define the correct, desired system behavior. Instead, we must resort to informal definitions. Intuitively, the FM is expected to provide a service that assists a user in accomplishing a specific task within a specific context. The lack of semantics, however, means we cannot formally assess the correctness or completeness of the FM output.

Despite these hurdles, we argue that distinct classes of obscure HFCs in FM components can be identified and prevented or mitigated. This paper undertakes such analysis for safety engineering tasks. We first formulate the problem of applying FMs to safety-relevant use cases. We then analyze recurring issues and commonalities across application scenarios, and propose five universal countermeasures to prevent or mitigate obscure HFCs. Finally, we present three real-world use cases and discuss practical experience in implementing these steps. The article aims to provide a pragmatic guideline for safety engineers approaching FM-based automation in critical systems. The referenced methods are selected illustrative examples rather than an exhaustive representation of literature in this field.

2 Problem Statement

A wide spectrum of diagnoses exists on the issues at hand. For example, [8] discusses safety assurance challenges of generic machine-learning systems, including aspects of uncertainty that effectively manifest as hazardous failure conditions. [10] dissects foundation model safety into the realms of trustworthiness, security, and responsibility. A comprehensive survey on LLM safety is provided in [24], along the dimensions of value misalignment, robustness, or misuse.

2.1 System Component Description

A generic FM system architecture and its essential components are shown in Fig. 1. We assume that the FM system is embedded in a larger *Safety Engineering System*, which could also involve human actors. At its core, the FM system includes one or multiple *FMs*, where we use *Foundation Model* throughout this article as the umbrella term for models operating on sequences of tokens that

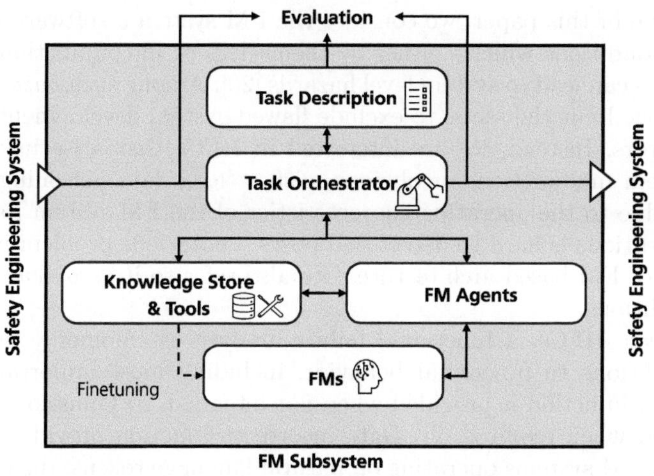

Fig. 1. Generic architecture of a typical FM-based subsystem embedded in a larger safety-engineering system.

represent natural language (in multimodal models also additional modalities such as images, after appropriate embedding), including large language models (LLMs), small language models (SLMs), or vision language models (VLMs), etc. Typical architectures integrate FMs into *FM agents* - functional wrappers that use FM processing to perform reasoning loops, autonomous decisions, or tool-driven actions [9,18]. On the agentic level, memory can be established to track conversation history. FM agents can choose to retrieve external knowledge from a *knowledge store* (i.e., documents, images, databases), or trigger *tools* such as calculators, web searches, SQL database searches. Information from an external knowledge store can also be integrated into the FM's internal knowledge through fine-tuning or re-training. A generic FM architecture includes some form of *task orchestrator* that coordinates an input task across one or multiple agents and aggregates their outputs into a final result. Orchestration may include additional information processing, for example, hard-coded or FM-based decomposition of a complex task into sub-tasks. Common aggregation schemes include ensemble voting and multi-agent debate (MAD) frameworks [25]. The task is defined in the form of a *task description* that can serve as the basis for system prompts and the design of the task orchestrator. Further, we determine task performance success given the FM system input and output during *evaluation*. Note that the above description of an FM system is generic: In individual system architectures, some of the components may be trivial identities, or merged together.

2.2 Hazardous Failure Conditions

Following established taxonomy [6,20], a *hazard* is a circumstance which can lead to damage, and is hence a key component of a system level safety analysis.

For the scope of this paper, we consider the FM system a software component, and study conditions which - either by themselves or in conjunction with other components - can lead to system level hazards [20]. Among such *hazardous failure conditions*, we limit the scope to exclude flawed system development or random software errors. Instead, we are interested in HFCs that arise in the absence of any system malfunctions or obvious misuse (e.g., data poisoning, LLM jailbreaking), due to the operating characteristics of the FM system itself. As this class of conditions is hard to detect and poses a notorious problem in the safety assessment of FM-based architectures, we also refer to it as *obscure* hazardous failure conditions.

To analyze HFCs, a functional failure analysis is commonly used to map failure conditions to functional behavior, including most importantly [20]; i) Commission: Function is provided when not intended, ii) Omission: Function is not provided when required, iii) Data incorrect: Function provided incorrectly. However, in FM systems operating on natural language tokens, the correct functionality can not be objectively defined due to a lack of formal semantics. As a consequence, the assessment about the presence or absence of a HFC is not given formally either. We propose to resort to the informal notion of an assumed functionality of the FM system, that is: *Deliver factual, complete information in an output format suitable to solve the task at hand.* This allows us to set up three categories of HFCs guided by the above principles: i) Factual, complete output is delivered, but not in a suitable format (e.g., information is confusing, masked by irrelevant details, of otherwise unusable quality, incorrectly formatted, etc.), ii) Output is factual but incomplete (e.g., missing information to solve the task), iii) Delivered information is not factual (e.g., hallucinations, inconsistent or counter-factual information).

In the following section, we set out to identify the most common and relevant sources of obscure HFC in generic FM setups. All discussed root causes trigger obscure HFCs in one or multiple of the above categories. Subsequently, we develop broadly applicable strategies in Sec. 3 to counter them. Our findings are derived from the empirical study of multiple safety-critical engineering tasks, where we have stress-tested FM systems in representative real-world scenarios from the domains of autonomous mobility, digital health, and industrial automation. Three of those practical use cases are briefly presented in Sec. 4.

2.3 Sources of Obscure Hazardous Conditions

1. **Experimental Design**

   a) **Choice of Model and Framework:** The safety engineering task may not be sufficiently represented in the knowledge of the selected (pretrained) FM. Poor experimental design choices can cause out-of-distribution input during deployment, with high risk of data misinterpretation. Further, well-chosen model reasoning capabilities are crucial for both FMs and FM agents. Agentic architectures can enforce overly open or narrow solution corridors, and must be selected carefully.

b) **Data Availability:** Relevant labeled data for individual safety engineering tasks is often scarce or not available, hindering large-scale testing and well-informed architecture design choices.

2. **Managing Input Diversity**
While many multi-modal approaches exist that can handle images, text, audio, etc., these do not necessarily balance information weights across modalities. VLMs seek to handle image and text input on equal footing in latent space [19], yet domain-specific information retrieval largely depends on the diversity and scale of annotated pre-training data. Information retrieved from different modalities may largely differ from a human observer's perception. Similarly, document converters for diagrams and other visuals embedded in text documents [27] may over- or under-represent visual content, causing imbalanced combined content.

3. **External Knowledge Integration**
a) **Retrieval Alignment:** Knowledge retrieval is typically based on a similarity search between the query and available text chunks in embedding space. However, relevant information often mixes with irrelevant or contradicting content in the same text passage. The retrieval alignment problem appears when embedding similarity does not reflect human understanding of similarity. This leads to specific information being missed, in particular, if chunk size is large compared to the desired information. As retrieved context quality directly shapes successful task delivery in RAG-based workflows, retrieval alignment is a prime root cause of HFC. Retrieval design becomes particularly important for extensive document pools, where semantic similarity search may return incomplete, mixed, or misaligned context.
b) **Fuzzy Tool Triggers:** Many agent frameworks make use of tools to retrieve external information. Similar to FM-based retrieval, the tool use is based on a similarity check of a provided tool trigger (typically a statement in natural language) and the query. Misaligned, imprecise, or conflicting trigger formulations can result in incorrect tool use, and thus in suboptimal context.
c) **Counterfactual Context:** Retrieved context - even if correctly aligned - can provide information that conflicts with the FM-internal knowledge. Such *counterfactual* context is challenging to process, and different models exhibit different behavior for staying faithful to internal knowledge or leaning to accept counterfactual context instead [21].

4. **Semantic Alignment of Generations**
While the alignment of generated answer semantics and user intent is a general challenge for FMs, we identify three common HFC causes in safety engineering:
a) **Completeness:** A recurring request in many safety-critical NLP tasks is to find all relevant items of interest, e.g., all hazards associated with a scenario, or all requirements derived from a work instruction. Ensuring an aligned understanding of completeness in such tasks is notoriously elusive.

b) **Usefulness:** Many scenarios pose specific requirements for the usefulness of an FM answer, which is closely related to the right level of technical detail in the answer.
c) **Semantic drift:** Many FMs show the tendency that the quality of generations in terms of factual accuracy and terminological coherence degrades over longer generation periods. Such drifts can compromise semantic alignment, e.g. over longer conversations or output windows [26].

5. **Syntactic Alignment of Generations**
FMs can be unsuccessful in providing output in a desired structure (e.g., list formats, jsons, etc.), even when urged to do so via prompting. If subsequent processing steps rely on the format structure, this can lead to HFCs. This may also include answers that mix correctly structured output with unsolicited FM prefaces or epilogues. In safety engineering workflows, syntactic deviations are particularly critical, as outputs must remain efficiently reviewable and auditable. Therefore, an important challenge is *schema-first* output (e.g., strictly schema-conform or otherwise machine-parseable responses).

6. **Evaluation**
Evaluating FM output against a given reference (both in natural language) requires deep semantic understanding of the two text items. FM judges are popular alternatives to often insufficient reference-free assessments (e.g., n-gram matching like Rouge score [22]). However, accurate similarity scoring from FM judges in safety engineering is challenging, requiring domain expertise which may not be included in the internal knowledge of typical FMs.
a) **Wording:** Technical terms in highly specialized safety contexts represent semantics that can deviate from broader interpretations as encoded in typical embedding models. For example, two terms may not be synonyms in this context, in contrast to their use in a colloquial setting.
b) **Structure:** A higher structural similarity of two text items increases their semantic similarity assessment for FM evaluators. However, texts with context-specific structural encoding can be misjudged, i.e., ambiguous sentence structures that are all correct within a specific context.
c) **Expert Dependency:** LLM judges largely benefit from trusted reference-matching to assess whether generated safety artifacts are correct, complete, and useful. However, creating such references requires domain expert knowledge and nuanced semantic judgment, which is why human expert review remains an important evaluation element in safety engineering. In niche applications, human expert reviews are typically expensive, scarce or absent.

3 A Five-Step Countermeasure Concept

To counter the above challenges, we present a collection of five generic measures, which can be adopted to mitigate or prevent obscure HFCs any FM-based safety engineering task. All concepts are derived heuristically from practical experience

Countermeasures

Sources of obscure HFC	I. Narrow domain	II. Setup diligence	III. Interpretability and uncertainty estimation	IV. Iterative workflows	V. Hybrid workflows
1. Experimental design	●	●	○	○	○
2. Managing input diversity	●	●	○	●	●
3. External knowledge integration	●	●	●	●	●
4. Semantic Alignment of Generations	●	●	●	●	●
5. Syntactic Alignment of Generations	●	●	○	●	○
6. Evaluation	●	●	●	●	●

Fig. 2. Mapping of obscure HFC sources and countermeasure approaches. A full circle indicates that the respective measure is generally suitable to address the corresponding obscure HFC. A blank circle means that the measure is likely not effective in countering the condition.

in real-world FM engineering, see the Sec. 4 for selected examples. Additional concepts exist, but the five measures presented here have been identified as the most relevant approaches in the studied use cases. Fig. 2 shows how the countermeasures map to the previously identified HFC sources.

I **Narrow Domain**
Many functional issues emerge from the FMs' inability to identify appropriate contextual windows - the more open-world a problem, the more room for ambiguity and misinterpretation. A precise specification of the NL task scope is the most fundamental, and arguably most important, strategy for the prevention of HFCs. Practical examples include adding detailed specifications to the NL task, guided reasoning with predefined guardrails representing the problem domain, or explicitly excluding certain operational scenarios from the task. Narrowing down the system scope can identify a better experimental design, stimulate more targeted retrieval or multi-modal processing, facilitate the semantic or structural alignment of FM generations, and help to achieve more accurate evaluations (see Fig. 2).

II **Setup Diligence**
a) Experimental settings that are not carefully prepared often lead to HFCs emerging from incompatible components. We recommend benchmarking-informed selection strategies to choose FM models, retrievers, reasoning agents, etc., that match best with the studied scenario. Fine-tuning can help to customize FMs to better suit specific safety-engineering tasks, while further enabling privacy-preserving local deployments with efficient resources.

b) Metrics for LLM judges must be aligned with the overall task goal. Test-driven development is important to detect and eliminate HFCs early in the production cycle.
c) Exploiting diverse input and retrieval data modalities can enrich the retrieved information, if balanced appropriately.
d) If data is scarce, augmentation techniques such as synthetic data generation should be considered. If retrieval data is abundant, on the other hand, a filtering and pre-selection is important to improve context retrieval quality.
e) Using stencils for generated output formats provides structural guidance and can help to reduce semantic misses.
f) In retrieval workflows, experimental parameters such as chunk size, chunk overlap etc. should be chosen such as to reflect the structure of provided data (e.g. documentation) appropriately.

III **Interpretability and Uncertainty Estimation**
With these techniques, the internal workings of FM models are monitored in order to identify patterns reflecting unusual behavior. This typically concerns intermediate activations, for example in attention heads [12].
a) Qualitative interpretations can support human review in semi-automated tasks and facilitate consistency in *auditable-by-design* safety engineering. For example, in retrieval procedures, word attributions in the source context help to plausibilize FM reasoning.
b) Quantitative activation monitoring enables uncertainty estimations in fully automated scenarios. A prime goal is the detection of hallucinations, after semantic interpretation of the monitored patterns [5,16,17]. Uncertainty scores can further be leveraged for certification guarantees based on statistical properties (derived during a calibration phase and checked at runtime). Some examples include approaches based on randomized smoothing or conformal prediction [23].

IV **Iterative Workflows**
Complex tasks are best tackled in multiple steps.
a) For a single agent, this can be a chain-of-thought (CoT) or reasoning loop, possibly guided by task-specific rules or orchestration guardrails. For example, agentic REACT [28] enforces a clear breakdown into distinct steps.
b) Harnessing multiple agents for ensemble voting or multi-agent debate frameworks can boost overall output quality [25], for example by iteratively refining completeness and usefulness or checking intermediate plausibility.
c) Multi-stage retrieval approaches help to improve context quality, especially in large corpora of documents [7], or reflect on retrieval alignment [11]. Robustness can be improved with techniques such as HyDE-style [13] query expansion, where a hypothetical, information-dense pseudo-document is drafted first, and subsequently retrieval is performed based on matching of the real and the pseudo-document.

V **Hybrid Workflows**
In many situations, non-AI-based tools can be harnessed to support or replace individual sub-tasks that are deterministic and well-defined in nature. We refer to such methods that feature a combination of FM-based and non-FM-based steps as *hybrid* approaches. Examples include FMs that reason in an interplay with knowledge graphs or other hard-coded rules [5,14]. Hybrid approaches can reduce inaccuracies by establishing effective reasoning guardrails. In retrieval setups, hybrid searches may use similarities in both dense and sparse embeddings to augment semantic search with a keyword-style matching or occurrence counting. Such methods can be more effective in identifying the most relevant text chunks.

The above concepts represent generic strategies that have proven effective in countering obscure HFCs in practical settings. Clearly, their impact depends further on the given experimental setups, and the specific implementation details. The above collection of measures is not exhaustive, and concepts may have mutual dependencies in specific problem instances. We therefore propose the above list of countermeasures as a principled starting point, which can guide the process of systematically eliminating or suppressing obscure HFCs in FM-based systems, whereas the context of a given problem instance needs to be considered to derive explicit design choices. For illustration, we describe in the following Section 4 three use cases from the fields of mobility, healthcare and industry automation, and demonstrate how the previously described five concepts can be applied in practice to address obscure HFCs.

4 Use Cases

4.1 LLM-Assisted Safety Requirements Engineering (Mobility)

Ensuring consistency and completeness in safety documentation remains a major challenge in safety-critical systems engineering. Safety artifacts - often comprising of hundreds of pages of natural language specifications, diagrams, and tabular data - must remain consistent across the system engineering lifecycle, despite fast iteration cycles, distributed international engineering teams, and evolving system architectures. Even minor inconsistencies can propagate into flawed product designs, making efficient and reliable review processes essential. Traditional manual safety requirement reviews imply high human effort, long lead times, and small inconsistencies that are easily overlooked.

In this use case, we leverage LLMs to support safety requirements engineering by automating elements of review, authoring, and validation, see Fig. 3. The scope is limited (I) to finding inconsistencies in safety case documentation of AUTOSAR Front Light Management [2]. We use carefully curated safety engineering artifacts (image and text) RAG sources for context retrieval, including safety concepts, item definitions, preliminary architectures, safety goals, functional safety requirements, as well as work instructions and templates for technical safety requirements (TSR) (II). The FM system generates and evaluates

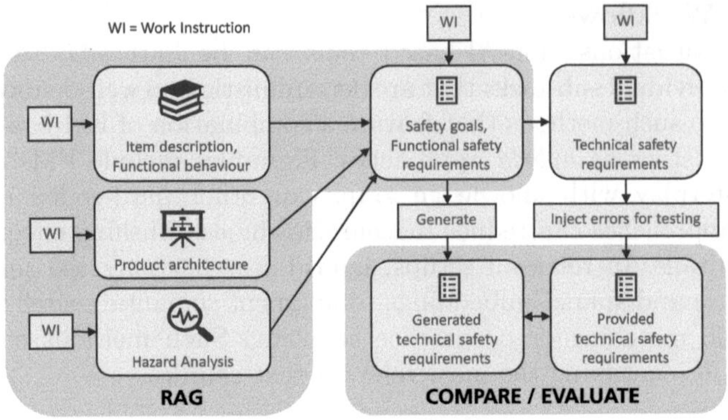

Fig. 3. Architecture of LLM-based requirements engineering use case.

TSRs against existing provided TSR to identify mismatches, inconsistencies, or missing content. Multi-stage retrieval can be used to identify the most relevant chunks (IV). Knowledge graphs help to understand the system hierarchy and component dependencies, glossary mapping combats synonym confusions and promotes domain-specific definitions (V).

The LLM evaluation framework categorizes requirement matches (*covered* or *partially covered*), detects unmatched requirements, and handles ill-structured items (e.g., requirements containing multiple clauses). Unmatched requirements are flagged as potential risks and subsequently reviewed by a human expert for a severity assessment. Attention monitoring can be applied during the TSR generation to discover early hallucinations (III). A score (1–5) systematically quantifies requirement quality for both generated and provided TSRs. Our approach demonstrates that LLM-assisted analysis can significantly enhance the efficiency and rigor of safety documentation workflows by enabling automated consistency checks and improved TSR quality.

4.2 LLM-based Multi-modal Symptom Analysis (Healthcare)

Accurate aggregation and interpretation of symptom data across sources, modalities, and time is critical for medical consultation and patient empowerment along monitoring, diagnosis and treatment. Patient-reported symptoms often remain unstructured, multilingual, and scattered across heterogeneous data sources (e.g., symptom diaries, wearables, medical notes). Manual review and analysis is tedious for time-pressured clinicians during appointments, and challenging for medically untrained patients. This creates gaps in longitudinal symptom tracking and analysis.

We study a multi-agent LLM system to automate symptom extraction, validation, analysis, and summarization from multi-modal patient data. The scope is narrowly defined to symptom extraction and terminology check from patient free-text, pattern identification from time-series data, context interpretation, and patient-friendly report generation (I). We exclude diagnostic or treatment recommendations to minimize semantic misalignment.

The FM system architecture orchestrates four specialized agents, adressing multiple HFC sources and enabling intermediate validation and interpretability (III, IV), see Fig. 4: *Agent 1* extracts symptoms and context from patient diaries using a locally deployed, fine-tuned language model (FLM) trained on a small proprietary expert-annotated dataset augmented with synthetic examples. The FLM addresses domain adaptation for lay terminology in patients' symptom descriptions, data scarcity, privacy risks through local deployment, and minimizes required compute resources (II). Text chunk highlighting enables patients and clinicians to verify extraction correctness, supporting FM system validation (III). *Agent 2* validates extracted symptom terms against trusted medical knowledge bases via RAG, ensuring correct mapping to standardized terminology (II, III). *Agent 3* applies deterministic statistical methods for time-series symptom pattern detection and correlation analysis, avoiding LLM hallucination risks in quantitative tasks (V). This hybrid workflow with deterministic correlation analysis and structured terminology checks is intended to constrain the LLM outputs to reduce HFCs (V). *Agent 4* consolidates findings from the previous agents into a structured, patient-friendly symptom report to support medical consultation using predefined template structure for syntactic alignment, with provenance links to source data (III). An *orchestrator* coordinates the four agents in a sequential pipeline with intermediate validation checkpoints (IV).

The system evaluation employs multiple complementary metrics such as exact-match precision for symptom extraction against ground truth, and LLM-as-a-judge assessments for symptom completeness, medical conciseness, and relevance (II). Validation using real-world expert-annotated data is important to ensure statistical guarantees and detect edge cases (II). Our approach demonstrates how obscure HFCs can be countered in a symptom extraction and analysis FM system embedded in an overall diagnostic and treatment journey.

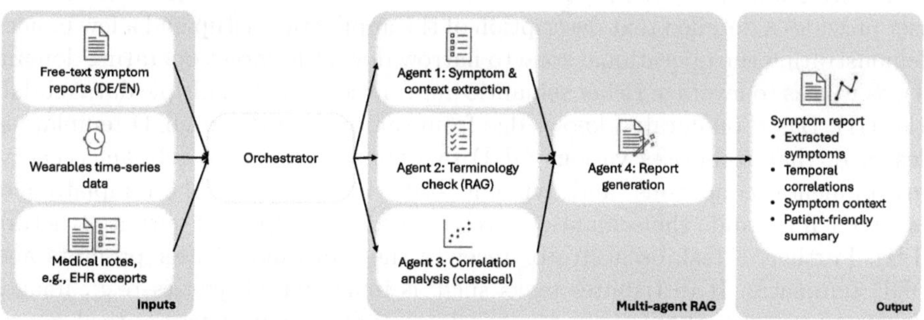

Fig. 4. Architecture of the LLM-based multi-modal symptom analysis system.

4.3 Generation of Operational Design Domain Description Using Multi-modal FM (Industrial Automation)

An Operational Design Domain (ODD) formally defines the conditions, actors, and environmental factors of an application. Standardizing ODDs promotes reusability and streamlines data management through consistent metadata and scenarios. This clarity is vital for safety-critical tasks to minimize operational risks. While sectors like automotive [4] and rail [3] use established standards, industrial ODD descriptions vary significantly across applications. To address this challenge, we develop *ODD Agent*, a LLM-assisted framework that simplifies ODD specification for industrial automation. It leverages FMs in combination with expert knowledge to extract ODD specifications from safety documents, using multi-modal input queries from text and images.

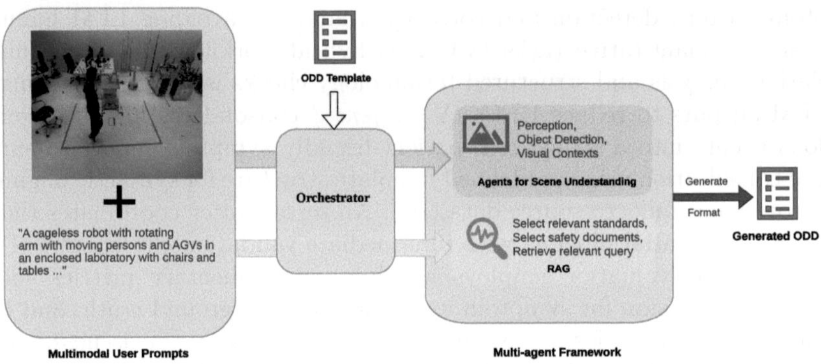

Fig. 5. Architecture for the multi-modal, FM-assisted generation of a structured description of Operational Design Domains for industrial automation applications.

Figure 5 illustrates the pipeline for the ODD Agent framework. The domain user provides a detailed text description of the application setup and a few images demonstrating the operational zone to narrow down the target operating domain (I), as well as to leverage richer semantic information across multiple data modalities (II). Based on curated knowledge from safety experts, an ODD template is designed, that comprises various ODD taxonomies related to relevant scenery, environmental conditions, and dynamic/static components. The template acts as a stencil to guide the semantic search towards the expected output structure (I,II). Further, a task orchestrator receives the multi-modal user prompts and ODD templates. It distributes tasks such as image-based queries (e.g., images showing persons, robots around a cageless robot arm in safety-critical zones, see Fig. 5) to various scene understanding agents based on VLMs (IV). This is followed by a customized RAG approach to query and retrieve template keys from relevant safety standards and documents. The prior organization of relevant domain safety documents and standards used for RAG and semantic search highlights the necessity of setup diligence (II). The orchestrator also collects the concurring outputs from different agents for each template key to combat misinterpretations. Finally, a ODD template is populated in standardized formats like the ASAM standard [1]. The generated ODD description can be iteratively enhanced by incorporating feedback from safety experts and subsequent user inputs (IV).

The evaluation framework includes an LLM judge that compares the generated ODD with a ground-truth ODD template. Quantitative metrics such as completeness and correctness scores (schema, actor counts, static elements) are validated, along with an analysis for critical errors based on visual contexts. This iterative workflow helps reduce uncertainty in the evaluation metrics (III, IV). In summary, the ODD Agent demonstrates FM support in the assessment of safety compliance in an industrial automation setup, and provides guidance about dataset coverage for setting up any ML training or testing pipeline.

5 Conclusion

In this article, we argue that the most critical failure conditions of FM systems performing safety-relevant tasks are best understood as *obscure HFCs*: Situations where model and infrastructure work as designed, but the delivered output is still hazardous, as it is factually or structurally unsuitable. Starting from a generic FM system architecture, we trace such HFCs back to recurring root causes, in particular poor experimental design and model choice, mismanaged input diversity, misaligned retrieval and tool use, semantic and syntactic misalignment between user intent and model outputs, and the difficulty of reliably evaluating safety-relevant results with FM judges. From this analysis, we distill five conceptual countermeasures that offer safety engineers a structured approach to design, integrate, and assess FM components: (I) narrowing the domain and task scope, (II) exercising diligence in the experimental setup (models, retrieval, data, metrics), (III) monitoring interpretability and uncertainty to detect abnormal or hallucinated behavior, (IV) using iterative workflows with multi-step reasoning, multi-agent or human-in-the-loop refinement, and (V) adopting hybrid workflows that combine FM capabilities with deterministic tools and rule-based systems. We demonstrate implementations of these five steps in three representative safety engineering use cases from different domains: LLM-assisted safety requirements engineering (mobility), LLM-based multi-modal symptom analysis (healthcare), and multi-modal ODD specification (industrial automation). This framework is a practical starting point, providing safety engineers across domains with shared vocabulary and practical levers to reason about hazardous failure conditions of FMs, to deliberately shape system behavior, and to design mitigation strategies aligned with safety assurance principles.

Acknowledgment. We thank Mario Trapp for insightful discussions. Parts of this work have been funded by the Free State of Bavaria in the DSgenAI project (RMF-SG20-3410-2-18-4).

References

1. ASAM OpenODD: Concept paper. https://www.asam.net/index.php?eID=dumpFile&t=f&f=4544&token=1260ce1c4f0afdbe18261f7137c689b1d9c27576
2. AUTOSAR. https://www.autosar.org/fileadmin/standards/R23-11/CP/AUTOSAR_CP_EXP_SafetyUseCase.pdf
3. DIN DKE SPEC 99004:2025-05, spezifikation von ODD im schienenverkehr; text Englisch. https://doi.org/10.31030/3610746, https://www.dinmedia.de/de/-/-/390496147
4. Operational design domain – implementation of BS ISO 34503. https://www.bsigroup.com/en-GB/insights-and-media/insights/brochures/operational-design-domain-implementation-of-bs-iso-34503/
5. Agrawal, G., Kumarage, T., Alghamdi, Z., Liu, H.: Can knowledge graphs reduce hallucinations in LLMs? : a survey http://arxiv.org/abs/2311.07914
6. Avižienis, A., Laprie, J.C., Randell, B., Landwehr, C.: Basic concepts and taxonomy of dependable and secure computing **1**(1), 11–33 (2004).https://doi.org/10.1109/TDSC.2004.2
7. Balu, B.V., et al.: Towards automated safety requirements derivation using agent-based RAG. https://doi.org/10.48550/arXiv.2504.11243, http://arxiv.org/abs/2504.11243
8. Burton, S., Herd, B.: Addressing uncertainty in the safety assurance of machine-learning **5**, 113258 (2023). https://doi.org/10.3389/fcomp.2023.1132580, https://www.frontiersin.org/articles/10.3389/fcomp.2023.1132580/full
9. Chase, H.: Langchain. https://github.com/langchain-ai/langchain
10. Chen, C., Gong, X., Liu, Z., Jiang, W., Goh, S.Q., Lam, K.Y.: Trustworthy, responsible, and safe AI: a comprehensive architectural framework for AI safety with challenges and mitigationhttps://doi.org/10.48550/arXiv.2408.12935, http://arxiv.org/abs/2408.12935, issue: arXiv:2408.12935
11. Chen, P.B., Zhang, Y., Cafarella, M., Roth, D.: Can we retrieve everything all at once? ARM: an alignment-oriented LLM-based retrieval method. https://doi.org/10.48550/arXiv.2501.18539, http://arxiv.org/abs/2501.18539
12. Chuang, Y.S., Qiu, L., Hsieh, C.Y., Krishna, R., Kim, Y., Glass, J.: Lookback lens: detecting and mitigating contextual hallucinations in large language models using only attention map. https://doi.org/10.48550/arXiv.2407.07071, http://arxiv.org/abs/2407.07071
13. Gao, L., Ma, X., Lin, J., Callan, J.: Precise zero-shot dense retrieval without relevance label. https://doi.org/10.48550/arXiv.2212.10496, http://arxiv.org/abs/2212.10496
14. Geissler, F., Roscher, K., Trapp, M.: Concept-guided LLM agents for human-AI safety codesign. https://doi.org/10.48550/arXiv.2404.15317, http://arxiv.org/abs/2404.15317
15. Graydon, M.S., Lehman, S.M.: Examining proposed uses of LLMs to produce or assess assurance arguments. https://ntrs.nasa.gov/citations/20250001849
16. Huang, L., et al.: A survey on hallucination in large language models: Principles, taxonomy, challenges, and open questions **43**(2), 1–5 (2025) . https://doi.org/10.1145/3703155, https://dl.acm.org/doi/10.1145/3703155
17. Kuhn, L., Gal, Y., Farquhar, S.: Semantic uncertainty: linguistic invariances for uncertainty estimation in natural language generation. https://openreview.net/forum?id=VD-AYtP0dve

18. Liu, J.: LlamaIndex. https://doi.org/10.5281/zenodo.1234, https://github.com/jerryjliu/llama_index
19. Masry, A., et al.: AlignVLM: bridging vision and language latent spaces for multimodal document understanding. https://doi.org/10.48550/arXiv.2502.01341, http://arxiv.org/abs/2502.01341
20. McDermid, J.: Software hazard and safety analysis. In: Damm, W., Olderog, E.R. (eds.) Formal Techniques in Real-Time and Fault-Tolerant Systems, vol. 2469, pp. 23–34. Springer Berlin Heidelberg. https://doi.org/10.1007/3-540-45739-9_2, http://link.springer.com/10.1007/3-540-45739-9_2, series Title: Lecture Notes in Computer Science
21. Ming, Y., et al.: FaithEval: can your language model stay faithful to context, even if "the moon is made of marshmallows"
22. Mohammadshahi, A., Scialom, T., Yazdani, M., Yanki, P., Fan, A., Henderson, J., Saeidi, M.: RQUGE: reference-free metric for evaluating question generation by answering the question. https://doi.org/10.48550/arXiv.2211.01482, http://arxiv.org/abs/2211.01482
23. Robey, A., Wong, E., Hassani, H., Pappas, G.J.: SmoothLLM: defending large language models against jailbreaking attacks. https://doi.org/10.48550/arXiv.2310.03684, http://arxiv.org/abs/2310.03684
24. Shi, D., et al.: Large language model safety: A holistic surve. https://doi.org/10.48550/arXiv.2412.17686, http://arxiv.org/abs/2412.17686
25. Smit, A., Duckworth, P., Grinsztajn, N., Barrett, T.D., Pretorius, A.: Should we be going MAD? A look at multi-agent debate strategies for LLMs. https://doi.org/10.48550/arXiv.2311.17371, http://arxiv.org/abs/2311.17371
26. Spataru, A., Hambro, E., Voita, E., Cancedda, N.: Know when to stop: a study of semantic drift in text generation (2024). https://arxiv.org/abs/2404.05411
27. Team, D.S.: Docling technical repor. https://doi.org/10.48550/arXiv.2408.09869, https://arxiv.org/abs/2408.09869, version Number: 1.0.0 _eprint: 2408.09869
28. Yao, S., et al.: ReAct: synergizing reasoning and acting in language models. https://doi.org/10.48550/arXiv.2210.03629, http://arxiv.org/abs/2210.03629

Practical Challenges in AI Risk Classification and Assessment: Insights from an AI Sandbox Pilot

Danilo Brajovic[1,2(✉)], Benjamin Fresz[1,2(✉)], Elena Dubovitskaya[3], Nadine Ferber[1], Theo Jacobs[1], Lena Lörcher[1], and Marco F. Huber[1,2]

[1] Fraunhofer Institute for Manufacturing Engineering and Automation IPA, Stuttgart, Germany
{danilo.brajovic,benjamin.fresz}@ipa.fraunhofer.de
[2] Institute of Industrial Manufacturing and Management IFF, University of Stuttgart, Stuttgart, Germany
[3] Justus Liebig University Giessen, Hesse, Germany

Abstract. The EU AI Act mandates regulatory sandboxes in each member state to support AI system development under regulatory supervision. However, concrete guidance on their design remains limited. This paper presents lessons from KIRR Real, a sandbox initiative that conducted Legal Quick Checks for 15 AI systems across manufacturing, automation, and human resources. The requirements assessed were based on the AI Act and, where applicable, the Machinery Regulation. Across 15 organizations, we identify recurring challenges in risk classification, the interpretation of unclear concepts such as "self-evolving behavior" under the Machinery Regulation, technical implementation, documentation, and resource allocation, indicating that additional guidance is needed. Current guidance inadequately addresses common practices like fine-tuning pre-trained models, and established AI development practices prove insufficient for AI Act compliance. We derive recommendations for future sandboxes emphasizing early-stage engagement, low-threshold access, and integrated legal-technical expertise, contributing to safe and compliant AI deployment.

Keywords: AI Act · Regulatory Sandbox · AI Safety · Compliance · Risk Classification

1 Introduction

The European Union's Artificial Intelligence Act (AI Act) is a landmark framework for AI safety and trustworthiness [8]. Articles 57–62 mandate the establishment of at least one regulatory sandbox per member state to provide controlled environments where AI providers can develop, test, and validate innovative systems under regulatory supervision while ensuring compliance. Despite

D. Brajovic and B. Fresz—Equally contributed.

© The Author(s), under exclusive license to Springer Nature Switzerland AG 2027
E. Troubitsyna et al. (Eds.): SAFECOMP 2026, LNCS 16868, pp. 260–275, 2027.
https://doi.org/10.1007/978-3-032-34867-8_17

considerable scholarly attention to the AI Act's provisions and philosophical approach [5,9,14], significant uncertainty remains regarding the practical implementation of regulatory sandboxes. Questions persist about organizational structure, operational procedures, and specific support mechanisms. This knowledge gap poses substantial challenges as member states begin establishing sandboxes.

1.1 The KIRR Real Project

The project name KIRR Real stands for *Reallabor für rechtskonforme KI und Robotik* in German, which translates to "Regulatory Sandbox for Compliant AI and Robotics" in English. As a precursor to formal AI Act sandboxes, the project explored practical compliance support approaches, with a particular focus on the interplay between the AI Act and the Machinery Regulation, which becomes applicable from 2027. KIRR Real offered so-called "Legal Quick Checks" (LQCs) to companies developing AI systems, providing structured assessments against the AI Act and, where applicable, Machinery Regulation requirements. Each LQC included the risk classification assessment required by the AI Act, conducted by a law firm, followed by the evaluation of the implementation via the corresponding documentation based on applicable requirements, conducted by technical experts. The project completed 15 assessments in various sectors including manufacturing, automation, and human resources management.

1.2 Contribution

This paper makes three contributions. First, it provides an early empirical account of AI Act-oriented regulatory support, based on 15 LQCs of AI systems across application domains. These cases cover different development patterns (from-scratch models, GPAI, fine-tuning, API integration) and regulatory constellations (AI Act alone and AI Act plus Machinery Regulation).

Second, it identifies cross-cutting implementation challenges that current guidance does not adequately address, including persistent uncertainty in risk classification, gaps in technical implementation and documentation practices, complexities arising from overlapping regulatory frameworks, and the specific difficulties of using pre-trained and third-party models under the AI Act.

Third, it derives design implications for future AI Act regulatory sandboxes, offering recommendations for sandbox operators, policymakers, and AI developers. These include principles for early-stage, low-threshold engagement, integrated legal and technical expertise, and mechanisms to feed sandbox practice into standards development and regulatory refinement.

This paper is structured as follows: Sect. 2 reviews regulatory sandboxes and existing implementations. Section 3 discusses AI Act provisions and their interplay with safety regulations. Section 4 describes the approach used in KIRR Real. Section 5 presents the key findings, condensed into lessons and recommendations in the discussion in Sect. 6. Section 7 concludes the paper.

2 Related Work

Regulatory sandboxes originated in the FinTech sector, with the UK Financial Conduct Authority pioneering this approach in 2016 [3]. The concept creates controlled environments for testing innovations with regulatory oversight while potentially relaxing certain requirements. This enables "regulatory learning"—a bilateral process benefiting both regulators, who gain insights into emerging technologies, and innovators, who receive compliance guidance [1,13]. The OECD has emphasized the role of sandboxes in promoting flexibility and innovation while maintaining regulatory objectives in the digital age [1].

Spain was the first EU member state to launch an AI sandbox pilot aligned with the AI Act draft in 2022–2023, highlighting challenges in risk classification and cross-sectoral coordination [10]. Since then, multiple member states have developed similar initiatives, sometimes originating from sandboxes focused on data protection, such as the Datatilsynet sandbox in Denmark or a sandbox pilot by the Swedish Data Protection Authority IMY. In Germany, the Federal Network Agency (Bundesnetzagentur) is currently establishing coordination mechanisms for AI supervision and sandbox operations [4]. While these initiatives provide valuable precedents, gaps remain regarding practical AI Act compliance challenges, the interplay of the AI Act with sector-specific safety regulations, and concrete methodologies for early-stage assessments.

3 The AI Act and Regulatory Sandboxes

Regulation (EU) 2024/1689—the AI Act [8]—entered into force on 1 August 2024 as one of the world's first comprehensive horizontal legal frameworks for AI. It takes a risk-based approach, calibrating obligations to the potential harm a system (or the use of its outputs) may cause across three tiers: (1) prohibited practices (Article 5), such as subliminal manipulation or social scoring; (2) high-risk systems (Article 6), subject to the strictest requirements; (3) minimal-risk systems, subject to no specific AI Act obligations. Additionally, some systems such as chatbots face transparency obligations under Article 50. Systems facing only transparency obligations are sometimes called limited-risk systems.

High-risk classification under Article 6 is triggered by two independent routes: Article 6(1) covers AI systems that are safety components of, or themselves constitute, products subject to third-party conformity assessment under EU harmonization legislation listed in Annex I (e.g., machinery, medical devices); Article 6(2) covers systems intended for Annex III use cases, in areas such as critical infrastructure, employment, and education. Article 6(3) provides a limited exception for certain Annex III systems that do not pose a significant risk of harm to health, safety, or fundamental rights of natural persons—assessed against four criteria concerning the limited scope of systems, e.g., systems for narrow procedural tasks. This exception proved highly contested in practice (see Sect. 5.1). Systems performing profiling of natural persons remain high-risk regardless.

High-risk obligations encompass risk management, data governance, technical documentation, logging, transparency, human oversight, accuracy, robustness, cybersecurity, quality management, conformity assessment, and post-market monitoring. Under the AI Act as originally adopted, Annex III obligations apply from 2 August 2026 and Article 6(1) obligations from 2 August 2027. Following the May 2026 political agreement on the Digital Omnibus, now subject to formal adoption, these dates are expected to shift to 2 December 2027 for stand-alone high-risk systems and 2 August 2028 for those embedded in regulated products.

Articles 57–62 of the AI Act establish the framework for regulatory sandboxes [8]. Each member state must ensure access to at least one sandbox providing controlled environments for developing, testing, and validating innovative AI systems before market placement. Key provisions include facilitating innovation while ensuring compliance, providing guidance on risk classification and requirements, enabling real-world testing with safeguards, and facilitating cross-border cooperation. The European Commission published a draft implementing act with further specifications [6], especially regarding the organization of such sandboxes. Under the AI Act as adopted, such sandboxes must be operational by 2 August 2026, possibly postponed by the Digital Omnibus to 2 August 2027.

3.1 Interplay with Safety Regulations

AI systems typically exist under multiple regulatory frameworks simultaneously. The AI Act establishes horizontal requirements, while sector-specific regulations address domain-specific safety and performance criteria. The AI Act explicitly references other legal acts in its risk classification scheme, particularly for high-risk systems based on Annex I [8]. This regulatory layering is intentional, but creates additional complexity that sandboxes must address.

In KIRR Real, the focus, in addition to the AI Act, was the EU Machinery Regulation. The new regulation will supersede the currently applicable Machinery Directive in January 2027 and introduces specific requirements for machinery incorporating AI, particularly regarding safety components with AI capabilities [7]. For the applicability of the Machinery Regulation, the degree of integration of an AI system into a machine is important. If machine data is processed without directly interfering with the operation of the machine, the Machinery Regulation is typically not applicable. This includes the aggregation of data for predictive maintenance or productivity analysis. If the AI system has an influence on the control system of the machine or how it is operated by personnel, it becomes part of the machine and is covered by the Machinery Regulation, for example, vision systems for detecting workpieces or a chatbot citing the machine's documentation. The strictest requirements from the Machinery Regulation apply if an AI system has an influence on the safety of the machine. This occurs if it is part of the safety-related control system, acting as a safeguard, for example for detecting persons and stopping the machine before a collision occurs. Safeguards are listed in Annex I of the Machinery Regulation, indicating that stricter requirements for conformity assessment apply. Particular attention is given to components with so-called "self-evolving behavior".

4 Methodology: The KIRR Real Approach

AI Act compliance requires integrated legal and technical expertise. Sandboxes must combine legal interpretation of the AI Act and related regulations, technical understanding to evaluate implementations beyond documentation, and sector knowledge of specific regulations and practices. Neither purely legal nor purely technical assessments suffice. As such, KIRR Real established a structured assessment process combining legal and technical expertise. The procedure followed the AI Act's risk-based approach through four stages:

1. Risk classification assessment determining whether the system falls under prohibited practices, high-risk or minimal risk categories and whether transparency obligations apply (supported by a law firm);
2. Identification of applicable requirements based on classification;
3. Documentation evaluation against AI Act requirements, conducted by two separate reviewers independently—with, based on the use case, either two AI experts or one AI expert and a robotic safety expert; and
4. Structured feedback on compliance gaps and recommendations, integrating the results of both reviewers.

4.1 Use Cases

Table 1 summarizes the 15 assessed use cases, including their AI Act risk type, the relevance of the Machinery Regulation, and whether they were assessed using the lightweight LQC format.

Companies were selected from four open calls throughout the project. The selection of companies was based on four main criteria: (1) novelty of the use case, i.e., whether a comparable case had already been assessed within the project; (2) expected complexity and generalizability, with high-risk or borderline cases being favored; (3) a preference for small and medium-sized enterprises (SMEs); and (4) funding-related constraints, limiting participation to one use case per company and to organizations located in the German federal state of Baden-Württemberg. LQCs were provided free of charge.

4.2 Iterative Adaptation and Cross-Case Synthesis

Halfway through the project, the team introduced the format "Legal Quick Check Light", responding to a key insight: compliance assessments should engage companies during early development stages rather than only when systems approach completion. This lightweight format provided preliminary risk classification, high-level guidance on requirements, compliance roadmaps throughout development, and overall lower resource requirements for the companies involved. Instead of reviewing documentation, the assessment process focused on the review of specification and concept papers as well as an open discussion with the companies, usually represented by engineers involved in the development of the AI system. Of the 15 LQCs conducted, four were Legal Quick Check Light.

Table 1. Overview of the 15 assessed use cases with AI Act risk category, Machinery Regulation (MR) relevance, and whether the lightweight LQC format was used (*light*). In the AI Act risk column, "+ T" denotes additional transparency obligations.

ID	Short description	AI Act risk	MR	light
1	Radar-based detection of persons and automated guided vehicles to adapt speed/stop of a collaborative robot	High	Yes	No
2	Reinforcement-learning library enabling users to build RL applications; no pre-trained models provided	n/a	No	No
3	Scene understanding for humanoid robots to detect unsafe situations and enable safe falling	High	Yes	No
4	Large Language Model/Retrieval-Augmented Generation-based, customer-specific maintenance chatbot for technical documentation	Low + T	Yes	No
5	Camera-based biometric access control for non-public areas with local face database	High	No	No
6	Speech-to-text GPAI model creating structured text for different domains (e.g., medical, inspection)	Low + GPAI	No	No
7	Radar-based detection of persons, humanoid robots and AGVs to enable safe human–robot collaboration	High	Yes	No
8	Embedding-based matching of employee profiles to project roles with LLM-generated summaries	High	No	Yes
9	Vision-based check whether the hazard zone of a machine is free before automated reset of a safety stop	High	Yes	Yes
10	On-board anomaly detection on an autonomous mobile robot with natural-language notifications to staff	Low + T	Yes	Yes
11	Platform incl. avatar agent for semi-automated applicant interviews and scoring in recruiting processes	High + T	No	No
12	Platform for developing and training vision models for optical quality inspection (e.g., injection needles)	Low	No	Yes
13	Prediction of remaining load cycles for plastic safety gears on a test bench to shorten durability tests	Low	No	No
14	Library/platform for data preparation, training and deployment of robotics vision models	n/a	No	No
15	Credit-risk assessment of natural persons with bias detection, explainability and unlearning features	High	No	No

Beyond adapting the assessment format, the project also established a structured, qualitative process for deriving cross-cutting findings. During each LQC, the two technical experts involved documented noteworthy observations as they arose—for instance, ambiguities in risk classification or implementation challenges not covered by existing guidance. Upon completion of all 15 assessments, a workshop brought together all five technical experts to compare observations, identify recurring patterns, and distinguish isolated from systematic challenges. The findings reported in Sect. 5 reflect only those challenges that emerged across multiple assessments or are grounded in direct legal interpretation.

5 Results: Key Findings

In the following section, the main findings identified via the LQCs are presented. They are split into findings regarding the risk classification and applicability of the AI Act, terminological uncertainties in the Machinery Regulation, specific problems with the use of LLMs and GPAI models, and uncertainties regarding the amount and specificity of the required documentation.

5.1 Risk Classification and AI Act Applicability

Risk classification emerged as a central uncertainty. The AI Act's risk-based approach makes classification the critical first step determining all subsequent obligations [8]. However, given the recent entry into force of the AI Act and the current non-application of high-risk requirements, no real precedents for risk classification are available at this time.

In practice, the main challenges in risk classification often crystallized around specific borderline questions rather than broad categories. A recurring issue concerned the notion of "self-evolving behavior" under the Machinery Regulation, as described in Sect. 5.2.

Another point of ambiguity concerned the exception clause in Article 6(3) AI Act, which allows certain AI systems not to be classified as high-risk where they do not "materially influence the outcome of decision-making". In several LQCs, participants struggled to determine when a contribution to a decision should be considered merely supportive rather than materially influential (e.g., ID 4 and 10). As the AI Act does not further operationalize this threshold, different actors may reasonably arrive at differing classifications for similar systems.

A further set of classification challenges arose from fine-grained distinctions with far-reaching legal consequences. One example is biometric systems: while biometric identification is listed as a high-risk use case in Annex III(1), biometric verification, "the sole purpose of which is to confirm that a specific natural person is the person he or she claims to be", is explicitly excluded from the high-risk category. Article 3(36) AI Act, however, ties this distinction to the technical difference between 1:1 and 1:n matching of biometric data (ID 5).

The AI Act fundamentally aims to safeguard EU citizens from harm. This objective suggests that AI systems whose failure could entail safety implications should fall within the high-risk category. However, the use cases analyzed indicate that this expectation is not consistently met, particularly regarding systems with indirect safety implications: Under current interpretation, AI systems performing quality checks during manufacturing of safety-critical components were not classified as high-risk, although errors could result in defective components slipping through, and thus endanger end users (ID 12 and 13).

Further uncertainties arose regarding the delineation of what constitutes an AI system. In two cases (ID 2 and ID 14), the AI Act was deemed inapplicable, contrary to the expectations of both the companies that submitted applications and the assessors who reviewed them. In both instances, the companies provided software libraries enabling their customers to develop AI solutions, to which the

AI Act would then apply. While this interpretation is logically consistent, it implies that library providers must incorporate sufficient information and documentation capabilities into their products to enable customers to develop AI Act-compliant solutions. Notably, as soon as the library includes pre-trained models, the provider is also classified as an AI provider under the AI Act.

5.2 Self-evolving Behavior in the Machinery Regulation

Pursuant to Article 25(2), read in conjunction with Annex I, Part A, Nos. 5 and 6 of the Machinery Regulation, an external conformity assessment is mandatory, in particular, for safety components with fully or partially self-evolving behavior using machine-learning approaches to ensure safety functions, as well as for machinery incorporating embedded systems with fully or partially self-evolving behavior using machine-learning approaches to ensure safety functions, where those systems have not been placed independently on the market; in the latter case, the conformity assessment relates only to those systems. At the same time, such safety components or embedded systems qualify as high-risk AI systems within the meaning of the AI Act, since they satisfy the conditions set out in Article 6(1)(a) and (b) of the AI Act.

Machinery incorporating embedded systems with fully or partially self-evolving behavior may itself qualify as an AI system within the meaning of Article 3(1) of the AI Act and simultaneously constitute a "product" within the meaning of Article 6(1)(a) of the AI Act. Given that such machinery falls within the scope of the Machinery Regulation and is subject to a third-party conformity assessment under that regulation, the cumulative requirements set out in Article 6(1)(a) and (b) of the AI Act for classification as a high-risk AI system are also met. Accordingly, the notion of "self-evolving behavior" becomes pivotal. With regard to safety components and embedded systems ensuring safety functions, it operates as the decisive trigger for third-party conformity assessment under the Machinery Regulation and, subsequently, for the qualification of such safety components and machinery as high-risk AI systems under the AI Act. This interaction is, however, likely to be modified by the Digital Omnibus on AI. Under the provisional political agreement, the Machinery Regulation is to be exempted from direct applicability of the AI Act's high-risk requirements; instead, the Commission would be empowered to adopt delegated acts under the Machinery Regulation introducing AI-specific health and safety requirements for AI systems classified as high-risk under the AI Act. The amendment thus does not remove AI-related safety concerns but shifts their treatment from the AI Act's Chapter III regime to the sector-specific machinery framework, with the aim of avoiding overlapping conformity-assessment and compliance obligations.

The notion of "self-evolving behavior" is not defined in the Machinery Regulation and does not correspond to a standard technical term. In practical terms, it may be understood to include online or continual learning systems, adaptive control approaches, and reinforcement learning systems that modify their policies through interaction with the environment. The key unresolved question is therefore how broadly that notion should be construed for safety-relevant machine-learning systems. A narrow reading would confine the term to systems that continue to learn and adapt their parameters during operation after being placed on the market, thereby excluding most industrial applications based on pre-trained and frozen models. A broad reading, by contrast, would also encompass systems whose functional logic is produced through a data-driven training process in which parameters are iteratively adjusted and typically evolve in a non-deterministic manner due to stochastic optimization and randomness. On this understanding, the adjective "self-evolving" describes the manner in which the model's behavior is formed during training, rather than requiring ongoing modification after deployment. Several considerations support a broad understanding of "self-evolving behavior". First, such an interpretation is consistent with the protective purpose of the Machinery Regulation, which generally favors an expansive reading of provisions designed to safeguard human health and safety. This is particularly relevant because machine-learning systems may be very complex, difficult to predict and can contain distinctive failure modes irrespective of whether they continue to learn in the field. Secondly, Annex I, Part A, Nos. 5 and 6 refer broadly to "fully or partially self-evolving behavior using machine learning approaches ensuring safety functions", without expressly limiting the concept to learning during operation or to adaptation after the product has been placed on the market. If the legislature had intended such a restriction, it could easily have made this explicit. Thirdly, Recital 55 supports the same reading. It explains that third-party conformity assessment is intended to cover systems with fully or partially self-evolving behavior based on machine learning, while excluding software that is neither capable of learning nor of further development and merely executes predefined automated functions. This contrast suggests that the distinction lies between adaptive or emergent behavior based on ML and static, purely rule-based automation, rather than necessarily between learning before and after the product has been placed on the market.

At the same time, certain considerations support a narrow construction of "self-evolving behavior". Linguistically, the term suggests continuing change over time and therefore more naturally points to behavior that evolves during operation than to parameter adaptation confined to the development phase. This reading also finds support in Annex III of the Machinery Regulation. In particular, the General Principles refer to hazards that are foreseeable at the time the machinery or related product is placed on the market as part of the "intended evolution" of its self-evolving behavior or logic. Similarly, Annex III, Sects. 1.1.6(f) and (g) and 1.2.1 address systems with fully or partially self-evolving behavior or logic operating with varying levels of autonomy and impose safeguards concerning, inter alia, interaction with persons, the communication of planned actions, and control-system constraints. Taken together, these provisions more readily suggest systems whose behavior may continue to develop in operation or after deployment than models whose parameters are fixed once training is complete. Moreover, such a reading would restrict third-party conformity assessment to systems that are especially difficult to predict and validate ex ante because their behavior may continue to change after deployment. While this proportionality consideration is not expressly stated in the Regulation, it nonetheless provides an additional policy argument in favor of a narrow interpretation.

Given the still limited doctrinal engagement with this notion, particularly in the specific context of the Machinery Regulation, it is noteworthy that the emerging literature already supports both a narrow [12] and a broad interpretation [2]. Whether in implementing guidance, harmonized standards, or case law, this interpretative divergence creates tangible legal uncertainty for manufacturers and supervisory authorities. Against that background, legal advisers may, as a matter of professional caution, be inclined to recommend that manufacturers proceed on the basis of the broader interpretation pending authoritative clarification of the legal position.

5.3 GPAI and LLMs

The distinction between general-purpose AI (GPAI) and AI system proved ambiguous in practice. Only one system in our sample was classified as GPAI (ID 6). The AI Act does not precisely define the degree of use-case specificity required to exclude a system from GPAI classification. For instance, consider a system that transcribes spoken language into text using a (domain-specific) ontology: Does such a system constitute GPAI due to its applicability across a variety of scenarios, or is the use case sufficiently specific to preclude this classification?

More commonly, (potentially high-risk) AI systems were developed by fine-tuning existing models or integrating AI functionality via APIs, rather than training models from scratch (e.g., ID 4). Available guidance largely assumes development from the ground up and offers limited practical advice for fine-tuning and integration scenarios. Participants also reported that, from the outside, it is difficult to assess which LLM or base-model provider is best suited to address specific technical, legal, and compliance challenges, or even whether a provider permits the use of its systems in high-risk applications, as publicly available information often remains high-level. This leaves uncertainty regarding the allocation of responsibilities (which party is responsible for which obligations), the assessment and documentation of base models, the extent of testing and validation required after fine-tuning, and how to maintain documentation when using third-party models. This guidance gap is problematic because fine-tuning and API integration are chosen precisely to reduce development effort. If the compliance implications of these approaches remain unclear or disproportionately complex, their practical attractiveness is significantly reduced.

5.4 Technical and Documentation Uncertainty

The absence of harmonized standards at the time of assessment introduced some of the greatest practical difficulties. Even where companies had identified the relevant requirements, they faced substantial uncertainty about how to implement them in concrete technical terms. Many AI Act obligations are formulated abstractly—for example, requirements on "accuracy, robustness, and cybersecurity" or "sufficiently transparent" systems—leaving wide room for interpretation.

This ambiguity was mirrored in the documentation we observed, which varied significantly across companies despite templates for documentation being available, such as in Annex IV of the AI Act. Companies expressed uncertainty regarding the required scope and level of detail (how much documentation is sufficient), how to balance transparency against protection of intellectual property, and what conformity assessment bodies and market surveillance authorities are likely to expect in practice. In the absence of experience and feedback from these entities, organizations struggle to assess whether their documentation is adequate. This complicates project budgeting, as companies currently cannot yet estimate the effort needed to achieve a compliant AI system.

Ultimately, this creates a risk that organizations investing heavily in compliance and documentation incur unnecessary costs, while competitors that invest fewer resources may gain a short-term competitive advantage.

6 Discussion

The 15 Legal Quick Checks indicate that the main implementation challenges of the AI Act are structural rather than idiosyncratic. Difficulties in risk classification, handling of pre-trained and third-party models, and operationalization of technical and documentation requirements appeared as recurring patterns across sectors and system types.

Strikingly, we did not observe systematic differences between small and medium-sized enterprises and larger companies in terms of their preparedness for the AI Act. Knowledge of the AI Act and the ability to translate its requirements into concrete processes and artifacts did not scale proportionally with organizational resources. This suggests that the main implementation challenges are rooted less in available capacity than in the complexity and novelty of the regulatory framework itself, and that support instruments such as sandboxes need to be designed for "typical" small teams rather than assuming that larger organizations will naturally cope on their own. It should be noted that selection bias may influence these findings: Participation in KIRR Real required companies to be aware of the project and actively apply. This likely excludes both companies that perceive no compliance challenges with their AI use cases and those unaware of the AI Act altogether.

6.1 Implications for Sandbox Design and Operation

The findings have direct implications for how AI Act regulatory sandboxes should be designed and operated.

First, early and iterative engagement is essential. Many of the issues identified in KIRR Real (e.g., the selection and integration of pre-trained models) are difficult to change once systems approach deployment. Following the idea of "regulatory learning" and iterative adjustment [1,13], we introduced the "LQC Light" format during the project. The uptake of this format showed that lightweight, early-stage assessments can meaningfully shape system design rather than merely documenting compliance ex post.

Second, AI Act sandboxes must explicitly address multi-regulation constellations. In several cases, the decisive questions emerged at the intersection between the AI Act and the Machinery Regulation. Sandboxes that focus solely on AI Act text, without integrating relevant sectoral regimes, risk providing incomplete or misleading guidance. Coordinated involvement of authorities responsible for AI and sector-specific regulation is therefore key, especially for high-risk use cases linked to safety components.

Third, integrated legal and technical expertise is not optional. KIRR Real showed that legal interpretation alone could not answer questions about model architectures, training strategies, or realistic documentation practices, and technical analysis alone could not resolve borderline classifications or allocate obligations between actors. Effective sandboxes will need multidisciplinary teams that understand both regulatory requirements and common AI development patterns (including fine-tuning and API-based integration).

Finally, sandboxes should be prepared to deal with recurring patterns rather than only idiosyncratic projects. In our sample, the same types of questions arose across different organizations: Treatment of software libraries, responsibilities in multi-actor chains, documentation of pre-trained models, and treatment of indirect safety functions. Systematically capturing and generalizing such patterns, and feeding them into guidance and standardization is a central part of the sandbox function rather than a by-product.

6.2 The Participation Dilemma and Governance Trade-Offs

KIRR Real also highlighted a structural challenge for AI Act sandboxes: The participation dilemma. Sandboxes promise early guidance and reduced uncertainty, but participation simultaneously increases visibility to regulators and can expose non-compliance that might otherwise remain undetected. This tension is particularly salient in regimes where the same or closely related authorities both operate the sandbox and later supervise or enforce compliance.

From a governance perspective, this creates a trade-off between learning and enforcement. If sandboxes are perceived as extensions of enforcement, risk-averse organizations—especially those uncertain about their risk classification or operating in borderline areas—may avoid participation. Conversely, if sandbox participation offers meaningful protection (for example, temporary relief from administrative fines during good-faith experimentation, as envisaged in the AI Act's sandbox provisions [8]), authorities must accept some limitation on immediate enforcement to gain long-term compliance benefits.

If the sandbox is not operated by a notified body, problems with confidentiality arise: In publicly funded projects, the publication of use cases and results might be required, as was the case in KIRR Real. Companies in early stage development of applications did not want their project idea to be publicly available, even when published anonymously, as their competitors might use that idea themselves or associate them with it. Additionally, if the third-party assessor is seen as a potential competitor (in our case, an applied research institute), companies might not be willing to provide enough information for a true assessment of their application, despite non-disclosure agreements being in force.

Our experience suggests that low-threshold, non-binding initial assessments can partially resolve this tension. In KIRR Real, the LQC Light format operationalized this idea: It allowed organizations to explore regulatory expectations with limited exposure and resource commitments before deciding whether and how to pursue their initially envisioned system concept. Additionally, if participation in a regulatory sandbox requires a high resource investment, as the current draft of the implementing act seems to suggest, many companies might not be willing to join. More generally, clear communication about the separation between advisory sandbox activities and formal enforcement, and about the concrete benefits of participation (e.g., documented guidance that can be referenced in future conformity assessments), will be crucial for uptake and for the legitimacy of AI Act sandboxes.

6.3 Limitations and Directions for Future Work

The insights from KIRR Real must be interpreted in light of several limitations.

Empirically, the sample comprised 15 cases from a specific region (Baden-Württemberg) and with a sectoral focus on manufacturing, automation, and human resources. Other domains, such as healthcare, finance, or public services, may face different constellations of risks and overlapping regulations. The assessments were conducted during a limited time window, in an early phase of AI Act implementation, before harmonized standards and further guidance had matured. Longitudinal effects—such as how organizations maintain compliance over time or respond to evolving interpretations—could not be observed.

Methodologically, the Legal Quick Check format provides point-in-time assessments. It is well suited to identifying current gaps and giving concrete feedback, but less suited to supporting continuous improvement or post-market monitoring. As a precursor initiative, KIRR Real also lacked formal regulatory powers: It could not issue binding decisions, grant derogation, or directly coordinate with market surveillance authorities. Furthermore, the resource intensity of detailed assessments limits scalability if similar formats were to be offered nationally.

These limitations suggest several directions for future work. Comparative studies of different sandbox models (e.g., continuous engagement, cohort-based programs, or infrastructure-oriented testbeds) could clarify which formats work best for which types of AI systems and organizations. Longitudinal research following systems from early development through deployment and operation would shed light on how AI Act compliance evolves in practice. Finally, systematic analysis of how harmonized standards and subsequent guidance affect the types of questions raised in sandboxes could inform both regulatory design and standardization priorities.

6.4 Practical Recommendations

Based on KIRR Real, we propose the following practical recommendations.
For regulatory bodies and sandbox operators.

- Provide staged support formats, from lightweight early screenings to in-depth reviews, to enable engagement throughout the development life cycle.
- Explicitly integrate sector-specific regimes (e.g., the Machinery Regulation) into sandbox procedures where relevant, and coordinate with responsible authorities.
- Focus guidance on common practices, such as fine-tuning of pre-trained models, API-based integration, and indirect safety implications.
- Capture and aggregate recurring issues and solutions, and feed them into guidance documents, standardization processes, and the Commission's implementing measures [6].

For policymakers.

- Clarify risk-classification criteria and borderline concepts (e.g., self-evolving behavior, GPAI boundaries, indirect safety use cases) through guidance and examples.
- Prioritize harmonized standards for common high-risk use cases and prevalent architectures, with particular attention to multi-actor development chains.
- Ensure that conformity assessment mechanisms and market surveillance practices foster a consistent application of AI Act requirements across member states, so that self-assessments and "AI Act compliant" labels have substantive meaning and do not degrade into unverified marketing claims.

For AI developers and organizations.

- Address AI Act requirements from the outset, aligning governance processes and technical design (e.g., using frameworks such as ISO/IEC 42001 [11]).
- Treat documentation as a continuous activity across the life cycle, not as an end-of-project exercise.
- Engage with sandbox and support programs early, particularly when using pre-trained models, APIs, or near high-risk boundaries.

7 Conclusion

This paper reports empirical insights from KIRR Real, an early AI Act sandbox precursor that conducted 15 Legal Quick Checks across manufacturing, automation, and human resources. The assessments revealed that organizations struggle most with context-dependent risk classification (especially where the AI Act intersects with the Machinery Regulation), with applying AI Act obligations to development patterns based on pre-trained and third-party models, and with translating abstract requirements into concrete technical measures and documentation. Existing AI development practices and documentation habits are therefore not sufficient to ensure AI Act compliance without additional support.

Building on these findings, we derived implications for the design of AI Act regulatory sandboxes. Effective sandboxes should support early and iterative engagement, integrate legal and technical expertise, explicitly address multi-regulation constellations, and systematically capture recurring patterns such as fine-tuning and API-based integration. We also highlighted a participation dilemma: sandboxes depend on voluntary uptake, yet participation increases regulatory visibility. Addressing this structural tension through low-threshold formats and clear benefits will be central to their success.

As member states implement Articles 57–62 AI Act, experiences from initiatives like KIRR Real can inform the design of sandboxes that genuinely bridge the gap between legal requirements and development practices. Future work should compare different sandbox models, observe compliance over longer periods, and examine how emerging harmonized standards and guidance change the types of questions that arise in practice.

Acknowledgements. KIRR Real was funded by the Ministry of Economic Affairs, Labor and Tourism Baden-Württemberg and conducted by the Fraunhofer Institute for Manufacturing Engineering and Automation IPA and the Active Research Environment for the Next generation of Automobiles (ARENA2036).

References

1. Attrey, A., Lesher, M., Lober, C.: The role of sandboxes in promoting flexibility and innovation in the digital age. Going digital toolkit policy note, OECD (2020). https://goingdigital.oecd.org/toolkitnotes/the-role-of-sandboxes-in-promoting-flexibility-and-innovation-in-the-digital-age.pdf
2. Balaire, P.: Decoding self-evolving behaviour in the EU machinery regulation: beyond continuous learning (2025). www.intertek.com/blog/2025/09-25-eu-machinery-self-evolving-behaviour/
3. Bromberg, L., Godwin, A., Ramsay, I.: Fintech sandboxes: achieving a balance between regulation and innovation. J. Bank. Finance Law Pract. **28**(4), 314–336 (2017)
4. Bundesnetzagentur: KI-Aufsicht und Reallabore (2024). https://www.bundesnetzagentur.de/DE/Fachthemen/Digitalisierung/KI/start.html
5. Ebers, M., Hoch, V., Rosenkranz, F., Ruschemeier, H., Steinrötter, B.: The European commission's proposal for an artificial intelligence act – A critical assessment by members of the robotics and AI law society (RAILS). J **4**(4), 589–603 (2021). https://doi.org/10.3390/j4040043
6. European Commission: Draft implementing act on AI regulatory sandboxes (2025). https://digital-strategy.ec.europa.eu/en/consultations/commission-seeks-feedback-draft-implementing-act-establish-ai-regulatory-sandboxes-under-ai-act
7. European Parliament and Council: Regulation (EU) 2023/1230 on machinery. Official Journal of the European Union, L 165/1 (2023). https://eur-lex.europa.eu/eli/reg/2023/1230/oj
8. European Parliament and Council: Regulation (EU) 2024/1689 Laying Down Harmonised Rules on Artificial Intelligence (AI Act). Official Journal of the European Union, L 2024/1689 (2024). https://eur-lex.europa.eu/eli/reg/2024/1689/oj

9. Floridi, L.: The European legislation on AI: a brief analysis of its philosophical approach. Philos. Technol. **34**, 215–222 (2021)
10. Future of Life Institute: AI regulatory sandbox approaches: EU member state overview (2025). https://artificialintelligenceact.eu/ai-regulatory-sandbox-approaches-eu-member-state-overview/
11. Information Technology - Artificial Intelligence - Management System. International standard, International Organization for Standardization (2023). ISO/IEC: ISO/IEC 42001:2023
12. Kittelmann, M., Mössner, T.: Maschinen mit Künstlicher Intelligenz: Anforderungen an Hersteller nach der neuen EU-MVO und Bezüge zur KI-VO. Arbeitsschutz in Recht und Praxis **6**(7), 217–221 (2025)
13. Ringe, W.G., Ruof, C.: Regulating Fintech in the EU: the case for a guided sandbox. Eur. J. Risk Regul. **11**(3), 604–629 (2020)
14. Veale, M., Zuiderveen Borgesius, F.: Demystifying the draft EU artificial intelligence act - Analysing the good, the bad, and the unclear elements of the proposed approach. Comput. Law Rev. Int. **22**(4), 97–112 (2021)

Quantitative Methods

Probabilistic Confidence In Assurance 2.0 Arguments

Robin Bloomfield[1] and John Rushby[2(✉)]

[1] City St George's, University of London, London, UK
R.E.Bloomfield@citystgeorges.ac.uk
[2] Computer Science Laboratory, SRI International, Menlo Park, CA 94025, USA
Rushby@csl.sri.com

Abstract. We present a new method for assessing probabilistic confidence in assurance case arguments that is simple, systematic and sound. It exploits the ways claims are decomposed in a structured argument and provides different approaches according to the different degrees of (in)dependence and diversity among subclaims, and how they eliminate the concerns that undermine confidence in their parent claims. The method uses only elementary probabilistic constructions that are well-known in other contexts (e.g., Fréchet bounds) but we interpret and apply them in a manner that is specifically focused on assurance arguments and requires no background in probabilistic analysis.

We show that the method is not susceptible to counterexamples that Graydon and Holloway exhibit for other approaches to quantifying confidence [12] and we recommend it as an additional tool and point of view in assessing Assurance 2.0 arguments that can contribute to dialectical examination and evaluation of cost/confidence tradeoffs for different risk levels and the overall balance of confidence across a structured argument.

1 Introduction

We have previously described overall assessment of confidence in an assurance case [7,9] within the methodology that we call Assurance 2.0 [6]. The primary positive assessment is *logical soundness*, and this should be subjected to dialectical challenge by exploration of potential *defeaters* [4]. For overall confidence, all defeaters must themselves be defeated or else be shown to pose negligible risk and accepted as *residual doubts*, and soundness must be established *indefeasibly* [16], meaning no credible new information would change the assessment.

Indefeasible assurance does not amount to certainty: there can always be some *aleatoric uncertainty* (uncertainty *in* the world, such as how many sensors fail during a mission) and some *epistemic uncertainty* (uncertainty *about* the world, such as whether evidence for some attribute is really conclusive). These uncertainties can be reduced (at increased cost) but not eliminated, so we set a threshold that balances cost and confidence bounds on indefeasibility. This balance is generally struck and stated informally (if stated at all), but sometimes it can be useful to attempt to quantify and express it numerically.

© The Author(s), under exclusive license to Springer Nature Switzerland AG 2027
E. Troubitsyna et al. (Eds.): SAFECOMP 2026, LNCS 16868, pp. 279–295, 2027.
https://doi.org/10.1007/978-3-032-34867-8_18

Quantification can sharpen thinking and assist decision making when used appropriately, though it can also convey false precision [17]. It can be particularly valuable in situations where a graduated treatment of confidence is needed. This arises in systems where different items pose different risks, and is exemplified by the Design Assurance Levels (DALs) A–E of DO-178C for software in commercial aircraft [14], and similar "integrity levels" in other standards and guidelines. In Assurance 2.0 we insist on logical soundness, even for items that pose less risk, but to reduce the cost of their assurance we may reduce some thresholds on evidence and rigor. To make reductions in a reasonable and justifiable manner, it is useful to assess some numerical estimate of confidence in an assurance case, and that is the topic of this paper, and of a more detailed technical report [10].

Subjective probability [13] provides a natural measure for numerical assessment of confidence; these subjective judgments, preferably achieved through consensus based on rational methods of assessment and calculation, should not be confused with frequentist or other objective measurements, but can be combined with such measures (e.g., observations from testing or initial deployment) to yield credible worst-case estimates of long-term reliability [18].

We have two ways of organizing assessment of probabilistic confidence [6]. First, there are circumstances where a probabilistic measure (e.g., probability of failure on demand, *pfd*) is an explicit part of the top claim or goal (some nuclear systems are like this). In this case, probabilistic measures will also appear in the evidence and in the internal claims of the argument and will be propagated by specific theories that are cited by the steps of the argument. We refer to this as *internal* probabilistic assessment. Alternatively, the top claim will be unconditional (e.g., "the system is correct"), and we develop a separate *external* probabilistic assessment of our subjective confidence (i.e., belief) in the truth of this claim and of those claims and evidence that support it. (It is plausible that these methods could be used in combination: for example, aleatoric uncertainty could be treated internally while epistemic uncertainty is assessed externally.)

Here, we are concerned with external assessments, where we and many others have proposed methods for calculating probabilistic or other numerical assessments for structured assurance arguments. Most assess logical soundness informally, so their techniques for numerical assessment carry much of the burden for justifying assessment. In Assurance 2.0, we assess logical soundness rigorously, separately, and prior to numerical assessment. This allows numerical assessment to provide a view that supports and complements the logical one, and enables probabilistic models that can be simple because they serve only a single purpose.

Our previous probabilistic treatments [9] used very elementary and conservative calculations, termed "the product calculation" and "sum of doubts," whose deficiencies we acknowledge [7, Section 3.5]. In Sect. 2, we introduce our new method, which is the main novel contribution in this paper. In Sect. 3 we consider Graydon and Holloway's critique of some other methods (of which six use rather complex constructions based on Dempster-Shafer or other forms of evidential reasoning, and five use Bayesian Belief Networks) [11, 12] and explain why ours is not susceptible to the flaws they identify. Section 4 provides conclusions.

We finish this introduction with a brief overview of Assurance 2.0 essentials [6,8] for those who have not yet read the earlier documents. An assurance case is modeled as a network of argument steps over propositions called *claims* and represented diagrammatically as *claim nodes*, connected by *argument nodes* or *steps*. Claims denote atomic propositions (typically expressed in controlled natural language) or a conjunction or disjunction of these. An argument step justifies a *parent claim* on the basis of a set of supporting *subclaims* that function as premises to the inference applied by the step, together with an optional *sideclaim* that records any explicit conditions required for the inference to be sound. The combination of an argument step with its parent claim, subclaims, and sideclaim is called an *argument block* (as in building block). Assurance 2.0 has just six different types of block [5].[1] The distinction between subclaims and sideclaims is methodological rather than logical—sideclaims could be represented as ordinary premises—but it is useful because these conditions on the argument step are often the main source of residual concerns[2] (for example, whether the subclaims consider all possible cases) and are best scrutinized explicitly.

2 The New Method

Our new method for probabilistic confidence in Assurance 2.0 works compositionally, claim by claim, on the argument of the case from bottom to top. Although our description is specific to Assurance 2.0, the general ideas are applicable to most assurance case frameworks.

The leaf nodes at the bottom of a completed assurance case argument can be references to external subcases, assumptions, evidence, or residual concerns. External subcase references need to be expanded in place (like a macro), or separately evaluated and represented by the confidence assessed for their top claim (like a subroutine). Assumptions state conditions on the overall argument, such as the maximum number of sensors that may be faulty. Their confidence, which may be based on historical experience or the judgment of Subject Matter Experts (SMEs), estimates the probability that each is true. The treatment of residual concerns is briefly described in Sect. 2.3. We begin with evidence.

2.1 Confidence in Evidence

Evidence is usually provided to an Assurance 2.0 argument in two steps, as portrayed in Fig. 1. In the first step, reference to the actual evidence is supplied to an *Evidence Incorporation* block that supports a claim about *something measured*: that is, it tells us what the evidence *is*. In Fig. 1 we see an example where the evidence is provided by requirements-based testing and the measurement claim states that this achieved some specific level of structural coverage (e.g.,

[1] Exact defeater (or negation) blocks [4] are an addition to the original five.

[2] Historically, we have spoken of residual *doubts*, but in a probabilistic context we prefer residual *concerns* as *doubt* is used for the probabilistic function $1 - P(a)$.

MC/DC). In a full assurance case the evidence incorporation node will reference a detailed justification that the evidence indeed supports the measured claim, which may cite an established theory for the test protocol involved with necessary assumptions referenced in a sideclaim.

In the second step, the measurement claim is transformed, using a *substitution* block, into a claim about *something useful* stating what the evidence *means* in the context of this assurance case. In Fig. 1 the useful claim is that the software contains no unreachable code. The reasoning is documented and justified in the substitution block, citing a suitable theory of testing and reachability.

In previous papers and reports [7, 9, 15] we describe in detail how claims should be evaluated against evidence by examining *confirmation measures* that help us assess the discriminating power of our evidence (i.e., its ability to justify this specific claim rather than others). These measures, such as Good's $\log \frac{P(E \mid C)}{P(E \mid \neg C)}$, where C represents the useful evidential claim and E the measurement claim, are constructed from various estimated conditional probabilities. But once they have been applied and their supported judgment accepted, we need a numerical assessment of confidence in the useful claim, given the evidence—and for this the simple posterior probability estimate $P(C \mid E)$, as estimated by assessors or other SMEs, is appropriate. This provides the first term in the following expression, where we use `conf` for confidence and `P` for probability:

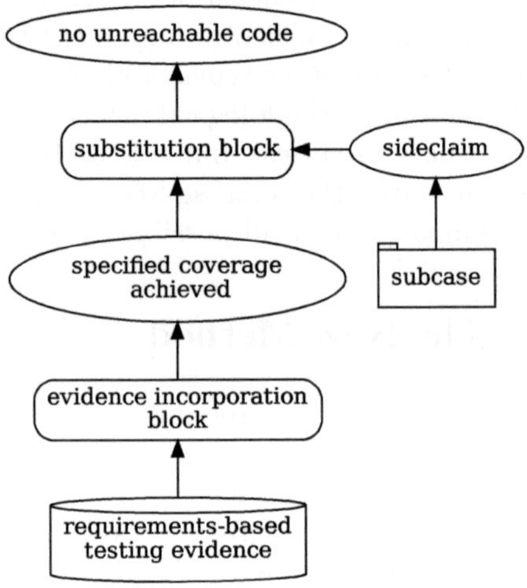

Fig. 1. Evidence Example

```
conf(useful claim) =
  P(useful claim measurement claim) × conf(sideclaim).
```

Argument blocks in Assurance 2.0 generally have a sideclaim recording conditions that may be necessary to ensure their justification is sound. Here, it might specify that coverage must be measured on executable object code (EOC), not source code. For reasons that are explained in the following section, confidence in the useful claim is then the product, shown above, of the conditional probability just described and confidence in the sideclaim, which is evaluated by recursive application of the techniques being developed here.

2.2 Confidence in Reasoning Steps

Above the leaf nodes, an Assurance 2.0 argument consists of reasoning steps that each justify a parent claim on the basis of one or more subclaims and, possibly, a sideclaim. The reasoning steps comprise *concretion*, *substitution*, and *exact defeater* (aka. negation) blocks, which each have a single subclaim, and *decomposition* and *calculation* blocks that each have two or more subclaims [5].

Blocks with a Single Subclaim

Substitution blocks used in combination with evidence incorporation have already provided a simple example of argument blocks with a single subclaim.

The general case for blocks with a single subclaim is portrayed in Fig. 2, where our task is to estimate probabilistic confidence in the truth of the parent claim given previously estimated confidence in the components of its supporting argument block, namely its subclaim, sideclaim, and the inference used in its argument. In logical assessment, the inference is generally assumed to be deductive and the sideclaim states conditions for it to be so. An argument block's logical interpretation is then `sideclaim` $\supset$ `(subclaim` $\supset$ `parent)`, which is equivalent to `sideclaim` $\wedge$ `subclaim` $\supset$ `parent`.[3]

As with logical assessment, we prefer that the inference in an argument step is considered deductive in probabilistic assessment also, and confidence in the parent claim is then directly derived from that in

`sideclaim` $\wedge$ `subclaim`.

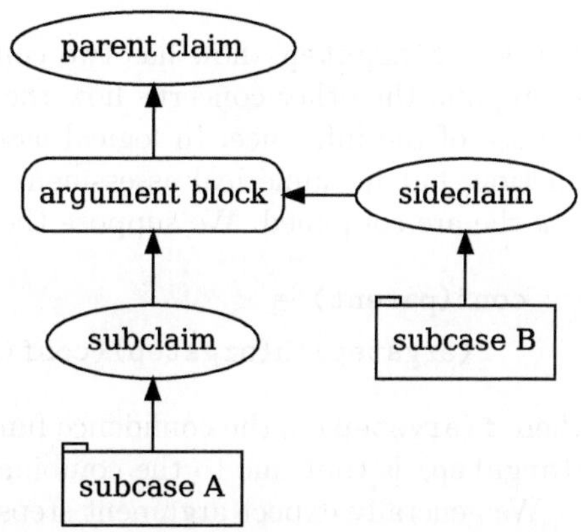

Fig. 2. Argument Block with a Single Subclaim

Confidence is a subjective probability, so we can apply probability laws and hence, by the chain rule, we obtain (1) below. As always, the first term in the product is confidence in the subclaim under the conditions stated in the sideclaim.

$$\texttt{conf(subclaim} \wedge \texttt{sideclaim)} = \texttt{conf(subclaim | sideclaim)} \times \texttt{conf(sideclaim)}. \quad (1)$$

But in Assurance 2.0, we evaluate arguments only when sideclaims are (believed to be) true, so we think of them as context and generally speak simply of `conf(subclaim)`, which is assessed by applying the methods described here to the subclaim's subcase. Hence,

$$\texttt{conf(parent)} = \texttt{conf(subclaim)} \times \texttt{conf(sideclaim)}.$$

However, we need to allow for the possibility that we may be less than certain in the inference performed in the argument step, and we do this by applying some

[3] We use $\supset$ for material implication, $\wedge$ for conjunction, and $\cap$ for set intersection.

function to `conf(subclaim)` as shown below, where the (higher-order) function `f` depends on the reasoning performed in the specific `argstep`.

$$\texttt{conf(parent) = f(argstep)(conf(subclaim))} \times \texttt{conf(sideclaim)}. \quad (2)$$

Applying this formula to Fig. 1: if our confidence in the measurement claim about test coverage, given its evidence is 0.8; confidence in the sideclaim, given its subcase, is 0.9; and confidence in the inference from coverage to reachable code—that is `f(argstep)(a)`—is `0.95` × `a`, then confidence in the parent (useful) claim of no unreachable code is the product of these three numbers: 0.684.

Blocks with Multiple Subclaims

The analysis for argument blocks having multiple subclaims proceeds as above, except that "subclaim" in (2) becomes plural, so we now have

$$\texttt{conf(parent) = f(argstep)(conf(subclaims))} \times \texttt{conf(sideclaim)}.$$

However, `f(argstep)` now has two components: one concerns the inference, as before, and the other concerns how the multiple subclaims combine to form the premise of the inference. In logical assessment, the combination is simply conjunction, but in numerical assessment we make fine-grained distinctions about *how* the are conjoined. We suppose the components can be factored so that

$$\texttt{conf(parent) =} \quad (3)$$
$$\texttt{f(argstep)(h(argstep)(conf(subclaims)))} \times \texttt{conf(sideclaim)}$$

where `f(argstep)` is the confidence function due to the inference, as before, and `h(argstep)` is that due to the combination of subclaims.

We generally expect argument steps to employ just one or the other of these components, so one of the functions will be the identity (or multiplication by a simple factor). For example, we often have a substitution block performing inference (but no combination) above a decomposition block performing combination (but no inference), rather than a single block that performs both of these.

In the following, we concentrate on combination and will suppose `f(argstep)` is some simple constant factor `k` that defaults to 1, and we will initially postpone consideration of the sideclaim, so (3) becomes

$$\texttt{conf(parent) = h(argstep)(conf(subclaims))} \quad (4)$$

and our task is to estimate `h(argstep)` for various kinds of reasoning steps.

We focus on decomposition blocks, which can be used in several different ways, possibly requiring different probabilistic interpretations and therefore different calculations for `h(argstep)`.

Decomposition Blocks, Diversity Case. Let us return to the evidential step of Sect. 2.1 and develop it further. Suppose requirements-based testing did not quite achieve the desired coverage. Rather than tinker with the testing evidence or the reasoning, we could decide to buttress them with static analysis: the idea being that testing and static analysis build on different foundations and are therefore "diverse" and may be supposed to fail independently [3]. This diversity argument is represented in the decomposition block shown in Fig. 3.

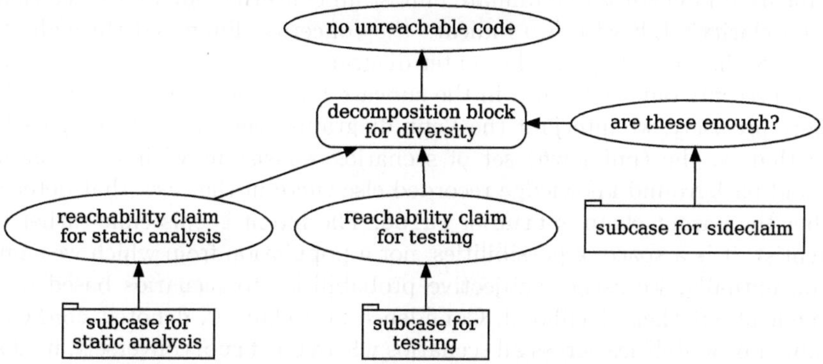

Fig. 3. Decomposition Block for Argument by Diversity

In some discussions of assurance cases, diversity arguments of this kind are regarded as disjunctions, the idea being that "if you don't like one subcase, you can use the other." This is not our interpretation: we think of one subclaim eliminating some concerns about the parent claim and the other subclaim eliminating other concerns, and we conclude the parent claim only when *both* subclaims have been assessed as valid. Hence, the logical interpretation of a decomposition argument has to be conjunction and the adequacy of any specific decomposition (e.g., are these specific subclaims sufficiently diverse and comprehensive?) is stated as a sideclaim and supported by its own subcase.

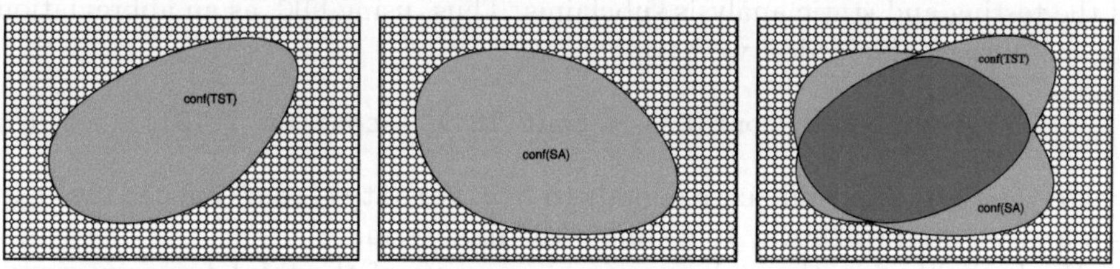

Fig. 4. The background circles represent concerns about the parent claim, and the colored shapes represent those eliminated (i.e., covered) by subclaims for testing (left), static analysis (center) and their combination (right)

Logical assessment is holistic, but when we assess confidence we are interested in the *quantity* of concerns that have been eliminated and their impact on the parent claim. This interpretation is illustrated in the Venn diagrams of Fig. 4. The blue shape in the left hand diagram indicates the probability mass of concerns in the parent claim eliminated by testing and the background circles indicate those concerns that remain. The area of the blue shape represents the confidence delivered by the testing subcase: `conf(TST)`. (Of course, in practice we want useful evidence to eliminate almost all concerns, but we reduce this for pictorial clarity.) Likewise, we indicate the concerns eliminated through static analysis by the pink shape in the center diagram of Fig. 4, where the area of the shape represents our confidence in the subcase for static analysis: `conf(SA)`.

More precisely, we interpret the Venn diagrams over an *epistemic possibility space*: that is, the (unknown) set of scenarios, consistent with scope assumptions and background knowledge recorded elsewhere in the case, that determine whether the parent claim is `true` or `false`. The intent is epistemic rather than frequentist: it is a space of possibilities, not a population from which we sample.

Conceptually, we assign subjective probabilities to scenarios based on our judgment about their likelihood. Confidence in a claim `C`, denoted `conf(C)`, is the sum of probabilities across all scenarios where `C` is `true`; conversely, its `doubt` is the sum across scenarios where it is `false`. When we remove $x\%$ of doubt, we gain $x\%$ in confidence.

A *concern* about claim `C` is a specific scenario in which `C` is `false`. In the diagrams, the small background circles represent such concerns and the rectangles correspond to the totality of these (i.e., doubt in the parent claim). Colored regions represent sets of scenarios (i.e. events), where region size corresponds to the sum of probabilities, the probability mass, associated with those scenarios.

The concerns eliminated by the combination of requirements-based testing and static analysis are shown in the diagram on the right of Fig. 4 where their combined probability mass corresponds to the union of the blue and pink shapes. Thus, confidence in the parent claim delivered by the two subclaims corresponds to the area of this union.[4] This area can be calculated as the sum of the areas for testing and for static analysis, less their overlap (shown in magenta), which would otherwise be counted twice. The overlap corresponds to the intersection of the testing and static analysis subclaims. Thus, using `NUC`, as an abbreviation for "no unreachable code," we have

$$\texttt{conf(NUC)} = \texttt{conf(SA)} + \texttt{conf(TST)} - \texttt{conf(SA} \cap \texttt{TST)} \qquad (5)$$

where the right hand side corresponds to `h(diversity)(conf(subclaims))` in (4), and we need some way to estimate the final term.

We know that static analysis covers a portion of the total "concern space" equal to `conf(SA)` so, assuming independence in the concerns eliminated by the two methods, it covers the same proportion of the space covered by testing,

[4] Notice how far we have departed from the standard interpretation of probabilistic logic: logical assessment *whether* the parent claim is true requires the *conjunction* of subclaims, whereas confidence assessment for *how strongly* we should believe it requires their union (traditionally associated with *disjunction*).

which itself covers conf(TST) of the total space. Hence, our estimate for their overlap conf(SA ∩ TST) is conf(SA) × conf(TST)[5] and therefore

$$\texttt{conf(NUC)} = \texttt{conf(SA)} + \texttt{conf(TST)} - \texttt{conf(SA)} \times \texttt{conf(TST)}. \quad (6)$$

Recalling that doubt(x) represents 1 - conf(x), this becomes

conf(NUC) = 1 - doubt(SA) × doubt(TST), and therefore

doubt(NUC) = doubt(SA) × doubt(TST).

We refer to this calculation as the *product of doubts* and it generalizes to arguments with n subclaims in the obvious way:

$$\texttt{doubt(parent)} = \prod_{i=1}^{n} \texttt{doubt(subclaim}_\texttt{i}\texttt{)}. \quad (7)$$

Unlike our previous methods, where confidence in a parent claim is always less than that in any of its subclaims, we see from (5) that a diversity case can *increase* confidence in the parent. For the specific formulas in (6) and (7), we assume concerns not eliminated by each subclaim are independent, and we believe, or hope, this will be so if the subclaims are "diverse," and the sideclaim to the argument node should therefore include a requirement for "adequate diversity."

For notational brevity we omitted the sideclaim in the calculations above but we recall from (3) that the propagated confidence is the product of confidence in the inference from subclaims to parent claim (i.e., the calculation described above) and confidence in the sideclaim. Additionally, a function f or constant factor k representing strength of the inference from combined subclaims to parent claim can be applied if those assessing the argument consider it desirable.

So, in the example above, if we are 95% confident that testing ensures no unreachable code and 90% confident that static analysis does the same, then the doubts are 5% and 10% respectively, and their product is 0.5%. Hence, confidence in the combination is 99.5%. However, if our confidence in the sideclaim for adequate diversity is only 90%, then overall confidence in the parent is reduced to 89.55% (as the product of 0.995 and 0.9). Thus, diversity increases overall confidence only when there is strong confidence in diversity and independence. Notice that consideration of defeaters to independence contributes to dialectical examination and can raise issues that cause logical assessment to be revisited.

Product of doubts is accurate when the independence assumption underlying diversity is valid and less so when it is not. There is more opportunity for inaccuracy when confidence in the individual subclaims is low. For example, if we are 49% confident in each of two subclaims, the product of doubts calculation gives 74% confidence in their combination, implicitly assuming independence, whereas

[5] These and several other formulas are well-known probabilistic tautologies; we spell them out so that these and later derivations are easier for nonspecialists to follow.

in reality it could range from 49% (positive correlation and total overlap) to 98% (negative correlation and completely disjoint). The latter may be a deliberate, alternative, strategy to diversity, and we consider it next.

Decomposition Blocks, Partitioned Case. In contrast to the previous example where, in the Venn diagram representation, the subclaims substantially overlap, we now consider the case where they are disjoint. This arises in decompositions over different hazards, or over the components of a larger system, or over different operating conditions, etc. The idea is that the subcases partition the "concern space" and the subclaims are each local to a separate partition.

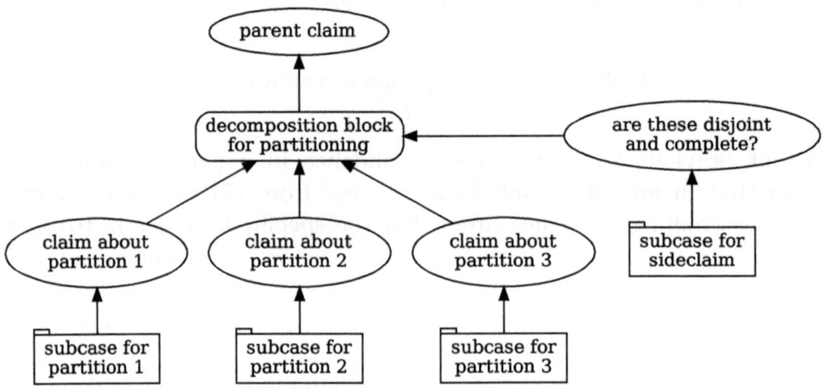

Fig. 5. Decomposition Block for Partitioned Argument

To be concrete, we could suppose we are decomposing the argument over two hazards. This might suggest two partitioned subcases, one for each hazard, but reflection suggests this is insufficient: we also need to consider possibilities such as a second hazard instance arising while the system is dealing with a first. We will combine these possibilities into an "everything else" partition, so that we have three in total. This case is portrayed abstractly in Fig. 5.

The Venn diagram corresponding to this argument is shown in Fig. 6 where the background rectangle represents the total concern space partitioned into the three subcases. The partitions are of different sizes because we may have more or less concern about them: `weight(x)` indicates the fraction of the total concerns or risks, depending on the claim, associated with partition `x` (so the `weights` sum to one) and is represented pictorially by the size of the partition. The colored shapes represent our confidence in the subcase for their associated partition, denoted `conf′(x)`; we use the prime ′ to indicate this confidence is assessed relative to its partition, not the whole space, which is denoted by plain `conf(x)` where `conf(x)` = `conf′(x)` × `weight(x)`. Our confidence in the overall decomposition, given the sideclaim, is then the weighted sum of these (this is

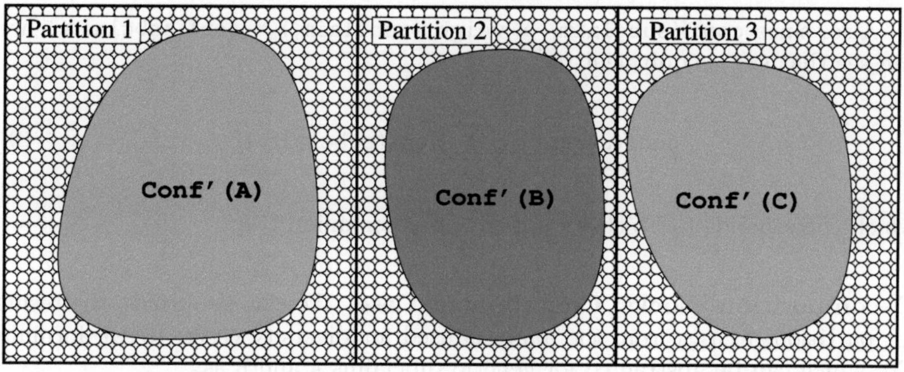

Fig. 6. Partitioned Subclaims

just a Total Probability calculation):

$$\begin{aligned}\texttt{conf(parent)} = \texttt{conf}'\texttt{(A)} &\times \texttt{weight(Partition 1)}\\ + \texttt{conf}'\texttt{(B)} &\times \texttt{weight(Partition 2)}\\ + \texttt{conf}'\texttt{(C)} &\times \texttt{weight(Partition 3)}.\end{aligned}$$

Notice that if the `weights` are equal, this is simply the arithmetic mean of the subclaim confidences; we call this the **averaging case**.

As in the previous cases, the confidence propagated from this argument step is the product of confidence in the combination of subclaims, calculated as above, and confidence in the sideclaim. Thus if we assess that 60% of the risks are associated with Hazard 1 alone, 30% with Hazard 2 alone, and 10% with other cases, and that our confidence in the subclaims associated with these are 90%, 95% and 80%, respectively, then basic confidence in the combination of subclaims is $90 \times 60 + 95 \times 30 + 80 \times 10 = 90.5\%$ and this needs to be multiplied by confidence in the sideclaim, say 99%, to yield an overall confidence of 89.6%. As in the previous sections, a function `f` or constant factor `k` can be applied to represent confidence in the inference from the combination of subclaims to the parent claim if those developing or assessing the argument consider it desirable.

The formula above obviously generalizes to decompositions with n partitioned subclaims as follows.

$$\texttt{conf(parent)} = \sum_{i=1}^{n} \texttt{conf}'(\texttt{subclaim}_i) \times \texttt{weight}(\texttt{subclaim}_i) \tag{8}$$

where the right side corresponds to `h(partition)(conf(subclaims))` in (4).

Despite its different derivation, this is essentially equivalent to Jeffrey's rule of combination, which is (implicitly) used in some Dempster-Shafer treatments of quantified confidence in assurance cases [1, Section 3.2].

Also, since $\texttt{conf}'(\texttt{subclaim}_i) \times \texttt{weight}(\texttt{subclaim}_i) = \texttt{conf}(\texttt{subclaim}_i)$, we can see that (8) is equivalent to

$$\texttt{conf(parent)} = \sum_{i=1}^{n} \texttt{conf}(\texttt{subclaim}_i), \tag{9}$$

which corresponds to Fréchet's upper bound for union [19].

Decomposition Blocks, Containment Case. We can also derive the result for partitioned cases from the analysis used for diversity cases. Recall formula (5), which can be abstracted for generic subclaims A and B as

$$\texttt{conf(parent) = conf(A) + conf(B) - conf(A} \cap \texttt{B)} \tag{10}$$

and imagine pulling apart the subclaim areas A and B in the concern space so that they no longer overlap—but then $\texttt{conf(A} \cap \texttt{B)}$ is zero. This corresponds to the partitioned case and we will obtain the same results under either the diversity or partitioned calculations, provided we take note that the subcase confidences are evaluated differently (cf. Formulas (8) and (9)).

We can reverse the mental experiment just described and imagine pushing the subclaims of a diversity case together until one is entirely within the other, as shown in Fig. 7. Ayoub and colleagues [1] call this a "containment" argument. Here, we analyze it as a special form of diversity case but consider their approach in our report [10, Section 3.2.5].

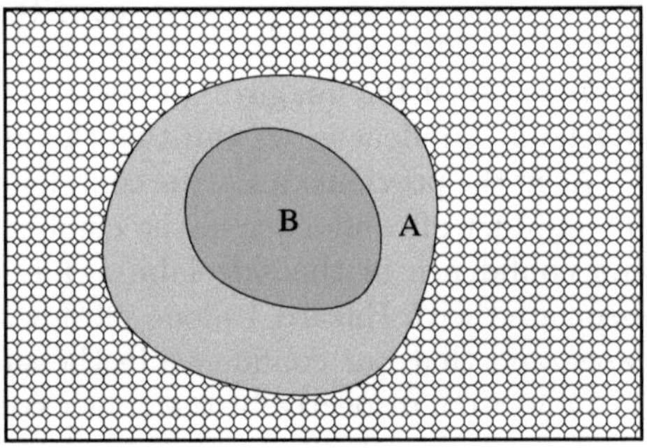

Fig. 7. Contained Subclaim

It is obvious in the diagram of Fig. 7 that the eliminated concerns are all covered by subcase A—and B adds nothing—so the value propagated should be simply conf(A). Formula (6) and its generalization (7) assume A and B are independent and give incorrect results in this case. However, they are derived from formula (10), which is true in all cases, and the challenge in applying it to special cases is to find good estimates for $\texttt{conf(A} \cap \texttt{B)}$. A conservative estimate is

$$\texttt{conf(A} \cap \texttt{B)} \leq \min\texttt{(conf(A), conf(B))},$$

which is the *Fréchet upper bound for intersection* [19]. In our containment case, the minimum is conf(B) and applying this in (10) gives conf(parent)

$\geq$ `conf(A)` as required.[6] More directly, (10) becomes

$$\texttt{conf(parent)} \geq \max\,(\texttt{conf(A)},\ \texttt{conf(B)}),$$

which corresponds to the *Fréchet lower bound for union* and the general case of containment among n subclaims is $\texttt{conf(parent)} \geq \max_{i=1..n}(\texttt{conf}(\texttt{subclaim}_i))$, which comes from the general form of the Fréchet bound [19].

Dually, the *Fréchet lower bound for intersection* provides an alternative derivation for partitioned cases as in (9); see [10, Section 3.2.5] for details.

Decomposition Blocks, Other Cases. We have seen that partitioned and containment cases can be reduced to diversity cases, so all are based on a single construction. We consider this uniformity a positive indication that our treatment is comprehensive and correct, and speculate that similar constructions can be developed for many other novel cases. However, we will now consider an example where this is not so.

Decomposition Blocks, Cumulative Case. Barrett and colleagues [2, page 32] consider a case where the parent claim that a dangerous AI function is taken out of service within one week is supported by the following three subclaims:

Detection: Incident monitoring detects all novel cyber attacks within one day,
Revision: A revision protocol ensures the safety case is updated (possibly invalidating the top claim) within five days of detection of a novel cyber attack.
Removal: The AI system will be taken offline within one day of the top-level safety case claim becoming false.

They represented the argument by a decomposition block similar to Fig. 5 and estimated confidence in the detection, revision, and removal subcases to be 48%, 81%, and 56% respectively. Using our earlier "product" and "sum of doubts" calculations [7,9], they obtained overall confidence of 21.772%, or 0, respectively, which they found disappointing.

This prompted us to consider whether our new methods might be applicable. However, although the argument resembles Fig. 5, the subcases are not independent (each is contingent on its predecessors) so the partitioned treatment is not appropriate, and neither is diversity. If we try to portray the concerns eliminated by each subcase in a Venn diagram, we obtain a three-subcase version of Fig. 7, suggesting containment. However, we can see that this example requires a different interpretation of the diagram: rather than concerns eliminated by some subclaims being contained within those of another, so that only that one is required, here the subclaims build support *cumulatively* and are all required. So, by the standard probabilistic interpretation of conjunction (the chain rule):

[6] Some might say that our confidence that each subclaim eliminates its indicated concerns could be weak, and so their overlapping combination is stronger and should recognize a contribution from `B`. We respond that in this case `B` is covering some "gaps" in the area supposedly covered by `A` and this is really a diversity case and `conf(A ∩B)` is therefore less than `conf(B)`.

```
conf(parent) = conf(detection & revision & removal)
             = conf(detection)                                  (11)
                 × conf(revision | detection)
                 × conf(removal | detection & revision).
```

Next, we notice that the definitions of the `detection` and `removal` subclaims stated earlier implicitly have this conditional form. Hence, the estimated subclaim confidences do correspond to the terms in formula (11), and this is the same as the product calculation mentioned earlier—but now we can see that for this case it is exact, rather than an approximation.

This example is a simple illustration of an argument where there are conditional relationships among subclaims. Because the relationships in this example are simple, we can easily calculate (as the product) the confidence propagated to the parent claim (subject, as usual, to adjustment by a function `f` or factor `k` and multiplication by confidence in any sideclaim). In arguments where the conditional relationships are more complex, it can be useful and appropriate to employ Bayesian Belief Networks (BBNs) and their tools, which are discussed in detail in our report [10, Section 4].

2.3 Residual Risks

An assurance argument may contain residual concerns: these are explicitly marked defeaters that we have been unable or have chosen not to eliminate or fully mitigate. Logical assessment checks that their consequences (i.e., residual *risks*) are insignificant and, on that basis, ignores them. Probabilistic assessment can follow this lead or, alternatively, we can attach a numerical assessment to each residual concern (i.e., our subjective confidence it will have no impact) and propagate this through the argument in the usual way. This assessment should consider the likelihood of harm due to the identified concern, its maximum severity, and number of instances if it may occur many times [10, Section 3.3].

3 Graydon and Holloway's Critique

There is much work on quantified assessment of assurance cases. Graydon and Holloway [11,12] provide a comprehensive description and critical evaluation of 12 different methods. For many of these methods, they reproduce the authors' examples and then construct variants that generate implausible results, showing the methods to be fallible and possibly flawed. We consider whether the underlying critiques apply to our method.

Many Subclaims Examples [11, pp. 36–39]. Abstracting from the details, the idea is that we have a decomposition step with many subcases, say 20. Of these, 19 have high confidence and one is very low. Most methods, including those introduced by us in the previous section, are not particularly sensitive to the degree of confidence in one subclaim among many, so confidence in the parent claim will be dominated by the majority and will therefore be high in this example. If acceptability of the argument is based on numerically high confidence, then this argument will be accepted despite having one very weak element, and Graydon and Holloway rightly regard this as a cause for concern.

But this is not how Assurance 2.0 works, and the example is no threat to our methods. The subclaim with low confidence would surely not have been adjudged `true` in logical assessment and so this assessment will fail and we will be unable to proceed to probabilistic assessment.

On the other hand, logical assessment will not highlight cases where subclaims are judged `true`, but with less certainty than others: it does not even provide a way to talk about this. Probabilistic assessment in Assurance 2.0 is a lens that gives a different view of the case and allows us to speak rationally of confidence, so that weakly `true` subclaims that might otherwise have gone unremarked will be identified and this can then lead to further investigation.

Graydon and Holloway also consider variants on this scenario where subclaims provide undermining or counter-evidence, as can arise in methods where quantification extends to disbelief as well as belief. But in Assurance 2.0 we examine disbelief (using defeaters) prior to probabilistic assessment, where the case is entirely positive and the proposed scenarios cannot arise.

Imperfect Examples [11, page 45]. Graydon and Holloway provide examples where one subclaim of slightly lower confidence than the others will sharply lower confidence in the parent claim; confidence in the parent claim is also very sensitive to the number of subclaims. Our older "product calculation" is similarly susceptible, and this is among the reasons that we now deprecate its use except in special circumstances (e.g., Sect. 2.2, where it is exactly the right method). Our new methods are more balanced in this regard (mainly because they are based on addition rather than multiplication).

Optimistic and Pessimistic Counterexamples [11, pp. 87–90]. Graydon and Holloway consider a trivial example where massive changes in what we would call a sideclaim (to establish completeness of a hazard list) produce only minor changes in overall confidence. In Assurance 2.0, sideclaim confidence is a multiplicative factor in the overall confidence for an argument step (recall definition (1)) and therefore has a large impact, generating what we consider to be correct and appropriate values in these examples.

Some of the methods susceptible to this criticism use BBNs, which we also endorse in suitable circumstances but, as we note in our report [10, Section 5.2], BBNs must represent the specifics of the case under consideration and cannot be derived directly from a traditional assurance argument because they represent different concerns. Many of the BBN-based methods examined by Graydon and Holloway do employ such derivations and are flawed for this reason.

4 Summary and Conclusion

We have described a new method for external probabilistic assessment of confidence in Assurance 2.0 arguments. Our method works compositionally, claim by claim, on the argument of the case from bottom to top, propagating confidence from leaf nodes such as evidence, assumptions, and residual concerns, up to the top claim. Our method has variants for the different ways that subclaims can be combined to form the premise of an argument step, most notably for diversity and partitioning, and further variants are summarized in our report [10, Table 1]. The method uses only simple probabilistic constructions that are easily implemented and are well-known in other contexts.

We recommend confidence assessment as a supplement to logical and dialectical assessments, and not as a deciding judgement. It is particularly useful in examining the balance of concern and effort across a large assurance argument, or one composed of several semi-independent parts, and in guiding cost and confidence tradeoffs for different risk or "integrity" levels. For example, at high integrity, we may require independent teams to perform each "leg" in a diversity case, whereas the same team could be used for both at lower integrity levels, thereby reducing cost but also confidence in the sideclaim for independence. Probabilistic quantification can support rational evaluation of these tradeoffs.

Because Assurance 2.0 assesses logical soundness prior to probabilistic assessment, and because its probabilistic basis is simple and robust, our method is not susceptible to the counterexamples that Graydon and Holloway exhibit for other approaches. We acknowledge user skepticism towards quantitative methods, but we hope that the role we propose for quantitative methods in Assurance 2.0, where they complement logical and dialectical assessments, and the intuitive and simple calculations that underlie our methods, can allay these concerns and will prove valuable when using Assurance 2.0.

Acknowledgments. We thank Peter Bishop, Bev Littlewood, and Andrey Povyakalo of City St George's for their challenging critique of some of our earlier description and derivations, and the SafeComp reviewers who provided detailed and constructive comments that greatly improved the presentation.

This material is based upon work supported by the United States Air Force and DARPA under Contract No. FA8750-23-C-0519 and is released under Distribution Statement "A" (Approved for Public Release, Distribution Unlimited). Any opinions, findings and conclusions or recommendations expressed in this material are those of the author(s) and do not necessarily reflect the views of the United States Air Force, DARPA, or the United States Government.

References

1. Ayoub, A.: Assessing the overall sufficiency of safety arguments. In: Proceedings of the 21st Safety-Critical Systems Symposium, pp. 127–144. Bristol, UK (2013)
2. Barrett, S.: Assessing confidence in frontier AI safety cases. arXiv:2502.05791 (2025)
3. Bloomfield, R., Littlewood, B.: Multi-legged arguments: the impact of diversity upon confidence in dependability arguments. In: DSN 2003, pp. 25–34 (2003)
4. Bloomeld, R., Netkachova, K., Rushby, J.: Defeaters and eliminative argumentation in assurance 2.0. Tech. Rep. SRI-CSL-2024-01. arXiv:2405.15800 (2024)
5. Bloomfield, R., Netkachova, K.: Building blocks for assurance cases. In: ISSRE ASSURE Workshop (2014). https://doi.org/10.1109/ISSREW.2014.72
6. Bloomfield, R., Rushby, J.: Assurance 2.0: a manifesto. In: Proceedings of the 29th Safety-Critical Systems Symposium, York UK, (2021). Preprint at arXiv:2004.10474
7. Bloomfield, R., Rushby, J.: Assessing confidence in assurance 2.0. Technical Report SRI-CSL-2022-02 (2022). arXiv:2205.04522
8. Bloomfield, R., Rushby, J.: Assurance 2.0 in a nutshell. Technical note (2024). https://www.csl.sri.com/users/rushby/papers/a2nutshell.pdf
9. Bloomfield, R., Rushby, J.: Confidence in assurance 2.0 cases. LNCS, vol. 14780, pp. 1–23 (2024), expanded version available at arXiv:2409.10665
10. Bloomeld, R., Rushby, J.: Quantifying confidence in assurance 2.0 arguments, Technical Report SRI-CSL-2025-01 (2025), arXiv:2604.00034
11. Graydon, P.J., Holloway, C.M.: An investigation of proposed techniques for quantifying confidence in assurance arguments. NASA/TM-2016–219195 (2016)
12. Graydon, P.J., Holloway, C.M.: An investigation of proposed techniques for quantifying confidence in assurance arguments. Saf. Sci. **92**, 53–65 (2017)
13. Jeffrey, R.: Subjective Probability: The Real Thing. CUP (2004)
14. RTCA: DO-178C: Software Considerations in Airborne Systems and Equipment Certification. RTCA, Washington, DC (2011)
15. Rushby, J.: Mechanized support for assurance case argumentation. In: JSAI-isAI Workshop AAA. LNCS, vol. 8417, pp. 304–318 (2013). Preprint available https://www.csl.sri.com/users/rushby/abstracts/aaa13
16. Rushby, J.: The indefeasibility criterion for assurance cases, postproceedings of a Shonan Workshop (2016). https://doi.org/10.1007/978-981-15-5054-6_12
17. Spiegelhalter, D.: The Art of Uncertainty: How to Navigate Chance, Ignorance, Risk and Luck. Random House (2024)
18. Strigini, L., Povyakalo, A.: Software fault-freeness and reliability predictions. In: SafeComp 2013: LNCS, vol. 8153, pp. 106–117 (2013)
19. Wikipedia contributors: Fréchet Inequalities. https://en.wikipedia.org/wiki/Frechet_inequalities

The Statistical Assessment of Bayes–"sub"Optimal Binary Machine Learning Classifier Risk

Abraham Chan[1(✉)], Ilir Gashi[2], Sathish Gopalakrishnan[1], Karthik Pattabiraman[1], and Kizito Salako[2]

[1] University of British Columbia, Columbia, Canada
{abrahamc,sathish,karthikp}@ece.ubc.ca

[2] Centre for Software Reliability, City St George's, University of London, London, UK
{ilir.gashi.1,k.o.salako}@city.ac.uk

Abstract. In many safety-critical applications, binary classifiers detect undesirable operational states. Optimal adjudication schemes may provide an affordable means of constructing *Bayes-optimal* hybrid classifiers (i.e., a classifier configuration with the lowest expected cost of classification errors) by combining the outputs from multiple classifiers. However, statistical uncertainty poses significant challenges when applying such schemes, so that their use cannot be guaranteed to produce truly optimal configurations. We present statistical methods for estimating the extent to which various adjudication schemes, including optimal adjudication, fail to produce optimal classifier configurations. Applying these methods to two critical contexts—pneumonia diagnosis and concrete-defect detection—classical statistical bounds reveal how training-sample uncertainty, validation-sample uncertainty, and the relative cost of classification errors jointly constrain both the efficacy of adjudication schemes and the confidence that can be placed in the configurations they produce. So-called "smoothed" hybrid classifiers consistently give the lowest risk with the tightest optimality confidence bounds, particularly when the relative cost of classification errors is very low/high.

Keywords: Bayes-optimal · Optimal Adjudication · Statistical Modelling

1 Introduction

In safety-critical applications, such as healthcare diagnosis [1] and infrastructure maintenance [2], correct binary classification (e.g., does an X-ray reveal a tumour or not?) is necessary for ensuring safety and minimizing risk. For example, misdiagnosis can lead to unnecessary medical treatment, placing patients' health at risk, while undetected structural defects in buildings can endanger lives. Classifiers must err infrequently and incur minimal *loss*—loss can encompass injury, fatalities, property damage, financial cost, etc.

© The Author(s), under exclusive license to Springer Nature Switzerland AG 2027
E. Troubitsyna et al. (Eds.): SAFECOMP 2026, LNCS 16868, pp. 296–311, 2027.
https://doi.org/10.1007/978-3-032-34867-8_19

Various adjudication schemes have been proposed for improving detection rates and reducing error costs by combining the outputs from multiple base classifiers into a *hybrid* classifier. Among such schemes, optimal adjudication has the potential of producing hybrid classifier configurations that achieve the lowest expected loss—these are *Bayes–optimal* hybrid classifiers. Constructing such configurations requires accurate statistical estimates (typically obtained from operational data) of the base classifiers' individual and joint failure probabilities. This reliance on operational data is principled: improvements in classifier performance are established based on what is actually observed. However, the accuracy of these estimates is limited by data quality and quantity, and the resulting statistical error may lead to hybrids that do not significantly reduce expected loss, with limited statistical evidence of improvement. By applying classical distribution-agnostic statistical bounds to estimate the "efficacy" of various hybrid classifier configurations, we highlight these challenges and how they impact hybrid classifier performance—an important gap in ML safety-critical applications and reliability evaluation. Here, "efficacy" is an estimate of the difference between a (hybrid) classifier's expected loss and the Bayes–optimal expected loss. We make the following research contributions:

- A statistical framework for assessing the efficacy of classifier adjudication schemes; in particular, for assessing attempted constructions of "optimal adjudication" hybrid classifiers and the extent to which hybrids deviate from true Bayes-optimality. This includes a probabilistic model of classifier failure occurrence, failure impact, and various demonstrations of the use of classical inequalities (e.g., Markov, Chebyshev, Hoeffding) to bound the probability of the hybrids being suboptimal. In the scenarios considered, Chebyshev inequalities provide the tightest and most useful bounds on sub-optimality;
- The framework is applied to scenarios from healthcare and construction, each involving 7 diverse Machine Learning (ML) architectures for base classifiers and 8 adjudication schemes for constructing 12 hybrids. We investigate the impact of training/validation sample uncertainty and classification-error losses on hybrid performance. The model and analysis show these factors can undermine the construction and evaluation of hybrids;
- The results also show a ranking among the 8 adjudication schemes, with so-called "smoothed" hybrids consistently achieving the lowest expected risk and the closest performance to Bayes-optimality. The apparent dominance of "smoothed" hybrids becomes more pronounced when false negatives incur vastly larger losses than false positives, or vice versa.

The paper is organized as follows. Related work is discussed in Sect. 2, and the model of classifier failure is detailed in Sect. 3. The methodology and experimental design are described in Sect. 4, with results and their implications in Sect. 5. The paper concludes with limitations and future work in Sect. 6.

2 Background and Related Work

Background. Safety standards emphasise quantifying and bounding the probability of failure and risks; e.g., IEC 61508 [3] advocates this for electronic systems, and ISO/TR 20417:2021 [4] for ML solutions used in healthcare.

Optimal adjudication was proposed by Blough et al. [5] and Di Giandomenico et al. [6] as a way of combining functionally redundant software in a fault–tolerant configuration that minimizes expected loss. The mathematical basis for this is Bayes–optimality, developed within the Bayes decision-theoretic framework; see Wald [7]. Decisions/actions are taken that minimize conditional expected loss, conditional on observations that inform the decision/action. Bayesian decision theory has been applied to optimize the performance of ensembles of ML classifiers [8,9]. In [10], Bayesian methods are used to design classifiers and estimate their risk when small amounts of training data are available and the true data-generating process is uncertain. They derive decision rules that, in principle, minimize expected risk under the chosen Bayesian model. There are also non-Bayesian methods that steer a classifier towards Bayes-optimality during training, such as using the "BOLT" training loss-function for neural networks [11]. In this paper, we focus on empirically estimating the extent to which classifiers designed to be Bayes-optimal are truly optimal.

Related Work. An alternative evaluation technique to those presented in this paper is cross-validation, which gives estimates of the performance, "on average", of classifiers produced by an algorithm. This contrasts with our focus, which is an estimation of the performance characteristics of "specific" classifiers/hybrids. Indeed, our evaluation concerns the extent to which each classifier/hybrid is sub-optimal. Seminal works on cross-validation are Stone [12] and Geisser [13].

Traditional Receiver Operating Characteristic (ROC) techniques—such as the computation of accuracy, sensitivity, specificity, area under the curve, etc.—are often used for evaluating classifier performance; see Swets [14,15]. However, they are not suitable in answering the concerns of this paper. One limitation is that they assume the classifiers' conditional rates remain unchanged when base rates change, and do not account for relative loss from errors.

The "Convex Hull" Theorem constructs hybrids from multiple base classifiers, using convex combinations of base classifier outputs; see Fawcett [16]. The "sense" in which these classifiers are optimal differs from Bayes–optimality.

3 Probabilistic Model of Classifier Failure and Risk

This section describes the probabilistic model for classifier error occurrence, adapted from Salako [17], used for the experiments and statistical assessment.

Hybrid vs. Base Classifiers. There is a (large but finite[1]) set $\boldsymbol{\Omega}$ of classification tasks assigned to a classifier. A classifier's response to each task is binary:

[1] Given a maximum amount of memory that the software can occupy, consider the number of binary words that can fit in this memory as an upper bound on $\boldsymbol{\Omega}$.

the classifier either issues an alert (denoted by 1") or no alert (0"). A *hybrid* classifier is one whose response to a task is constructed from the responses of other classifiers on the same task. A *base* classifier is one that is not hybrid. Base classifiers are deterministic: they always produce the same response to a given task. However, the required response on a classification task may differ from a classifier's actual response, either because the classifier is flawed or because of limitations in the classification approach (e.g., reliance on techniques that are thwarted by code obfuscation [18,19], or techniques that only approximate correct behaviour, such as trained neural networks [20]). Consequently, a classifier may produce *false positive* errors (FPs) (raising an alert when it should not) or *false negative* errors (FNs) (failing to raise an alert when it should).

Syndromes. A collection of k classifiers partitions $\boldsymbol{\Omega}$ into at most 2^k subsets, where each subset contains all tasks triggering the same k-bit response pattern. Each such pattern is called a *syndrome*, indexed $i \in \{0, 1, \ldots, 2^k - 1\}$, with the j-th classifier's response on syndrome i given by the j-th bit of i's binary representation. For example, $i = 0 = (0\ldots0)_2$ has no alerts and $i = 3 = (0\ldots101)_2$ has an alert from the first and third classifiers.

The probability that a randomly chosen task is from subset i and it should not trigger an alert is p_i, while q_i is the probability that the task is from subset i but should trigger an alert instead. So, the probability of an FN is, at most, $\sum_{i=0}^{2^k-1} q_i$, and the FP probability is, at most, $\sum_{i=0}^{2^k-1} p_i$. It also follows that the j-th classifier makes an FP error with probability $p^j = \sum_{i\,:\,\text{base-2 } j\text{th component is } 1} p_i$, and an FN error with probability $q^j = \sum_{i\,:\,\text{base-2 } j\text{th component is } 0} q_i$.

Expected Loss. When a classifier makes an error, an associated loss L is incurred. The conditional expected losses, due to an FN or FP error, are l_{fn} and l_{fp}, respectively. The expected loss $\mathbb{E}[L]$ from a classifier failing is then $\mathbb{E}[L] = l_{fn}q^j + l_{fp}p^j$. This expression shows how the expected loss is determined by the sizes of the failure probabilities and the sizes of the conditional expected losses. We may rewrite this as $\mathbb{E}[L] = l_{fn}(q^j + \gamma p^j)$, where $\gamma = \frac{l_{fp}}{l_{fn}}$; γ is the relative cost of FPs to FNs. Choosing units in which l_{fn} is 1, we have

$$\mathbb{E}[L] = (q^j + \gamma p^j)\,. \tag{1}$$

Using Syndromes to Build Hybrid Classifiers. Syndromes can be used to construct hybrid classifiers, since a response — alert or no alert — can be associated with each syndrome. To minimise $\mathbb{E}[L]$, associate with syndrome i the response r that incurs the smallest expected loss when this is the incorrect response when syndrome i occurs. That is, solve

$$\arg\min_{r} \mathbb{E}[L\mathbf{1}_i \mid \text{response } r \text{ for syndrome } i] \tag{2}$$

for r, where $\mathbf{1}_i$ is an indicator function—it equals 1 when a random task from $\boldsymbol{\Omega}$ triggers the i-th syndrome, and equals 0 otherwise. If the proposed hybrid response r is "alert" when syndrome i occurs, then only FPs are possible and

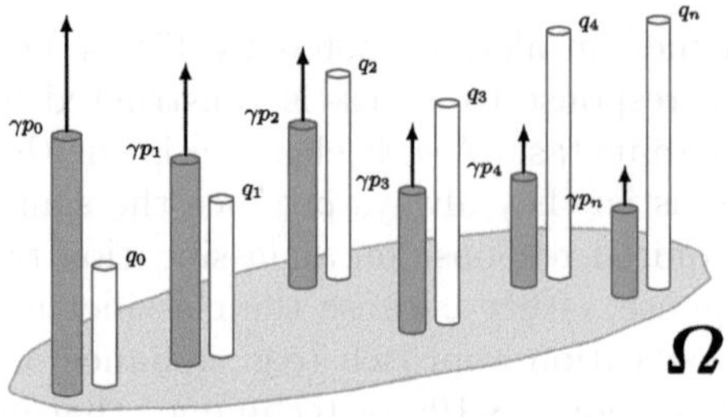

Fig. 1. Expected losses from FPs (i.e., γp_i) and FNs (i.e., q_i) associated with each syndrome i as illustrated by grey and white pillars, for each subset of $\boldsymbol{\Omega}$ defined by a syndrome. As γ increases, the expected loss due to an FP associated with syndrome i (i.e., γp_i) increases. Longer arrows indicate greater marginal increase in the conditional loss per unit increase in γ, due to relatively larger p_i values.

the probability of this error is p_i. If, instead, "no alert" is the response, then only FNs are possible and this error occurs with probability q_i. Consequently, $\mathbb{E}[L\mathbf{1}_i \mid \text{"alert" for syndrome } i] = \gamma p_i$ and $\mathbb{E}[L\mathbf{1}_i \mid \text{"no alert" for syndrome } i] = q_i$. Consequently, for the ith syndrome, the solution to (2) is:

$$\arg\min_r \mathbb{E}[L\mathbf{1}_i \mid \text{ response } r \text{ for syndrome } i] = \begin{cases} \text{"no alert", if } q_i \leqslant \gamma p_i \\ \text{"alert", } \quad \text{if } q_i > \gamma p_i \end{cases} \tag{3}$$

If $p_i > 0$ then, for all sufficiently large γ, "no alert" should be raised for syndrome i. And, if $q_i > 0$ then, for all sufficiently small γ, an "alert" should be raised. (If $p_i = 0$ and $q_i > 0$, "alert" is optimal for all γ; if $q_i = 0$, "no alert" is optimal for all γ.) Figure 1 illustrates how conditional expected losses change with γ.

Solving (2) for each syndrome's optimal response yields a syndrome-level decision rule whose expected loss is no larger than that of any individual base classifier used in isolation, or any other deterministic hybrid classifier whose response is a function only of syndromes. The hybrid classifier whose responses satisfy (3) is called the *Bayes-optimal* classifier. The number of comparisons required to solve for all of the optimal responses is exponential in the number of classifiers, i.e., at most 2^k comparisons for k classifiers. This uses a comparison table with 2^{k+1} entries – i.e., 2^k rows with two entries, q_i and γp_i, per row.

4 Methodology

This section details the experimental setup and statistical tools used to examine the impact of these construction challenges on classifier reliability.

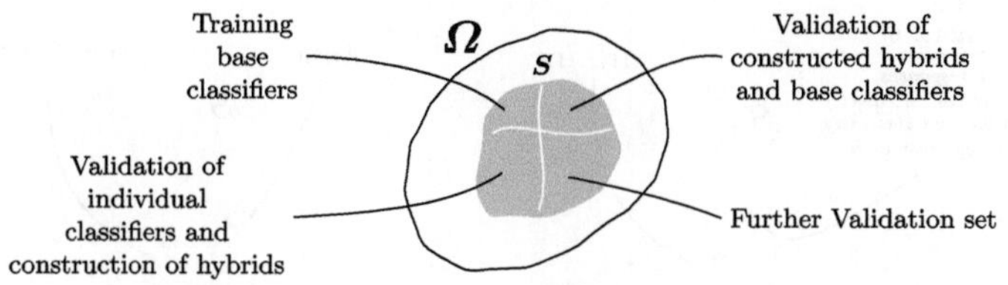

Fig. 2. Sample subset $\boldsymbol{S}$ from the set of all possible classification tasks $\boldsymbol{\Omega}$.

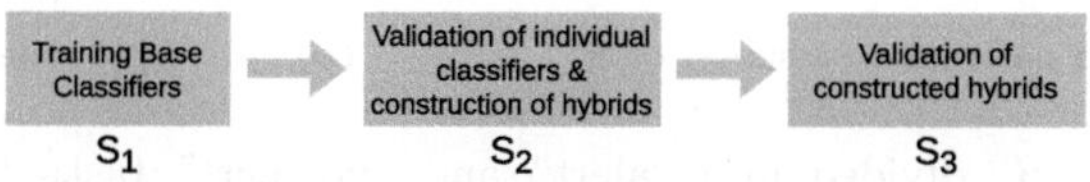

Fig. 3. Training and validation pipeline for the subsets of $\boldsymbol{\Omega}$ making up $\boldsymbol{S}$.

4.1 Experimental Setup

To assess classifier risk, consider a subset $\boldsymbol{S}$ of tasks from $\boldsymbol{\Omega}$. The experimental pipeline for this paper uses subsets of $\boldsymbol{S}$ (see Fig. 2) as follows (see Fig. 3): **i)** k classifiers are trained on a subset of $\boldsymbol{S}$; **ii)** the classifiers are validated on a second subset of $\boldsymbol{S}$. The risk-related statistics from this validation sample are used to configure various hybrid classifiers (using (3)); **iii)** the estimated risk from the hybrids and base classifiers are compared on a third and fourth subset of $\boldsymbol{S}$. The p_i and q_i probabilities in (3) characterise the random occurrence (across all of the $\boldsymbol{\Omega}$ subsets in Fig. 2) of ground-truth "no alert" and "alert" tasks that trigger syndrome i. In the rest of this section, we denote the subsets of $\boldsymbol{S}$ as $\boldsymbol{S}_1, \ldots, \boldsymbol{S}_4$

There are two datasets: Pneumonia [21] and Concrete [22]. Since we partition the dataset into custom subsets as later explained, we use the default training splits of each dataset. The Pneumonia dataset contains 5,239 X-ray images of paediatric patients from China. The Concrete dataset contains 40,000 images of concrete segments from Turkey. Both Pneumonia and Concrete are binary safety-critical ML datasets in the domains of medical image diagnosis (i.e., Pneumonia vs Benign) and concrete defect detection (i.e., Crack vs No Crack) respectively.

Each dataset has its own separate experimental pipeline for $\boldsymbol{S}$ and the construction of its subsets – no data is shared between datasets. For training, 61% of the images are used. The remaining classification tasks are divided into three subsets, each comprising of 13% of their datasets, for validating the base classifiers, constructing hybrids and comparing the risk from these classifiers respectively. $\boldsymbol{S}$ is split into four subsets, indexed $j = 1, ..., 4$.

In total, $k = 7$ base binary classifiers are trained from the ConvNet, DeconvNet, MobileNet, ResNet18, ResNet50, VGG11 and VGG16 architectures. Each binary classifier is independently trained, following the experimental pipeline for $\boldsymbol{S}$. Each classification determines whether an image corresponds to one of two binary classes in Pneumonia and Concrete.

Figure 4a illustrates how the 2^k syndromes divide $\boldsymbol{S}_j$ into (at most) 2^k further subsets — one further subset for each syndrome triggered by a task in $\boldsymbol{S}_j$.

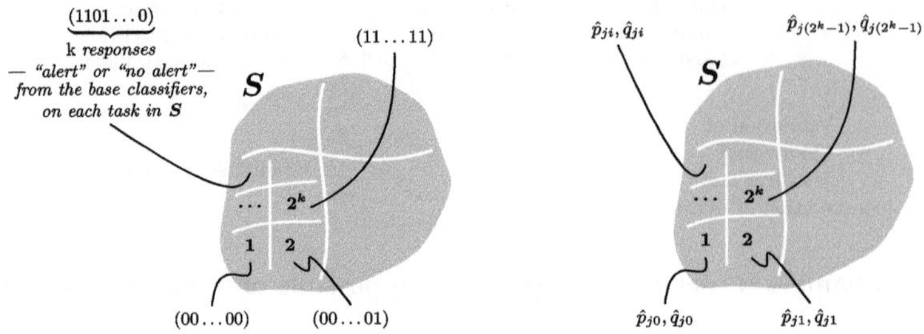

(a) Induced partition of a subset of $\boldsymbol{S}$ **(b)** "alert"/"no alert" tasks

Fig. 4. Subsets of $\boldsymbol{S}_j$ divided into "alert" and "no alert" tasks that trigger each observed syndrome. Estimates are normalised w.r.t. $\boldsymbol{S}_j$, $\sum_{i=0}^{2^k-1}(\hat{p}_{ji} + \hat{q}_{ji}) = 1$.

The responses of the base classifiers on the validation set are used to construct syndromes which form the basis of constructing hybrid classifiers. For each syndrome i triggered by a task from $\boldsymbol{S}_2$, the probabilities p_i and q_i are estimated as the proportion of the validation subsets that trigger the syndrome and have ground-truth classifications "no alert" (denoted $\hat{p}_{2i}$) and "alert" (denoted $\hat{q}_{2i}$), respectively. Figure 4b depicts the same idea as Fig. 4a, but in terms of these proportions. Using these proportions in optimisation solution (3), 7 hybrid classifiers are constructed, each differing by how the response is chosen for syndromes that are *not* seen when constructing the hybrid:

1. hybrid 0: the response is 0;
2. hybrid 1: the response is 1;
3. hybrid 2: a random response of either 0 or 1;
4. hybrid 3: the response is the most frequent response to syndromes that have been seen during construction;
5. hybrid 4: the response is the least frequent response to syndromes that have been seen during construction;
6. hybrid 5: the response is the majority response in the unseen syndrome;
7. hybrid 6: the response is the minority response in the unseen syndrome;

In addition, 5 hybrids are constructed from the following class of "smoothed" hybrids. For each syndrome i, if c_{0i} and c_{1i} are the counts of ground-truth "no alert" and "alert" tasks in $\boldsymbol{S}_2$, then a smoothed version of (3) is

$$\text{response} = \begin{cases} \text{"no alert"}, \text{ if } c_{1i} + \alpha_1 \leqslant \gamma(c_{0i} + \alpha_0) \\ \\ \text{"alert"}, \quad \text{ if } c_{1i} + \alpha_1 > \gamma(c_{0i} + \alpha_0) \end{cases} \tag{4}$$

for chosen $\alpha_0, \alpha_1 > 0$. The smoothed decision–rule (4) is equivalent to using a Beta prior distribution (with parameters (α_1, α_0)) to model not knowing what the conditional probability of an alert task is (conditional on each syndrome), and using the posterior mean from $\boldsymbol{S}_2$ to define the decision–rule.

The smoothed hybrids determine responses on unseen syndromes so that, as $\gamma \to \infty$ (resp. $\gamma \to 0$), the chosen responses match the limiting behaviour of the Bayes rule under the assumptions $p_i > 0$ for all syndromes (resp. $q_i > 0$), yielding "never alert" (resp. "always alert"). The other "non-smoothed" hybrids do not satisfy this asymptotic property. The 5 smoothed hybrids differ in the choice of (α_0, α_1): $(0.001, 0.001)$, $(0.1, 0.1)$, $(0.1, 1.0)$, $(1.0, 0.1)$, and $(1.0, 1.0)$.

In total, 7 base and 12 hybrid classifiers are validated and compared.

4.2 Statistical Assessment of Classifiers

Using $\boldsymbol{S}_3$, the empirical losses from all classifiers are compared, and confidence bounds are estimated for the sub-optimality of each classifier (i.e., estimates of the excess expected loss over that produced by the theoretical Bayes–optimal classifier, and estimates of how likely these excesses are).

Expected Loss. The following formalises classifier risk as expected loss; the empirical losses from our experiments estimate this loss. Consider a random vector $\mathbf{X}$ that is a one-hot encoding representing the ground-truth classification and syndrome triggered by a randomly chosen task from $\boldsymbol{\Omega}$:

1. $\mathbf{X}$ is a binary column vector of length 2^{k+1}, where only one non-zero component in the vector indicates the task is chosen from the subset;
2. a value of 1 for the "$2i$+1"-th component of $\mathbf{X}$ represents the occurrence of a ground-truth "no alert" task from $\boldsymbol{\Omega}$ that triggers syndrome i, while a value of 1 in the "$2i$+2"-th position represents an "alert" task occurring instead.

Hence, by the definitions of p_i, q_i, we have

$$\mathbb{E}[\mathbf{X}]^T = \langle p_0, q_0, ..., p_{(2^k-1)}, q_{(2^k-1)} \rangle \tag{5}$$

If a classifier issues "alert" for tasks that trigger the i-th syndrome, the only possible errors on these tasks are FPs – with associated loss γ. And, if "no alert" is issued, the only errors on these tasks are FNs – with associated loss 1. Let $\{r_i\}_{0 \leqslant i \leqslant (2^k-1)}$ be the responses of a classifier (either 1 or 0) on each syndrome. And, let the response vector $\mathbf{C}$ be the column vector of length 2^{k+1}, defined as

$$\mathbf{C}^T := \langle \gamma r_0, 1 - r_0, \ldots, \gamma r_{(2^k-1)}, 1 - r_{(2^k-1)} \rangle$$

Then, the loss L experienced by a classifier with response vector $\mathbf{C}$ on a random task is $\mathbf{C}^T\mathbf{X}$ and the expected loss for the classifier is

$$\mathbb{E}[L] = \mathbb{E}[\mathbf{C}^T\mathbf{X}] = \mathbf{C}^T\mathbb{E}[\mathbf{X}] = \mathbb{E}[\mathbf{X}]^T\mathbf{C}. \tag{6}$$

This measure of risk generalises expected loss (1) for base classifiers.

In our experiments, expected loss (6) is estimated by an empirical loss, derived as follows using validation set $\boldsymbol{S}_3$ of size N. Let $\{\mathbf{X}_n\}_{1 \leqslant n \leqslant N}$ be a sequence of N independent, identically distributed (i.i.d.) tasks that constitute $\boldsymbol{S}_3$, where all of the $\mathbf{X}_n$ have the same distribution as $\mathbf{X}$. Then, empirical loss is

$$\widehat{L} = \frac{1}{N}\sum_{n=1}^{N} \mathbf{C}^T\mathbf{X}_n = \mathbf{C}^T(\sum_{n=1}^{N}\frac{1}{N}\mathbf{X}_n) = \mathbf{C}^T\widehat{\mathbb{E}}_{\boldsymbol{S}_3}[\mathbf{X}]\,, \tag{7}$$

where

$$\widehat{\mathbb{E}}_{\boldsymbol{S}_3}[\mathbf{X}]^T := \langle \hat{p}_{30}, \hat{q}_{30}, ..., \hat{p}_{3(2^k-1)}, \hat{q}_{3(2^k-1)}\rangle \tag{8}$$

and the proportions $\hat{p}_{3i}, \hat{q}_{3i}$ are used as plug-in estimates for unknown probabilities p_i, q_i. Using (8) in (7), empirical loss for (base or hybrid) classifiers is

$$\widehat{L} = \sum_{i=0}^{2^k-1} r_i\gamma\hat{p}_{3i} + (1-r_i)\hat{q}_{3i}\,, \tag{9}$$

where r_i is the response of the classifier on a task that generates syndrome i.

Denoting the Bayes–optimal classifier risk as μ^*, an estimate of the gap between μ^* and a classifier's risk is the expected bias $\boldsymbol{B} := \mathbb{E}[\widehat{L} - \mu^*] = \mathbb{E}[\widehat{L}] - \mu^* = \mathbb{E}\mathbf{C}^T\mathbb{E}\mathbf{X} - \mu^*$. The last equality holds because the training, construction, and validation tasks are i.i.d.

The variance $\mathbb{V}(\widehat{L})$, used in the following analysis, is defined as:

$$\mathbb{V}(\widehat{L}) = \frac{1}{N}\mathbb{E}[\mathbf{C}^T\mathbb{V}\mathbf{X}\mathbf{C}] \;+\; \mathbb{E}\mathbf{X}^T\mathbb{V}\mathbf{C}\mathbb{E}\mathbf{X} \tag{10}$$

where $\mathbb{V}\mathbf{X}$ and $\mathbb{V}\mathbf{C}$ are the respective *covariance matrices* for $\mathbf{X}$ and $\mathbf{C}$.

Confidence Bounds on Sub-optimality. The following are confidence bounds on the gap in expected losses, between classifiers and the Bayes–optimal classifier. Throughout, K is the training/construction set size, N is the validation set size, ϵ is the deviation from the mean, while γ is the relative cost as previously defined in Sect. 3. For small $\epsilon > 0$, we bound $\mathbb{P}(|\widehat{L} - \mu^\star| \geqslant \epsilon)$ in four ways.

Inequality 1. ***Markov Absolute-Deviation (AD) Bound:***

$$\mathbb{P}(|\widehat{L} - \mu^*| \geqslant \epsilon) \leqslant \frac{1}{\epsilon}\left(\sqrt{\mathbb{V}(\widehat{L})} + |\boldsymbol{B}|\right) \tag{11}$$

Hence, for hybrid classifiers that use (3) *for their construction on observed syndromes,* $\widehat{L} \to \mu^*$ *in probability, as* $K, N \to \infty$.

Inequality 2. ***Chebyshev Mean Squared-Error (MSE) Bound:***

$$\mathbb{P}(|\widehat{L} - \mu^*| \geqslant \epsilon) \leqslant \frac{1}{\epsilon^2}\left(\mathbb{V}(\widehat{L}) + \boldsymbol{B}^2\right) \tag{12}$$

Hence, for hybrid classifiers that use (3) *for their construction on observed syndromes,* $\widehat{L} \to \mu^*$ *in probability, as* $K, N \to \infty$.

Inequality 3. ***Chebyshev Mean-Centered (MC) Bound:***

$$\mathbb{P}(|\widehat{L} - \mu^*| \geqslant \epsilon) \leqslant \frac{\mathbb{V}(\widehat{L})}{(\epsilon - \boldsymbol{B})^2} \qquad (\text{for } \epsilon > B) \tag{13}$$

Hence, for hybrid classifiers that use (3) *for their construction on observed syndromes,* $\widehat{L} \to \mu^*$ *in probability, as* $K, N \to \infty$.

Inequality 4. ***Hoeffding-type Bound:***

$$\mathbb{P}(|\widehat{L} - \mu^*| \geqslant \epsilon) \leqslant \psi(b_*) \tag{14}$$

where

$$\psi(b) := 2 \exp\left(-2\frac{N(\epsilon - b)^2}{(\max\{\gamma, 1\})^2}\right) + \mathbb{P}(|\mathbf{C}^T\mathbb{E}[\mathbf{X}] - \mu^*| > b)$$

for any $b \in (0, \epsilon)$ *and* $b_* = \arg\min_{b \in [0,\epsilon]} \psi(b)$.

When computing bounds, μ^* acts as a substitute for the unknown Bayes-optimal risk. For each γ, we use the smallest empirical loss among the classifiers.

5 Results

We report the empirical losses and sub-optimality bounds obtained by applying the methodology of Sect. 4. We examine classifier risks under typical relative costs (Sect. 5.1), then under extreme relative costs (Sect. 5.2), and finally identify the tightest confidence bounds on sub-optimality (Sect. 5.3).

5.1 What are the Classifier Risks Under Typical Relative Costs?

We compute the empirical losses from the classifiers from $\boldsymbol{S}_3$ and compare them in Fig. 5. We find that the 7 base classifiers tend to give the largest empirical losses for most values of the relative cost γ, followed by the empirical losses from non-smoothed hybrids, then the smoothed hybrids. Precise orderings of these empirical losses change as γ changes; the empirical loss curves cross at certain values of γ. For example, in Fig. 5b, MobileNet in Concrete performs better than ResNet50 and VGG16 when γ is small (less than 10^{-1}) but gives greater losses when γ becomes large (greater than 10). To a lesser extent, "crossings" are also observed among base classifiers and hybrids in Pneumonia (Fig. 5a).

In general, the order of the curves in Fig. 5 is not preserved as γ increases; within each figure, curves cross at various points. Two effects explain many of these crossings: **i)** the relative cost of FPs to FNs increasing in the plots (i.e., γ increasing) and **ii)** differences, between $\boldsymbol{S}_2$ and $\boldsymbol{S}_3$, in the estimated value at which the expected frequency of FPs becomes more costly than the frequency of FNs. The first effect—increasing γ—means that base classifiers that alert

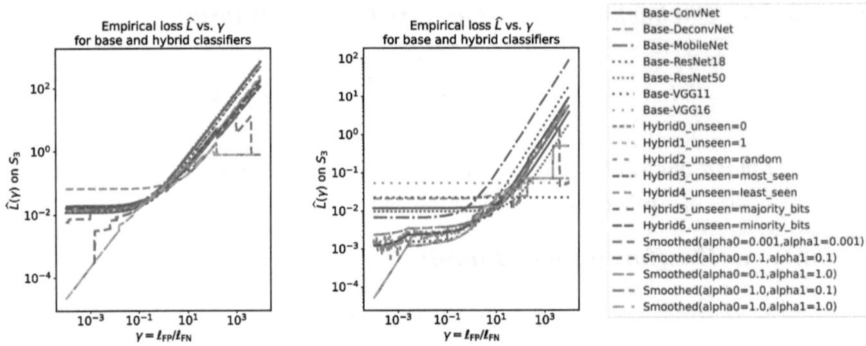

(a) Pneumonia (b) Concrete

Fig. 5. Empirical loss $\widehat{L}$ vs. relative FP-to-FN cost γ for classifiers from the Pneumonia and Concrete datasets—each dataset was divided across four subsets $\boldsymbol{S}_1$ to $\boldsymbol{S}_4$.

often will initially benefit from FPs not being too costly when γ is initially small. However, as γ increases and FPs become too costly, those base classifiers eventually become more risky than other base classifiers that are less likely to issue alerts. The second effect—uncertainty in determining when FPs become too costly as γ increases—explains the sharp "drops" and "rises" that sometimes cause a hybrid curve to cross other curves. These occur whenever there is a mismatch between the estimated "switch over" points for seen syndromes where raising an alert for the syndrome becomes too costly (see the heights of the pair of costs for each syndrome in Fig. 1). For syndrome i, let γ^*_{ji} denote the value of γ at which the estimated conditional expected losses for "alert" and "no alert" are equal, based on the proportions observed in subset $\boldsymbol{S}_j$. On the one hand, as dictated by $\boldsymbol{S}_2$, this "switch over" occurs at $\gamma^*_{2i}\widehat{p}_{2i} = \widehat{q}_{2i}$ while, on the other hand, the proportions from $\boldsymbol{S}_3$ estimate it at $\gamma^*_{3i}\widehat{p}_{3i} = \widehat{q}_{3i}$. Thus, if $\widehat{q}_{2i}/\widehat{p}_{2i} > \widehat{q}_{3i}/\widehat{p}_{3i}$, a hybrid curve based on (3) or (4) eventually exhibits a sharp "drop" as γ increases, otherwise it exhibits a sharp "rise".

5.2 What are the Classifier Risks Under Extreme Relative Costs?

We examine the risk of classifiers at extreme values of γ. For applications where the cost of FNs are more significant, low values of γ are of greater interest. Where the cost of FPs are more significant, high values of γ are of greater interest.

For extreme γ, smoothed hybrids give the smallest empirical losses among these classifiers, exemplified by the smoothed hybrid ($\alpha_0 = 1, \alpha_1 = 1$) curves in Fig. 5. Away from these extremes there is a range of γ, roughly $10^{-1} \leqslant \gamma \leqslant 10$, over which the classifier with the smallest empirical loss could be a base or hybrid; there is no "always best performing" classifier among these classifiers over this limited range of γ. For Pneumonia, at $\gamma = 1$, in Fig. 5a, the lowest empirical loss is 5.9×10^{-2} from hybrid 0 (where the response to "unseen during construction" syndromes is "no alert") and hybrid 3 (where the response

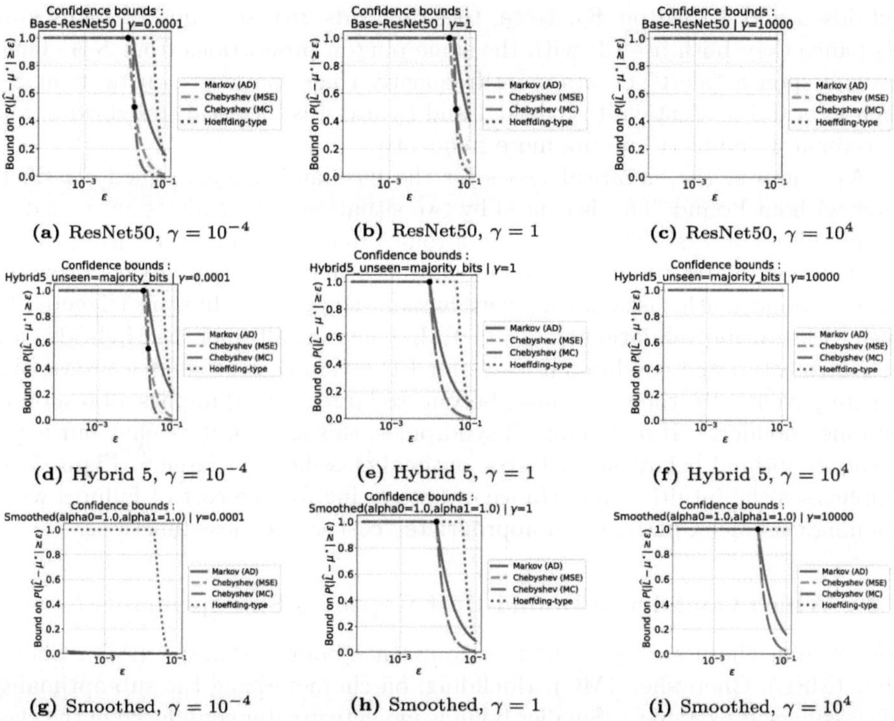

(a) ResNet50, $\gamma = 10^{-4}$ **(b)** ResNet50, $\gamma = 1$ **(c)** ResNet50, $\gamma = 10^4$

(d) Hybrid 5, $\gamma = 10^{-4}$ **(e)** Hybrid 5, $\gamma = 1$ **(f)** Hybrid 5, $\gamma = 10^4$

(g) Smoothed, $\gamma = 10^{-4}$ **(h)** Smoothed, $\gamma = 1$ **(i)** Smoothed, $\gamma = 10^4$

Fig. 6. Estimated confidence-bound curves on `Pneumonia` comparing `ResNet50`, `Hybrid 5`, and `Smoothed` (α_0=1.0, α_1=1.0) across $\varepsilon \in [10^{-4}, 10^{-1}]$ at three γ values.

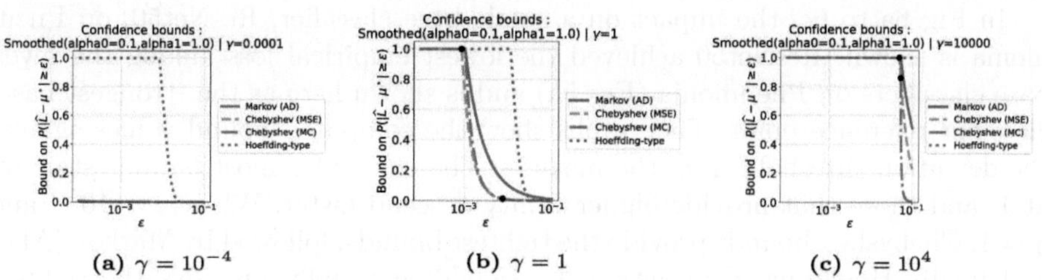

(a) $\gamma = 10^{-4}$ **(b)** $\gamma = 1$ **(c)** $\gamma = 10^4$

Fig. 7. Estimated confidence-bound curves for `Smoothed` (α_0=0.1, α_1=1.0) on `Concrete` across $\varepsilon \in [10^{-4}, 10^{-1}]$ at three γ values.

to "unseen during construction" syndromes is the most frequent response to syndromes "seen during construction").

The fact that hybrid's 0 and 3 have the same risk in the last example is not coincidence — in this instance, these hybrids *are* identical. In general, there is redundancy that occurs at random among certain hybrids. Precisely, hybrids 0 and 1 must equal either hybrid 3 or hybrid 4, with hybrids 0 and 1 necessarily being different. Which of these hybrids are redundant depends on the responses hybrids 0 and 1 give on $\boldsymbol{S}_2$. Note, these hybrids give *the same* responses on $\boldsymbol{S}_2$ (since they both use (3) with the same plug-in proportions from $\boldsymbol{S}_2$). Thus, if the response "alert" occurs more frequently than "no alert" on tasks in $\boldsymbol{S}_2$, then hybrid 3 is identical to hybrid 1 and hybrid 4 is identical to hybrid 0, and vice-versa if "no alert" occurs more frequently.

As γ grows, the empirical losses for the base and non-smoothed classifiers grow without bound. This is caused by two situations where alerts are issued: **i)** issuing alerts on tasks that trigger syndromes that were not seen during hybrid construction, or **ii)** issuing alerts on syndromes that are never triggered by "no alert" ground-truth tasks during construction (so $\widehat{p}_{2i} = 0$), but are triggered by such tasks during validation (so $\widehat{p}_{3i} > 0$). In contrast, all smoothed hybrids have non–zero estimated conditional losses for FP errors that grow as γ grows; this ensures, eventually, these estimates become so large that (4) implies all responses become "no alert". If $p_i > 0$ for all syndromes, this asymptotic behaviour is the eventual optimal behaviour of Bayes–optimal classifiers for large γ. Figure 5 re-emphasises the intuitive importance of accounting for the cost of failures when training classifiers, perhaps via appropriate "cost-aware" loss-functions.

5.3 Which Confidence Bound Best Captures Sub-optimality?

We examine the usability of the four confidence bounds [Markov (AD), Chebyshev (MSE), Chebyshev (MC), Hoeffding] on characterizing the sub-optimality of classifiers at extreme γ. Smaller bounds indicate greater confidence in the classifier being close to optimal. We analyse the bounds for the configurations with the lowest losses in each classifier category in Fig. 5: base classifiers, unsmoothed hybrids, and smoothed hybrids.

In Fig. 6a to 6c, the impact on a single base classifier, ResNet50, on Pneumonia is shown. ResNet50 achieved the lowest empirical loss among the seven base classifiers on Pneumonia (Fig. 5a) and is shown here as the strongest base-classifier reference point. The graphs show the computed bound values against the deviation threshold from the mean, ϵ. The curves for most bounds start off at 1, and those that provide higher utility descend faster. When $\gamma = 10^{-4}$ and $\gamma = 1$, Chebyshev bounds provide the tightest bounds, followed by Markov (AD), and finally Hoeffding. Conventionally, Hoeffding usually provides the tightest bounds for large sample sizes [23]. However, since datasets in safety-critical domains are smaller than that of general purpose ML, Chebyshev bounds may provide more utility over the other bounds. When $\gamma = 10^4$, all of the bounds equally collapse, meaning they are all highly sensitive to FPs. The propensity of the bounds to better capture high FN cases compared to high FP may be acceptable for safety-critical applications such as Pneumonia and Concrete where the cost of non-detection (potential loss of life) is much higher compared to false alarms (increased financial cost).

Hybrid 5 and the smoothed ($\alpha_0 = 1, \alpha_1 = 1$) were the best unsmoothed and smoothed hybrids for Pneumonia. Chebyshev bounds remain the most useful for hybrid classifiers. While the general pattern previously observed on ResNet50 also holds for (unsmoothed) hybrid 5 in Fig. 6d to 6f, smoothed hybrid ($\alpha_0 =$

$1, \alpha_1 = 1)$ in Fig. 6g to 6i show differences at extreme γ. When $\gamma = 10^{-4}$ in Fig. 6g, all bounds except for Hoeffding are extremely tight - this shows that smoothed hybrids can effectively reduce the loss from FNs. When $\gamma = 10^4$ in Fig. 6i, all bounds except Hoeffding still offer usefulness, avoiding excessive losses from FPs. Overall, smoothed hybrids improves robustness to both FNs and FPs. A similar pattern appears for Concrete in Fig. 7, where the best smoothed hybrid for Concrete ($\alpha_0 = 0.1, \alpha_1 = 1$), is also most robust across γ.

6 Conclusion, Limitations, and Future Work

Safety-critical systems require classifiers that minimize the risk associated with failures. While a *Bayes–optimal* classifier achieves the theoretical minimum, they are challenging to demonstrably construct. Within a statistical framework, this paper examines the construction of hybrids from multiple diverse base classifiers and demonstrates how smoothed hybrids best approximate Bayes-optimality across two safety-critical applications. The techniques employed are useful for supporting safety claims that bound the risks from classification errors.

While our bounds are derived at the classifier level, the classifiers represent only one component of broader system safety. Real-world systems comprise many elements beyond the ML model, including sensors, actuators, and human operators. Some classifier failures may be masked, while others may propagate and contribute to system-level hazards. Classifier-level bounds can therefore serve as evidence of residual safety risks associated with perception modules, in system-level standards such as IEC 61508 [3] and ISO 21448/SOTIF [24]. Bounding hazardous misclassifications can support quantitative safety assessment and fault tree analysis. Integrating classifier-level assurances into compliance arguments for system-level standards remains an important direction for future work.

Base/hybrid classifier outputs are treated as binary "alert"/"no alert", and base classifiers are modelled as deterministic functions of classification tasks. Additionally, datasets are assumed to be correctly labelled. Our analysis concerns the risk contribution of a classifier's generalization error, rather than adversarial or robustness failures, which may require separate assessment and modelling. Future work can include multiclass classifiers and the impact of uncertainty (i.e., label noise, distribution shifts) in a dataset's ground-truth labels.

The number of syndromes is exponential in the number of classifiers, which can impose significant training and runtime overhead when using (3) or (4), especially compared to individual classifiers. However, hybrids can be deployed in parallel to limit runtime overhead. For hybrids constructed using seen syndromes (hybrids 0–4), overhead is mainly limited to training.

Useful confidence bounds for small ϵ differences and extreme γ values might require large training and validation sets, which may be difficult to obtain.

We employ classical statistical methods; an advantage of this is their simplicity makes the important factors in our analysis clear. The trade–off is that these methods ignore effects such as distribution-shift or adversarial correlations across classification tasks.

While the Bayes–optimal classifier has the lowest expected loss, it does not necessarily have the lowest empirical loss. When computing the bounds, our estimate of the optimal expected loss introduces some statistical error, because the true Bayes–optimal expected loss μ^* is not guaranteed to be smaller than the empirical losses on a randomly chosen validation set. In future work, more accurate bounds could use estimates of a useful upper bound on μ^* computed from an independent data set (i.e., not from $\boldsymbol{S}_3$). Cross-validation is an alternative to estimating these bounds, albeit more computationally demanding.

Acknowledgements. We would like to thank the anonymous SAFECOMP reviewers for their feedback. This work was supported by a Discovery Grant and an Alliance Grant from the Natural Sciences and Engineering Research Council of Canada (NSERC), and a Four Year Fellowship from UBC.

References

1. Rasheed, K., et al.: Explainable, trustworthy, and ethical machine learning for healthcare: a survey. Comput. Biol. Med. (2022)
2. Li, K., et al.: Automatic crack detection in civil infrastructure based on a hybrid fine tuned MnasNet and adaptive patching. Sci. Rep. (2025)
3. IEC 61508: Functional safety of electrical/electronic/programmable electronic safety-related systems (2010)
4. ISO/TR 20417:2021: Medical device software — Guidance on the use of artificial intelligence (2021)
5. Blough, D., Sullivan, G.: A comparison of voting strategies for fault-tolerant distributed systems. In: Proceedings of SRDS (1990)
6. Di Giandomenico, F., Strigini, L.: Adjudicators for diverse-redundant components. In: Proceedings of SRDS (1990)
7. Wald, A.: Statistical decision functions (1950)
8. Li, B., Zhang, R.: Long-tailed classification from a bayesian-decision-theory perspective. In: Symposium on Advances in Approximate Bayesian Inference (2023)
9. Ichijo, N., et al.: Bayesian decision-theoretic prediction with ensemble of meta-trees for classification problems. In: Proceedings of ISITA (2024)
10. Dalton, L.A., Yousefi, M.R.: On optimal bayesian classification and risk estimation under multiple classes. EURASIP J. Bioinform. Syst. Biol. (2015)
11. Naeini, M.T., et al.: Universal training of neural networks to achieve bayes optimal classification accuracy. In: Proceedings of ICASSP (2025)
12. Stone, M.: Cross-validatory choice and assessment of statistical predictions. J. Royal Stat. Soc. (1974)
13. Geisser, S.: The predictive sample reuse method with applications. J. Am. Stat. Assoc. (1975)
14. Swets, J.A.: The relative operating characteristic in psychology. Science (1973)
15. Swets, J.A.: Measuring the accuracy of diagnostic systems. Science (1988)
16. Fawcett, T.: An introduction to ROC analysis. Pattern Recogn. Lett. (2006)
17. Salako, K.: Loss-size and reliability trade-offs amongst diverse redundant binary classifiers. In: Proceedings of QEST (2020)
18. Christodorescu, M., et al.: Semantics-aware malware detection. In: Proceedings of S&P (2005)

19. Schrittwieser, S., et al.: Protecting software through obfuscation: can it keep pace with progress in code analysis? ACM Comput. Surv. (2016)
20. Goodfellow, I.J., Bengio, Y., Courville, A.: Deep Learning (2016)
21. Kermany, D., et al.: Labeled optical coherence tomography (OCT) and chest x-ray images for classification. Mendeley Data, v2 (2018)
22. Özgenel, C.F., et al.: Performance comparison of pretrained convolutional neural networks on crack detection in buildings. In: Proceedings of ISARC (2018)
23. Hoeffding, W.: Probability inequalities for sums of bounded random variables. J. Am. Stat. Assoc. **58**(301), 13–30 (1963)
24. ISO 21448:2022: Road Vehicles – Safety of the intended functionality (2022)

Real-Time Aspects

MISRust: Mapping MISRA-C++ Coding Guidelines to the Rust Programming Language

Marius Molz[1(✉)], Niels Schneider[1], Sven Lechner[2], Stefan Kowalewski[1], and Alexandru Kampmann[1]

[1] Chair of Embedded Software, RWTH Aachen University, Aachen, Germany
{molz,schneider,kowalewski,kampmann}@embedded.rwth-aachen.de
[2] RWTH Aachen University, Aachen, Germany
sven.lechner@rwth-aachen.de

Abstract. The Rust programming language is increasingly being considered for safety-critical system development. However, established safety standards such as ISO 26262 require the use of coding guidelines that do not yet exist for Rust. This paper systematically examines each of the 179 MISRA C++ 2023 coding guidelines and classifies them into 6 categories based on their applicability to Rust. Our approach analyzes the rationale behind each MISRA rule to determine whether it remains valid in the Rust programming context. We find that 47.75% of the 111 as-is applicable MISRA rules are automatically enforced by Rust's language design, eliminating the need for explicit guideline enforcement. Furthermore, our analysis explicitly distinguishes between `safe` and `unsafe` Rust. We find that 69 guidelines are still relevant and still require either direct application or adaptation for Rust. Importantly, 36 of these rules are automatically satisfied when only using the `safe` subset of the Rust language. However, they are required again if `unsafe` Rust features are introduced. We also identify specific areas where new Rust-specific guidelines are needed. Where a guideline does not directly translate, we propose Rust-specific adaptations that preserve its intent.

Keywords: Rust programming language · Coding guidelines · Safety-critical systems · MISRA C++ Guidelines · Unsafe code · Automotive software engineering · Software certification

1 Introduction

Rust safeguards against entire classes of programming errors through its ownership system, type safety, and memory management model, such as use-after-free, data races, null pointer dereferences, buffer overflows, and uninitialized memory access. Hence, there is growing interest in using Rust for the development of safety-critical systems. For example, the Ferrocene project has achieved ISO

© The Author(s), under exclusive license to Springer Nature Switzerland AG 2027
E. Troubitsyna et al. (Eds.): SAFECOMP 2026, LNCS 16868, pp. 315–329, 2027.
https://doi.org/10.1007/978-3-032-34867-8_20

26262 tool qualification for a Rust compiler, while automotive initiatives such as Eclipse SDV are incorporating Rust components. [1,6,14]

Multiple safety standards across domains mandate the use of coding guidelines during software development, including ISO 26262 for automotive, DO-178C for aerospace, IEC 62304 for medical devices, and EN 50128 for railway systems. These guidelines limit the usage of specific language features and enforce specific program structures to address typical bugs that occur using a particular language. Various safety-related coding guidelines exist for C/C++ systems, including MISRA C/C++, AUTOSAR C++ Guidelines, or JSF++ for aerospace. However, no equivalent coding guidelines exist for Rust, despite growing interest in safety-critical domains. To date, the MISRA Consortium has only published an addendum to MISRA C:2025 that addresses the applicability of MISRA C:2025 to the Rust programming language, rather than providing a dedicated and comprehensive set of Rust-specific guidelines [5,6].

The lack of established safety-related coding guidelines is one of the obstacles to achieving compliance with standards when using Rust to develop safety-critical systems. We attempt to address this gap by investigating the transferability of the 179 existing MISRA C++ coding guidelines to Rust. We focus on MISRA C++ because it is widely adopted as a coding standard for automotive safety-critical systems and is explicitly recognized in ISO 26262. In particular, our mapping may aid developers in safety-critical domains migrating from C++ to Rust, as well as those writing new Rust code for safety-critical applications.

Rust distinguishes two language modes: in the default `safe` Rust language subset, the compiler statically enforces its ownership model and borrow checker, preventing undefined behavior at compile time. In `unsafe` Rust, activated by the `unsafe` keyword, some of these guarantees are suspended, allowing operations such as raw pointer dereferencing or foreign function calls, for which manual verification by the programmer becomes necessary again.

Our methodology is based on studying the underlying *rationale* of each MISRA rule that motivates its existence. Using the rationale, we investigate whether the rule still holds in Rust. We propose a mapping framework comprising 6 categories to express the degree to which a particular rule is applicable in the Rust context. Then, for all 179 MISRA rules, we propose a mapping to the 6 categories, ranging from *rule still required in* (`safe`) *Rust to rule not required, even in* `unsafe` *Rust.* Using this methodology, we pursue the following research questions:

- **Which MISRA C++ guidelines remain applicable to Rust?** We provide the first comprehensive mapping of all 179 MISRA C++ 2023 guidelines, identifying 69 guidelines still relevant to Rust. Of them, 58 are directly applicable, while the remaining 11 rules require some form of adaptation for Rust.
- **What rules can be dropped based on Rust's inherent guarantees?** We demonstrate that 47.75% of applicable MISRA rules are automatically satisfied by Rust's language design.
- **How does `unsafe` Rust affect guideline compliance?** We identify that 36 guidelines automatically satisfied in `safe` Rust become unsatisfied in `unsafe`

blocks, providing critical guidance for safety certification of Rust code containing `unsafe` sections.

The remainder of this paper is structured as follows. Section 2 reviews related work. Section 3 presents our methodology and classification framework. Section 4 details our mapping results and analysis. Section 5 discusses implications for safety certification, and Sect. 6 concludes.

2 Related Work

The increasing interest in Rust for safety-critical systems development has already led to a few exploratory studies, partial ruleset investigations, and early certification efforts. However, the landscape of coding guidelines regarding Rust in critical domains remains fragmented, with no complete systematic mapping between established safety standards and Rust's unique language design.

In 2019, Pinho et al. [8] provided one of the first works on the applicability of already established safety-focused coding guidelines to Rust. They motivate adopting Rust in safety-critical domains by emphasizing its ownership model, borrow checker, and lifetime system. As part of this investigation, a subset of the MISRA-C guidelines is examined for their relevance to Rust. The study provides analysis results on 9 selected MISRA-C rules. For each rule, the authors explain why it may be simplified when translated to Rust. However, the analysis remains limited in scope and does not attempt to generalize the applicability of rules across the entire rule set [8].

More recently, the MISRA Consortium explicitly addressed the use of Rust in safety-critical contexts. The MISRA C:2025 Addendum 6 "Applicability of MISRA C:2025 to the Rust Programming Language" [5] provides a high-level overview of how MISRA-C guidelines may translate to Rust. Applicability is evaluated separately for both `safe` and `unsafe` Rust code. The addendum also addresses the application of the MISRA C:2025 rule to Rust's C Foreign Function Interface (FFI), treating it as part of the `unsafe` category. Nevertheless, as it is only an addendum to the full MISRA-C ruleset, it does not constitute a detailed, rule-by-rule technical mapping, as, e.g., Rust code examples do, nor does it explain the rationale behind the classification decisions [5].

We map the MISRA C++:2023 rules to Rust instead of the MISRA C:2025 rules. The C++ ruleset takes features into account that are not available in C, such as classes and objects (structs and impl blocks), virtual function dispatch (traits and the `dyn` keyword), exceptions (`Result` and `panic!`), templates (generics and macros), and more, of which some features exist in similar forms within Rust. The addendum classifies the rationale of each MISRA C rule into four categories, namely *Undefined Behaviour*, *Implementation-Defined Behaviour*, *Code Quality*, and *Developer Confusion*, and provides a *yes*/*no*/*partial* applicability assessment for each rule in the context of Rust [5].

Beyond partial rule mappings, Munch et al. [6] investigated Rust based on the published MISRA-C Addendum. They analyzed how the Rust programming language differs from C/C++, established safety principles, and identified both enabling factors and open issues. A central challenge highlighted is dynamic memory allocation. The presence of memory allocation, both heap and stack, raises certification concerns, particularly in systems that require predictability [6].

In parallel with academic work, industry has begun developing certification-oriented Rust toolchains. Proprietary solutions, such as those from Ferrocene or HighTec, aim to qualify Rust compilers and tools for use in safety-critical system development. However, while Vector and HighTec claim to "[...] have removed the last hurdles for the use of Rust in the automotive sector" [16], it is still unclear how existing established coding guidelines methodically map onto Rust's unique language design. [6]

3 Methodology and Mapping Framework

Rather than evaluating a small subset of rules or focusing solely on conceptual alignment, we aim to provide a broader methodical mapping that supports rigorous reasoning about compliance, applicability, and necessary adaptations when using Rust in safety-critical contexts. We outline the ruleset classification process and discuss the resulting mappings.

All mapping results and supporting artifacts are publicly available as open-source materials at [https://github.com/embedded-software-laboratory/MISRust].

3.1 Approach

To ensure sound guideline mappings, we conduct a manual review of each of the 179 MISRA-C++ guidelines. For each guideline, we assess whether its underlying rationale remains applicable in Rust. Our analysis is limited to mapping the MISRA-C++ ruleset to Rust and does not extend to other safety standards. Where a guideline does not directly translate, we propose Rust-specific adaptations that preserve its intent.

If applicable, we present code examples for rules relevant to the mapping decision, which intentionally may not be fully compilable or executable, demonstrating compiler-enforced guarantees. We analyze how the guidelines apply to `safe` Rust and assess whether these guidelines remain necessary in the context of `unsafe` Rust. We also propose that runtime panics in Rust, i.e., unrecoverable program terminations triggered by operations such as the `panic!` macro, out-of-bounds access, or integer overflow, can be considered as `unsafe` behavior. Classification utilizes an analysis template as shown in Fig. 1:

The fields for the identifier and the specifier (if the guideline is a rule or directive) are taken directly from [4]. The categorization of each guideline into 1 out of the total 6 classes is given in the field *Analysis Result*. Furthermore, the discussion to justify the chosen class is contained in the field named *Comment*, and the code examples are included in the template.

Rule or Directive and guideline identifier [4]

Analysis Result
1 of 6 classes that are defined in this work

Comment: The comment section explains the rationale behind assigning a specific class in the context of Rust.
To support our claims and provide counterexamples where necessary, we may directly include illustrative code examples.

```
fn main() {
    // This is an example for the guideline analysis template
    println!("Template for guideline analysis");
}
```

Listing 1.1. Rust Code Example

Fig. 1. Template used for guideline analysis.

As far as possible, we provide references from the Rust reference [11] and the programming handbook [2] for every guideline in the discussion. The specific Rust version being evaluated is the current language edition 2024, compiler version `rustc 1.92.0 (ded5c06cf 2025-12-08)`.

Furthermore, the analysis is confined to Rust language constructs and the standard library. Cross-language interoperability through Rust's FFI is excluded. FFI introduces cross-language interactions that follow C semantics, thereby partially bypassing certain safety guarantees of the Rust compiler and borrow checker. We refer to MISRA C:2025 for an evaluation of the applicability of rules within the context of the Rust FFI.

3.2 Classification Process

Our classification procedure is depicted in Fig. 2. We define 6 main classes **C1 - C6** into which all 179 guidelines can be assigned. The classification begins by determining whether each guideline is applicable to the Rust programming language:

(C1) Not applicable - C++ Standard Library Usage Restriction: Guidelines whose rationale is exclusively tied to deficiencies or design issues in the C++ standard library. One such example is Guideline 25.5.3 "The pointer returned by the C++ standard library functions `asctime`, `ctime`, `gmtime`, `localtime`, `localeconv`, `getenv`, `setlocale`, or `strerror` must not be used following a subsequent call to the same function". This classification however does not imply that the Rust standard library is free of safety-relevant concerns. Safety implications of Rust standard library usage are discussed in Sect. 5 [4].

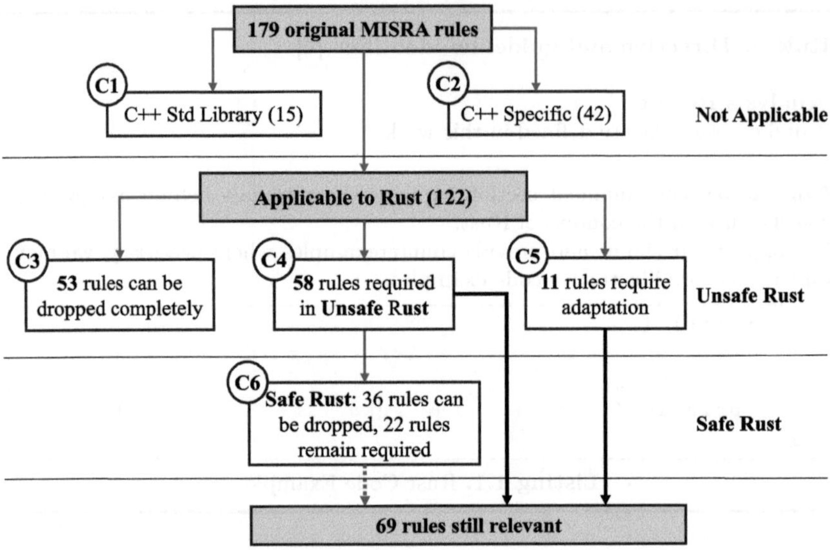

Fig. 2. Overview of the classification procedure and resulting rule count for each class. Class **C6** is a strict subset of **C4**.

(C2) Not applicable - C++ Feature does not exist in Rust: Guidelines that address language features available only in C++. These features are generally considered undesirable or `unsafe`, and the corresponding guidelines are therefore unnecessary in Rust. In the case of Guideline 12.2.1 "Bit-fields should not be declared", there simply does not exist an equivalent feature to *Bitfields* in Rust. Similar missing features in Rust are `goto`-Statements (Guideline 9.6.1), or virtual pointer-to-member functions (Guideline 13.3.4) [4].

After filtering out rules that are not applicable to Rust, we systematically analyze the remaining guidelines to determine their appropriate placement within the final classes **C3 - C6**. In addition, we identify a dedicated subset, **C6**, consisting of selected guidelines from **C4** that require special consideration. We consider these categories to represent the most substantive portion of the ruleset - the guidelines that remain necessary in Rust to preserve the MISRA-C++ ruleset's underlying rationale:

(C3) `Unsafe` Rust - Rule can be dropped: Guidelines in this category are applicable, but can be completely dropped. Rust enforces these Guidelines by design, even in `unsafe` contexts. One such example is Guideline 19.2.1. The rationale is to prevent confusing behavior and unexpected redefinitions caused by collisions in complex include structures. Due to strict import rules and compile-time checks against redefinition, the problems addressed by this rule are solved at compile time in Rust.

(C4) `Unsafe` Rust - Rule still required: These Guidelines are still required when using `unsafe` Rust. In general, these are rules that describe broader concepts, such as forbidding recursion or managing dynamic memory. The broadest category **C4** includes all rules that remain required when writing Rust code that uses `unsafe` language features. This applies in particular to scenarios where Rust's safety guarantees no longer hold automatically, such as manual memory management, manual pointer interaction, or FFI utilization. Such a rule regarding pointer interaction is Rule 8.2.7, "A cast should not convert a pointer type to an integral type": Interpreting a pointer type as an integer yields that pointer's address in Rust. Accessing a pointer itself must be handled in an `unsafe` block [11].

(C5) `Unsafe` Rust - Rule requires adaptation: Guidelines that address C++-specific features that do not exist in Rust, but whose underlying intent would be valuable if expressed in a Rust-specific form. We will present some rule modifications in Sect. 4.3.

(C6) Rule still required (`safe` Rust): Guidelines that are still required, even in a `safe` context. Within category **C4**, subcategory **C6** captures the subset of rules that remain necessary even in fully `safe` Rust. This is especially interesting if one can restrict oneself to only utilizing `safe` Rust language features. Again, this set also includes rules that address potential programming mistakes at the conceptual level, regardless of the implementation language. In the case of rule 0.3.1, Rust does not provide suspicious (imprecise) use warnings. Signaling for infinities exists, but the rationale behind the directive is not satisfied. Issues with IEEE 754 floating-point representation are independent of the Rust programming language, and developers must ensure that they prevent unintended software behavior when using Rust.

As shown in Fig. 2, a strict hierarchy exists among all classification categories. We identify classes **C4** and **C5** as the sets of guidelines that remain relevant to preserve the rationale of the MISRA-C++ ruleset within Rust.

4 Results and Analysis

In this section, we first give insight into the quantitative results of our classification. Afterwards, we propose possible adaptations for rules that are not applicable as-is, but still required in Rust.

4.1 Quantitative Overview

Of the original 179 MISRA rules, we classify 15 as not applicable because they focus solely on compliance with the C++ standard library (**C1**). Additionally, 42 rules that address features available only in C++ are considered not applicable and are classified into class **C2**. We therefore identify 122 rules to be applicable to Rust, of which 53 rules are automatically enforced, for example, by Rust's

Table 1. Distribution of the 179 guidelines across our 6 categories.

Category	Guidelines	[%]
Not applicable:	57/179	31.84
(**C1**) C++ Standard Library Usage Restriction	15/179	08.38
(**C2**) C++ Feature does not exist in Rust	42/179	23.46
Applicable:	122/179	68.16
(**C3**) Rule already satisfied in Rust language	53/179	29.61
Rule still relevant:	69/179	38.55
(**C4**) Required in `unsafe` Rust	58/179	32.40
(**C6**) From them, also required in `safe` Rust	22/179	12.29
(**C5**) Rule requires adaptation	11/179	06.15

unique ownership model or borrow checker concept. The remaining 69 applicable rules are still required to fulfill the rationale of the MISRA-C++ guidelines.

As depicted in Table 1, 38.55% of the 179 MISRA-C++ rules are still relevant. Only 6.15% of the original ruleset requires adaptation such that the rationale of the rule can be validated in the Rust language (class **C5**).

Due to the Rust language design, in total, 89 of the 111 as-is applicable rules are automatically satisfied within the `safe` subset of the language. However, guideline compliance in `unsafe` Rust, is especially interesting. Of the 89 rules automatically satisfied by `safe` Rust, 36 rules (**C4 - C6**) need to be reintroduced in `unsafe` Rust.

Table 2 shows how the 69 rules identified as still relevant (classes **C3** and **C4**) are distributed across the original 179 MISRA-C++ Rules, grouped by their respective topic identified by the MISRA Consortium. The distribution reveals highly variant topic relevance. Entire topic sections like these regarding language libraries fall largely into categories **C1, C2**, which are not applicable to Rust. Likewise, object-oriented design topics such as *Classes* (25.00%), *Derived classes* (16.70%), and *Special member functions* (12.50%) are only marginally relevant. This can be attributed to substantial differences in language design, particularly in inheritance and polymorphism. Additionally, Rust does not provide class-based inheritance, classes or class-like structures holding member functions and internal state, or inherited internal variables. While Rust allows behavior similar to classes through the use of structs and associated `impl` blocks, and supports polymorphism via traits acting as interface-like abstractions, these techniques do not permit inheritance of state or class behavior. Especially concepts central to C++, such as accessing inherited members, overwriting base-class functionality, or invoking base-class methods, are not supported in safe Rust. Therefore, many MISRA-C++ Rules addressing class concepts or inheritance are not applicable in Rust [10].

Table 2. Distribution of the 69 rules still relevant across the original 179 MISRA-C++ Ruleset. Grouped by MISRA-C++ topics.

MISRA-C++ topic	Rules still relevant	Topic relevance [%]
Language independent issues	5/10	50.00
General principles	4/4	100.00
Lexical conventions	1/12	08.30
Basic concepts	13/21	61.90
Standard conversions	5/9	55.60
Expressions	17/24	70.80
Statements	1/11	09.10
Declarations	2/8	25.00
Declarators	3/5	60.00
Classes	1/4	25.00
Derived classes	1/6	16.70
Member access control	1/1	100.00
Special member functions	1/8	12.50
Exception handling	2/8	25.00
Preprocessing directives	6/16	37.50
Language support library	4/12	33.30
Diagnostics library	1/2	50.00
Algorithms library	1/5	20.00

4.2 Unsafe Rust

While Rust enforces strong guarantees in a `safe` context, these guarantees are relaxed when using `unsafe` code. It should be emphasized that the `unsafe` keyword does not disable borrow checking or all other safety checks, but a few selected relaxations become possible: raw pointer dereferencing, `unsafe` function calls, interaction with static mutable variables, `unsafe` trait implementations, and union field access (see [13]). Of the 69 guidelines classified as still relevant to Rust, 36 are violated exclusively in `unsafe` code contexts.

These violations arise primarily from issues related to lifetimes, access sequencing, and (raw) pointer access. In `safe` Rust, such properties are statically verified by compiler checks.

The majority of these affected rules originate from foundational language features such as *Basic concepts*, *Expressions*, or *Standard conversions.* We conclude that `unsafe` Rust primarily reintroduces risks at a core language level, such as raw pointer arithmetic, and not at a higher level of abstraction. We identify these rule violations to closely resemble classic C or C++ safety issues, such as use-after-free, access sequences, shared resource access, or aliasing violations.

This motivates additional research into tools that specifically validate `unsafe` Rust sections.

4.3 Rules Requiring Adaptation

This category consists of guidelines that require adaptation when applied to Rust. MISRA guidelines [4] that fall into this category often refer to features available in Rust that do not map one-to-one to the C++ version. Although the original rationale might still apply, applying the guideline in Rust requires modifications.

Rule 4.1.1: A program shall conform to ISO/IEC 14882:2017(C++17). At the time of this publication, the Rust programming language does not yet have an ISO/IEC standard. Ongoing efforts working towards a formal specification already exist [11,12]. Should such an ISO/IEC standard be created, this rule should be amended to require conformance to it.

Rule 6.0.3: The only declarations in the global namespace should be `main`, namespace declarations and `extern "C"` declarations. Part of the rationale for this rule is the influence of include ordering on program behavior. Due to strict rules of unambiguous names when using crates in Rust, this rule is partially satisfied. Prevention of cluttering during lookup still requires following this rule in Rust code (see example in artifacts). A potential adaptation could involve the usage of ‘prelude’ modules that are often used within the Rust ecosystem.

Rule 6.4.2: Derived classes shall not conceal functions that are inherited from their bases. While Rust supports some form of inheritance using traits, Rust does not allow arbitrary method concealing. Traits fundamentally work by exposing all methods that they define on their implementation type. Structs can only shadow trait methods but cannot fully conceal them from consumers. Rust does not allow two functions with the same name and different parameter types. However, in a strict sense, the same behavior is present in Rust as in C++, because traits work the same way in both.

Rule 6.5.1: A function or object with external linkage should be introduced in a header file. The rule rationale indicates that external function or object linkages should be put into a header file and included everywhere to ensure that the underlying type declaration is the same everywhere. The same argument can be extended to Rust. An adaptation of the Rule could involve putting FFI declarations into a dedicated module at the root of a crate named *ffi*. A common approach within the ecosystem involves defining a sister crate with the *-sys* suffix that contains raw FFI bindings, and including the sister crate as a dependency in every project that requires these bindings.

Rule 12.3.1: The `union` keyword shall not be used. Rust provides `union` support. However, it suffers from the same drawbacks that the MISRA rule points out. An adaptation of this rule could involve recommending Rust `enums` as an alternative to `unions`. These are type-safe, support `match` statements, and do not require extensive tracking of `union` enum variants to prevent undefined behavior.

Rule 15.0.1: Special member functions shall be provided appropriately. Rust facilitates similar semantics using its `Copy`, `Clone`, `Drop`, and `Pin` traits. An adaptation of a rule should describe the implications of each of these traits in more detail. Additionally, the composition of structs that implement these traits should be noted.

Rule 18.5.1: A `noexcept` function should not attempt to propagate an exception to the calling function. Rust's error handling does not rely on exceptions. It uses the `Result` enum to provide type-safe access to error and success variants. However, functions returning a `Result` are not required to check and handle all error paths within their function body. An adaptation of the rule could require ensuring that all error variants are sufficiently handled within functions or are forwarded to the caller. The adaptation could naturally extend to the usage of panics, restricting them in favor of explicit, recoverable error handling.

Rule 19.0.1: A line whose first token is # shall be a valid preprocessing directive. While Rust does not use preprocessor directives, it supports conditional inclusion of code via function attributes. As shown in the example file, the rule can be adapted to require that all attributes be valid.

Rule 19.2.2: The `#include` directive shall be followed by either a `<filename>` or `"filename"` sequence. Because Rust does not use filenames, but rather crate names, and instead of the include directive, the `use` directive, the guideline is not directly applicable. In general, some restriction on the valid import syntax in Rust is required.

Rule 21.6.2: Dynamic memory shall be managed automatically. While all Rust standard library structs include proper *Drop* trait implementations to free all allocated resources, the rule should be adapted to include guidelines on how to correctly implement the *Drop* for custom structs relying on manual memory allocations.

Rule 22.3.1: The `assert` macro shall not be used with a constant-expression. While C++ offers a `static_assert` macro to execute the assertion at compile-time, Rust's assert macro works in a non-`const` and `const` context. This duality can lead to confusion among developers. A potential adaptation is to always wrap assertions intended to be executed within a `const` context inside a `const` block, making the intent clear.

5 Discussion

This section discusses the consequences of adapting our MISRA-C++-motivated Ruleset to Rust with respect to safety certification practice and places our results in context by identifying their validity limitations. We address the observed reduction in applicable required guidelines, the resulting emphasis on governing unsafe code, and the current lack of mature tools and widespread industrial adoption. The discussion also outlines factors that may affect the validity of our analysis, including ongoing language evolution, the subjectivity of manual guideline classification, and the utilization of only a few analysts.

5.1 Implications for Safety Certification

Our investigation carries several implications for the use of Rust in safety-critical domains and for establishing a coding guideline similar to the MISRA-C++ guidelines. Most notably, the resulting rule mapping suggests a substantial reduction in the guideline burden. Due to the Rust language design, approximately 61.45% of the original MISRA ruleset are considered unnecessary, as they are automatically fulfilled by the language intrinsics, or not applicable (categories **C1 - C3**). This strongly indicates that a future Rust-specific safety standard may be significantly more compact, potentially reducing the effort to reach safety compliance.

We identify a critical need for well-defined governance of `unsafe` Rust features, especially `unsafe` code blocks. Since `unsafe` disables some memory safety checks of the Rust compiler, such as raw pointer access [13], explicit certification guidelines must be defined. Only with such governance extending existing safety concepts can applications written in `unsafe` Rust be reliably safety-certified.

This paper also identifies a notable safety-related coding guideline deficit in Rust, similar to MISRA-C++. The Rust toolchain already utilizes expressive compiler diagnostics and strict compile-time code checks. Additional tools like the *Clippy* linter claim to further improve code quality [9]. However, some MISRA-C++ guidelines have no direct equivalent in the Rust ecosystem. These tools may be a good starting point in extending their static analysis system for future certification frameworks, as new linting rules, compile-time code checks, or even complete third-party analysis tools may be needed to achieve parity with the MISRA-C++ implied safety certification.

Additionally, the inapplicability of **C1** guidelines to Rust does not imply that the Rust standard library is free of safety-relevant concerns. The Rust standard library introduces its own features that are not addressed by the MISRA-C++ guidelines, such as the `Result` and `Option` types for error handling, or the `panic!` macro for unrecoverable errors. These features may introduce safety issues if not used properly, and therefore require original Rust-specific guidelines, as we identify in Sect. 4.3.

While interest in Rust for safety-critical systems is increasing, widespread industrial adoption is still limited [1,8,14]. Recent work on using Rust for safety-critical systems reports benefits from the language's integrated memory-safety guarantees, but highlights challenges with legacy code compatibility, the Rust ecosystem's maturity, and the steep learning curve involved in migrating to the new language [6–8,14].

In practice, partial integration of Rust and component replacements in a small, limited scope appears to be the most feasible adoption strategy. Small scope changes may allow organizations to leverage Rust whilst reducing risk in the re-certification effort. One prominent example is the introduction of Rust into the Linux Kernel, where it is integrated as a framework for developing device drivers in Rust. The Rust-For-Linux framework extends the existing architecture rather than replacing the C kernel code [3]. Seidel et al. [14] also recommend gradual integration of Rust in the context of safety-critical embedded space systems, noting that partial Rust rewrites should be explored in the context of highly critical system components [14].

Institutional research also reflects this incremental adoption pattern. The Eclipse SDV group, for example, has founded a Special Interest Group on Rust, focusing on the adoption in the automotive industry and safety-critical systems [1]. The DLR Institute of Software Technology has investigated the applicability of Rust in spacecraft systems, highlighting potential benefits and limitations for existing software frameworks, development processes, and standard compliance [15].

We therefore conclude that at present Rust is still more likely to complement than substitute established languages, such as C or C++, in safety-critical systems, meaning that widespread adoption is still hindered by a lack of available tooling, standardization, and certification.

5.2 Threats to Validity

Many factors may affect the validity of the analysis presented. Both C++ and Rust are evolving languages; future changes in their specifications or common best-practice guidelines may affect how individual MISRA-C++ rules map to similar guidelines for Rust. The classification results presented in this paper only reflect the state of the Rust language at compiler version `1.92.0`, edition 2024. Therefore, future revisions may be required if new language features emerge.

The classification of the ruleset involved manual interpretation of each individual MISRA-C++ Guideline. Even with our explicit classification template, decisions about whether a guideline is not applicable, automatically satisfied, or needs to be modified to be applicable to Rust (and is still required) depended on the analyst's subjective understanding of the Rust language and its safety concept and semantics. Inconsistent judgments or interpretation bias are therefore possible.

Our study relies on MISRA-C++ as a baseline for deriving an equivalent safety ruleset for Rust. While this provides a sensible starting point, we implicitly assume that the MISRA-C++ design priorities are compatible with the Rust architecture design to a certain level. As the baseline is targeted at C++, and given Rust's unique ownership and borrowing model, some guidelines (classes **C1 - C3**) are redundant or not applicable per design.

Finally, the classification was conducted by a small group of analysts ($n = 4$). While all four analysts independently classified the full set of 179 guidelines prior to a joint consensus discussion, no external verification was performed.

This introduces the residual risk of shared interpretation bias or misclassifications within the author group. Future work, therefore, should incorporate a multi-reviewer approach with independent verification to reduce subjectivity and reinforce the robustness of the resulting classifications.

6 Conclusion

We have provided a systematic rule-by-rule mapping of all 179 MISRA-C++:2023 guidelines to Rust, introducing a 6-category classification framework to assess each guideline's applicability and necessity in the context of Rust, distinguishing between `safe` and `unsafe` Rust. Although `safe` Rust substantially reduces the guideline burden, `unsafe` Rust is a critical topic in future safety-focused coding standards.

We identified 11 rules (6.15% of the original ruleset) that require adaptation to Rust, and proposed possible migrations to preserve the original intention of MISRA-C++.

Beyond the rules addressed in this mapping, we identify several Rust-specific features with no direct C++ analog that would require original guidelines in a future Rust safety standard. Most notably, the implicit panic behavior of `unwrap` and `expect` on `Result` and `Option` values, the hygiene and code generation properties of declarative and procedural macros, and the risk of panicking within `Drop` implementations during stack unwinding each represent safety-relevant concerns that the MISRA-C++ ruleset does not and cannot address. Furthermore, manual implementations of the `Send` and `Sync` marker traits can invalidate compiler-enforced thread safety guarantees, while additional areas including interior mutability patterns, integer overflow behavior diverging between debug and release builds, and misuse of `Pin` may introduce safety-critical issues that are equally beyond the scope of the present mapping. We therefore conclude that a comprehensive Rust safety standard would require not only the adapted MISRA-C++ rules identified in this work, but also a dedicated set of original guidelines addressing these Rust-specific concerns. A systematic treatment of these areas remains future work.

Overall, we hope that our presented mapping provides a concrete foundation for future standards and tool support, and that it extends the results of the MISRA-C:2025 Addendum. All mapping results and artifacts are made available as open source and accessible at [https://github.com/embedded-software-laboratory/MISRust].

References

1. Eclipse Foundation: Rust - Software Defined Vehicle | The Eclipse Foundation. https://sdv.eclipse.org/special-interest-groups/rust/
2. Klabnik, S., Nichols, C., Krycho, C., Community, R.: The rust programming language. https://doc.rust-lang.org/stable/book/

3. Li, Z., Narayanan, V., Chen, X., Zhang, J., Burtsev, A.: Rust for linux: understanding the security impact of rust in the linux kernel. In: 2024 Annual Computer Security Applications Conference (ACSAC), pp. 548–562 (2024). https://doi.org/10.1109/ACSAC63791.2024.00054
4. MISRA: MISRA C++:2023 Guidelines for the use of C++17 in critical systems. The MISRA Consortium Limited, Norwich, UK (2023)
5. MISRA: MISRA C:2025 – Addendum 6 (MISRA C:2025 ADD6) (2025). https://misra.org.uk/app/uploads/2025/03/MISRA-C-2025-ADD6.pdf
6. Munch, M., Lindner, M., Eriksson, J., Dzialo, P., Lindgren, P.: Rust for safety and security critical systems. In: 2025 IEEE Nordic Circuits and Systems Conference (NorCAS), pp. 1–7 (2025). https://doi.org/10.1109/NorCAS66540.2025.11231205
7. Müller, R., Nehlich, P., Klinkner, S.: Leveraging the rust programming language for space applications. In: 2024 IEEE Space Computing Conference (SCC), pp. 40–50. (2024). https://doi.org/10.1109/SCC61854.2024.00011
8. Pinho, A., Couto, L., Oliveira, J.: Towards rust for critical systems. In: 2019 IEEE International Symposium on Software Reliability Engineering Workshops (ISSREW), pp. 19–24. (2019). https://doi.org/10.1109/ISSREW.2019.00036
9. rust-lang: Introduction - Clippy Documentation. https://doc.rust-lang.org/clippy/
10. rust-lang: Object Oriented Programming Features - The Rust Programming Language. https://doc.rust-lang.org/book/ch18-00-oop.html
11. rust-lang: the rust reference. https://doc.rust-lang.org/reference/
12. rust-lang: tracking issue for the rust specification · Issue #113527 · rust-lang/rust. https://github.com/rust-lang/rust/issues/113527
13. rust-lang: unsafe rust - the rust programming language. https://doc.rust-lang.org/book/ch20-01-unsafe-rust.html
14. Seidel, L., Beier, J.: Bringing rust to safety-critical systems in space. In: 2024 Security for Space Systems (3S), pp. 1–8 (2024). https://doi.org/10.23919/3S60530.2024.10592287
15. Sommer, J., Gutierrez Rojo, T., Lund, A., Abdelmaksoud, H.I.E.: Rust for space applications and rtems-the good, the bad and the ecss. In: 18th ESA Workshop on Avionics, Data, Control and Software Systems (ADCSS2024). European Space Agency (ESA) (2024)
16. Vector Informatik GmbH: vector and HighTec bring functional safety applications with rust and AUTOSAR classic to the road. https://www.vector.com/int/en/news/news/safety-applications-with-rust-and-autosar-classic/

On the Timing Determinism of CUDA and Vulkan SC for Safety-Critical AI Workloads on Integrated SoC GPUs

Yousif M. Elsheikh(✉), Adrian Raiser, Fabian Frommherz, and Zamira Daw

Institute of Aircraft Systems, University of Stuttgart, Stuttgart, Germany
{yousif.elsheikh,zamira.daw}@ils.uni-stuttgart.de,
{st155111,st188446}@stud.uni-stuttgart.de

Abstract. AI-enabled functions are increasingly considered for aerospace and other safety-critical systems, where predictable execution is as important as acceleration. Standard GPU programming models such as CUDA do not inherently provide the timing guarantees required for certification, since runtime jitter and limited worst-case predictability remain problematic. Vulkan SC has been proposed as a more deterministic alternative for safety-critical GPU execution. This work compares CUDA and Vulkan SC on the NVIDIA Jetson Orin Nano SoC platform with a focus on timing determinism. We apply fixed clocking, warm-up runs, and process-level optimizations. We then analyze timing contributions from memory management, data movement, and kernel execution. The evaluation covers neural-network-related synthetic primitives, composite SAXPY-ReLU chains, and a fused neural-network workload. To relate the synthetic primitives to the NN workload, we compare an operation-count-equivalent unfused primitive baseline with the measured fused implementation across input sizes. The results show that Vulkan SC provides more stable kernel execution and consistently low kernel-side jitter. The growth analysis shows that CUDA kernel time is nearly input-size independent for small and medium input sizes, while Vulkan SC shows less uniform growth. For both APIs, the unfused baseline remains above the fused NN kernel time, while the relative gap decreases with increasing input size. H2D/D2H transfer timing scales almost linearly with transferred data size for larger transfers, but remains a major source of end-to-end variability, especially for Vulkan SC. Overall, Vulkan SC improves kernel-level timing determinism, while system-level predictability remains constrained by transfer overhead and shared-memory effects.

Keywords: GPU · Safety-critical systems · CUDA · Vulkan SC

1 Introduction

The integration of complex and computationally intensive machine learning (ML) algorithms into safety-critical systems requires specialized hardware and

Y. M. Elsheikh, A. Raiser and F. Frommherz—Equal contribution.

© The Author(s), under exclusive license to Springer Nature Switzerland AG 2027
E. Troubitsyna et al. (Eds.): SAFECOMP 2026, LNCS 16868, pp. 330–345, 2027.
https://doi.org/10.1007/978-3-032-34867-8_21

software solutions. In non-safety-critical domains, graphics processing units (GPUs) have emerged as an effective and widely available alternative to central processing units (CPUs) for parallel computation.

However, standard application programming interfaces (APIs) do not inherently provide deterministic runtime guarantees, introducing timing uncertainties and jitter that are unacceptable in safety-critical systems. In the aerospace industry, software must undergo rigorous certification processes where determinism is not just a performance goal, but a strict requirement for safety-critical components. Meeting these certification standards, such as Design Assurance Level (DAL) A, requires that the worst-case execution time (WCET) of all operations must be predictable with high confidence. Compute Unified Device Architecture (CUDA), despite being a dominant API, suffers from limited timing predictability, making it difficult to use in certified environments without extensive mitigation [6]. To address this, the Khronos Group proposed version 1.0 of Vulkan Safety-Critical (Vulkan SC) based on Vulkan 1.2 in 2022. According to the Khronos Group [3], Vulkan SC in contrast to Vulkan is characterized by being streamlined, deterministic and robust. For that they eliminated non-essential runtime functionality, such as sparse memory and descriptor update templates, ensured predictable execution times through offline pipeline compilation, static memory allocation, and explicit synchronization and removed undefined behaviors and ignored parameters while enhancing fault handling and aligning with Motor Industry Software Reliability Association (MISRA) C standards (C programming standard from the automotive industry). The viability of Vulkan SC in this sector is bolstered by industry solutions such as CoreAVI's (now Lynx) VkCore SC, which provides an implementation of Vulkan SC certified up to DAL A for safety-critical hardware. Such implementation-specific certification is independent of the benchmarking performed here.

This work does not provide certification evidence. It provides an empirical timing analysis of CUDA and Vulkan SC and identifies which execution phases are stable and which phases require further bounding, giving insights into the timing-related certification bottlenecks of using system-on-a-chip (SoC) integrated GPUs in safety-critical systems.

2 Background and Benchmark Workloads

2.1 Related Work

Perez-Cerrolaza et al. introduce a comparison between different GPU vendors and standards with respect to different certification standards in their survey [6]. They conclude that GPU hardware is rapidly evolving into safety-critical applications but is not yet matured. In addition, challenges with respect to programming languages and frameworks need to be addressed. One approach to resolve the framework challenge is Brook Auto [7]. A comparable evaluation was performed by Marc Benito et al., benchmarking OpenGL SC 2.0 against Brook Auto/BRASIL in an aviation-inspired case study [2]. Their assessment incorporates the ability to comply with certification standards, specifically DO-178 and ISO 26262, and conclude that both languages meet the performance requirements of advanced functionalities. Aleman et al. demonstrate the use of Vulkan SC in [1], showing how GPU code can achieve code coverage in testing.

Because Vulkan SC is relatively recent compared to the other GPU APIs, the existing body of literature does not provide a direct comparison between CUDA and Vulkan SC with respect to their performance. Current research efforts instead emphasize the certifiability and qualification aspects of GPU-based software development rather than conducting comprehensive empirical performance evaluations between these two frameworks, which this work aims to undertake.

2.2 CUDA and Vulkan SC Execution Models

The APIs compared here differ fundamentally in their execution models. In Fig. 1, the operation pipelines of CUDA and Vulkan SC are compared side by side. Besides the different pipelines, the terminology also differs. Therefore, Table 1 compares the main terms used in CUDA and Vulkan SC.

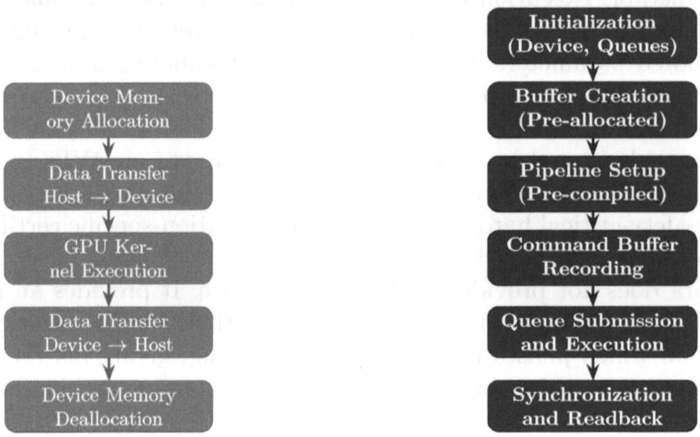

(a) CUDA operation pipeline (b) Vulkan SC operation pipeline

Fig. 1. Comparison of CUDA and Vulkan SC operation pipelines

CUDA. is NVIDIA's proprietary general-purpose GPU computing (GPGPU) platform and follows an execution flow based on dynamic device memory allocation, host-device transfers, kernel launch, result transfer, and memory deallocation. Even on the NVIDIA Jetson Orin Nano Developer Kit (Jetson) unified memory architecture (UMA) platform, this model still requires logical data orchestration and synchronization, which can introduce timing variance.

Vulkan SC. Developed by the Khronos Group, Vulkan SC provides a low-level interface with explicit control over memory, synchronization, and execution [4]. In contrast to CUDA, resources and pipelines are prepared in advance and work is submitted through command buffers, shifting the execution model from implicit stream handling to an explicit and more deterministic workflow.

Table 1. Differences in CUDA and Vulkan SC terminology

CUDA	Vulkan SC
Thread	Invocation
Thread Block	Workgroup
Grid	Dispatch Group
Warp	Subgroup
Kernel	Compute Shader / Pipeline
Shared Memory	Workgroup Storage (Shared)
Global Memory	Storage Buffer (SSBO)
Constant Memory	Uniform Buffer / Push Constants
Stream	Queue
Device	Logical Device / Physical Device
Context	Instance
Texture / Surface	Image / Sampler
__syncthreads()	barrier() / Memory Barrier
cudaMalloc()	vkAllocateMemory
cudaMemcpy()	vkCmdCopyBuffer

2.3 Neural-Network Primitive Operations

Table 2. Evaluated synthetic NN-related workloads

Workload	Expression
SAXPY	$y[i] = a\,x[i] + y[i]$
ReLU	$\max(0, x)$
Sigmoid	$\sigma(x) = 1/(1 + e^{-x})$
Tanh	$\tanh(x)$
Softmax	$\text{softmax}(z)_i = e^{z_i} / \sum_{j=1}^{K} e^{z_j}$
Chain	$f_{\text{chain}}(x) = (\text{ReLU} \circ \text{SAXPY})^n(x)$, $n \in \{10, 100, 1000\}$

The evaluation uses synthetic NN-related workloads that cover linear arithmetic, activation functions and composite SAXPY-ReLU chains, which represent increasing computational depth. The evaluated workloads are summarized in Table 2.

2.4 Fused Neural-Network

Motivated by the ADIMA peripheral-detection use case [5], we evaluate a fused feed-forward neural network (NN) workload and compare it with an operation-count-equivalent unfused primitive baseline based on Sect. 2.3. As shown in Fig. 2, each sample requires 380 SAXPY elements, 20 ReLUs, and 4 sigmoids. Thus, the unfused kernel baseline is defined as
$\texttt{K\_unfused}(N) = 380 \cdot \texttt{K\_SAXPY}(N) + 20 \cdot \texttt{K\_ReLU}(N) + 4 \cdot \texttt{K\_sigmoid}(N)$, where N is the number of input samples processed in parallel and `K_*` is the mean.

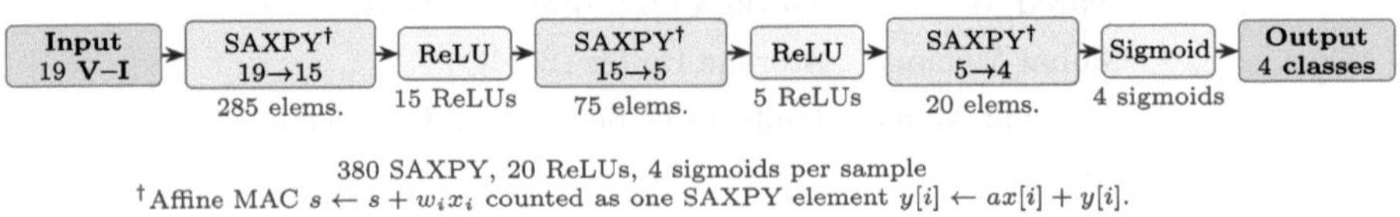

Fig. 2. Primitive-level structure of the fused NN workload. Each affine-layer MAC is counted as one SAXPY element multiply-add primitive.

The fused NN is implemented as a single kernel/shader. The NN weights and biases are fixed during inference. In CUDA, they are declared with the `__constant__` memory-space specifier and stored in device-side constant memory. In Vulkan SC, they are declared as `const` arrays in the compute shader.

2.5 Identified Computational Bottlenecks

Timing non-determinism is usually caused by three main sources: dynamic memory management, host-device data orchestration (host-to-device (H2D)/deviceto-host (D2H)), and frequency variability caused by power-management mechanisms such as dynamic voltage and frequency scaling (DVFS). These sources form the basis for the measures applied to improve overall runtime determinism before switching to Vulkan SC.

3 Experimental Setup

3.1 Platform

As platform for the experiments, we use the Jetson, a compact embedded SoC widely adopted in ML applications due to its favorable size, weight, power and cost (SWaP-C) characteristics. The device used in this study is the NVIDIA Jetson

Orin Nano Developer Kit (Model No. P3766), equipped with a 1024-core NVIDIA Ampere architecture GPU with 32 Tensor Cores, a 6-core Arm Cortex-A78AE v8.2 CPU, and 8 GB of 128-bit LPDDR5 memory. Since the platform follows a UMA, the CPU and GPU share the same physical random access memory (RAM), while the memory controller is integrated on the same chip. All relevant settings are summarized in Table 3.

The platform is operated in maximum performance mode and accessed via secure shell (SSH) and is run in headless-mode to reduce interference from graphical system activity. In addition, the CPU, GPU, and memory clocks are fixed using `jetson_clocks` in order to reduce variability caused by DVFS, while thermal throttling remains enabled as a safety mechanism. Since the first executions after boot show noticeable timing anomalies, the first 200 runs are excluded to ensure reproducible measurements. To characterize timing behavior, pure GPU kernel execution time is measured in CUDA via `cuda_runtime` hardware timers, excluding host-side API and synchronization overhead, while end-to-end runtime is measured using the `std::chrono` library for CPU tasks and Vulkan SC execution. To better trace timing anomalies back to their source, system-wide monitoring is additionally performed using `tegrastats`, system log files, and self-created dashboards. Input data are generated using a fixed random seed to ensure repeatability, and all experiments are repeated 1000 times. Related configuration setup/results are discussed in Sect. 3.3.

Table 3. Benchmarking configuration

Parameter	Value
CUDA version	12.6.68
Vulkan SC version	1.0.15
Power mode	Max. performance
Access	SSH
Clocks	Fixed
Threads per block	256
Number of blocks	Workload-dependent
Warm-up exclusion	First 200 runs
Random seed	42
Vector size (N)	10^7 and $\in \{10^1, ..., 10^7\}$ (for the timing growth analysis)
Data range	$x \in [-5, 5]$
Repetitions	1000

3.2 Metrics

The analysis focuses on five metrics: mean runtime μ, mean absolute jitter (MAJ), measured WCET, relative measured WCET, and skewness. Mean runtime captures average execution time. MAJ, defined as $\text{MAJ} = \frac{1}{n}\sum_{i=1}^{n} |t_i - \mu|$, quantifies the typical absolute deviation from the mean timing behavior. Measured WCET denotes the maximum observed runtime, while relative measured WCET measures its percentage overhead over μ. Skewness characterizes the direction of variability, with positive values indicating a tendency toward latency spikes and negative values indicating deviations toward shorter runtimes.

3.3 Parametrization

Thread Configuration. To select the number of threads per block, a strong-scaling study was performed in which the problem size remained fixed while the thread-block size was varied. As shown in Fig. 3, runtime decreases up to a configuration of 256 threads per block and increases again beyond that point. Therefore, all following experiments use 256 threads per block.

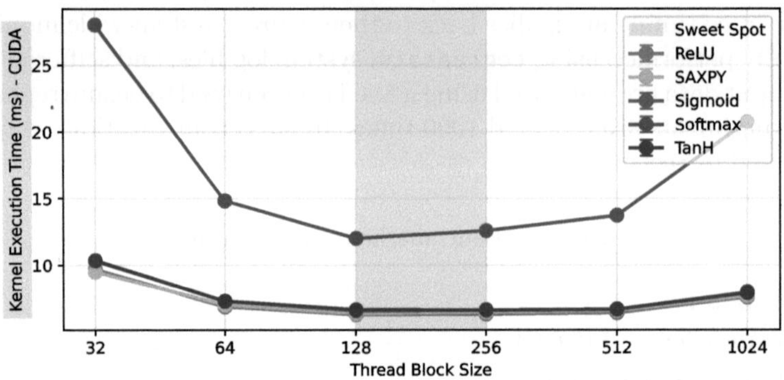

Fig. 3. Kernel execution time (ms) on Jetson hardware for CUDA kernels implementing ReLU, SAXPY, Sigmoid, Softmax, and Tanh, plotted against thread-block size (32–1024 threads per block)

Clock Configuration. To decide the clock configuration, we compared dynamic and fixed clocking. As shown in Fig. 4, fixed clocking reduces runtime variability and noticeably decreases the number of outliers. This motivated the use of fixed clocks in all subsequent experiments.

Warm-up Phase. Early iterations after reboot exhibited noticeable timing anomalies. This behavior is illustrated in Fig. 5 for the example of SAXPY memory allocation. To ensure steady-state and reproducible measurements, the first 200 runs after boot were excluded from the reported results.

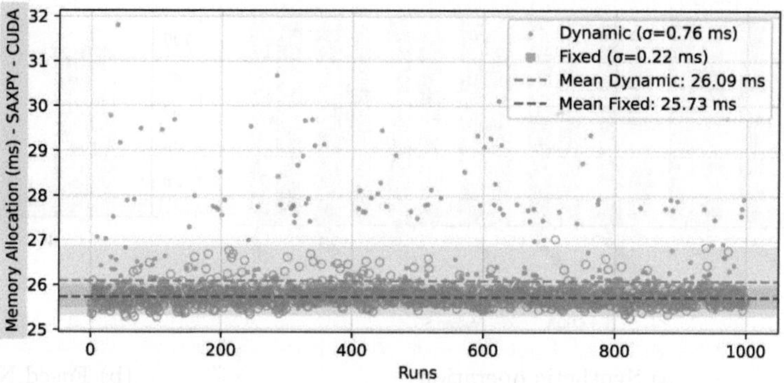

Fig. 4. Memory allocation time (ms) for SAXPY across 1000 runs under dynamic and fixed system clocking

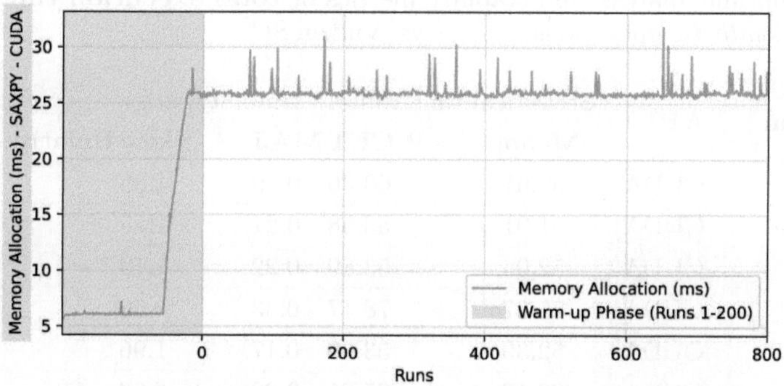

Fig. 5. Warm-up period for **memory allocation time** measurements over SAXPY 1000 runs. The shaded region marks runs 1–200, which are excluded from steady-state statistics

4 Results

Tables 4 and 5 summarize the synthetic operations experiment results, covering total execution time and kernel execution time, respectively.

4.1 Runtime Performance

In Fig. 7, the percentage contribution of all operations to the total runtime, comparing CUDA and Vulkan SC, is illustrated.

Regarding memory management, two observations can be made. First, memory allocation overhead in Vulkan SC accounts for a negligible share of the total runtime across all tests. Second, memory deallocation is not present in the Vulkan SC results. Furthermore, the kernel execution time for Vulkan SC is lower for individual operations and NN.

However, as the complexity increases, the kernel share of the runtime grows to a level similar to CUDA. It is also observed that data transfer times (H2D and D2H)

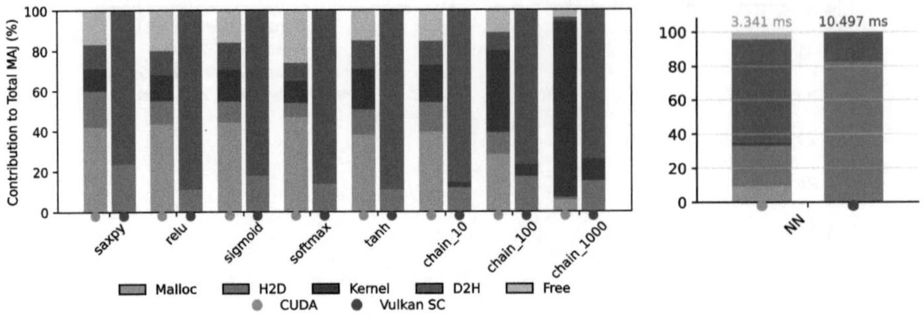

(a) Synthetic operations (b) Fused NN

Fig. 6. Relative contribution to MAJ for CUDA and Vulkan SC execution.

Table 4. Runtime analysis and stability metrics of **total execution time** over 1000 runs for the *synthetic* operations: CUDA vs. Vulkan SC

Operation	API	Statistical time values		Variability/jitter		Stability
		Mean	**WCET**	**MAJ**	**Skew**	**Relative WCET**
SAXPY	CUDA	58.61	60.36	0.19	1.35	2.98
ReLU	CUDA	51.91	53.58	0.21	1.55	3.22
Sigmoid	CUDA	52.05	53.89	0.22	1.79	3.55
Softmax	CUDA	75.67	78.17	0.33	0.60	3.32
Tanh	CUDA	52.39	53.99	0.17	1.96	3.06
$Chain_{10}$	CUDA	89.17	95.21	0.22	6.06	6.77
$Chain_{100}$	CUDA	395.26	400.04	0.41	1.44	1.21
$Chain_{1000}$	CUDA	3453.26	3462.02	2.23	0.08	0.25
SAXPY	Vulkan SC	68.29	70.69	0.71	-1.66	3.52
ReLU	Vulkan SC	46.92	49.16	0.96	-1.22	4.78
Sigmoid	Vulkan SC	47.12	49.36	0.99	-1.09	4.76
Softmax	Vulkan SC	53.00	55.49	0.87	-0.54	4.70
Tanh	Vulkan SC	47.26	49.03	0.73	-1.69	3.74
$Chain_{10}$	Vulkan SC	105.40	109.97	0.96	-1.08	4.33
$Chain_{100}$	Vulkan SC	482.63	484.95	0.83	-1.30	0.48
$Chain_{1000}$	Vulkan SC	4251.58	4255.24	1.22	-0.55	0.09

represent a larger proportion of the runtime in Vulkan SC than in CUDA. In contrast, for individual operations in CUDA, memory allocation represents the largest component of the total runtime. As visible in Table 4, Vulkan SC requires approximately 17% more total execution time for SAXPY, 18% for $Chain_{10}$, 22% for $Chain_{100}$, and 23% for $Chain_{1000}$. In contrast, for the activation functions, Vulkan SC exhibits lower total execution times. ReLU and Tanh are executed roughly

Table 5. Runtime analysis and stability metrics of **kernel execution time** over 1000 runs for the *synthetic* operations: CUDA vs. Vulkan SC

Operation	API	Statistical time values		Variability/jitter		Stability
		Mean	WCET	MAJ	Skew	Relative WCET
SAXPY	CUDA	6.66	7.03	0.03	1.61	5.54
ReLU	CUDA	6.59	6.95	0.04	1.25	5.42
Sigmoid	CUDA	6.74	7.17	0.05	0.84	6.44
Softmax	CUDA	12.88	13.32	0.05	0.62	3.43
Tanh	CUDA	6.72	6.99	0.06	0.24	4.01
$Chain_{10}$	CUDA	37.31	43.20	0.06	24.92	15.77
$Chain_{100}$	CUDA	343.30	347.59	0.24	2.80	1.25
$Chain_{1000}$	CUDA	3401.40	3409.76	2.12	0.08	0.25
SAXPY	Vulkan SC	1.99	2.04	0.00	4.26	2.54
ReLU	Vulkan SC	1.66	1.68	0.00	0.28	1.53
Sigmoid	Vulkan SC	1.71	1.73	0.00	0.04	1.40
Softmax	Vulkan SC	7.80	7.84	0.00	2.77	0.52
Tanh	Vulkan SC	1.72	1.74	0.00	-0.30	0.94
$Chain_{10}$	Vulkan SC	39.68	40.94	0.03	9.49	3.16
$Chain_{100}$	Vulkan SC	416.00	416.88	0.07	1.94	0.21
$Chain_{1000}$	Vulkan SC	4182.78	4185.61	0.17	1.66	0.07

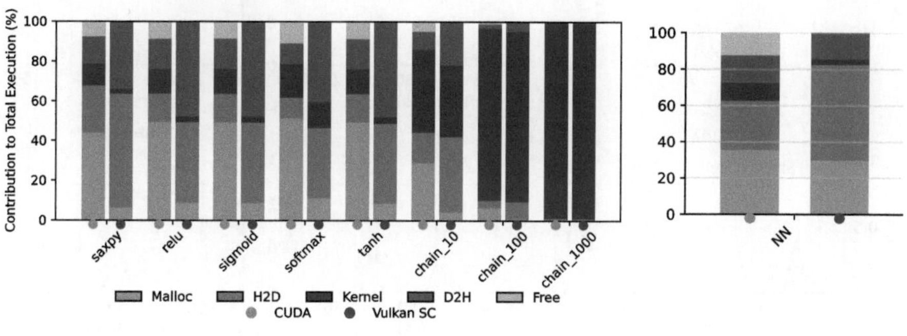

(a) Synthetic operations (b) Fused NN

Fig. 7. Relative contribution of each execution step to total execution time for CUDA and Vulkan SC.

10% faster, Sigmoid about 9% faster, and Softmax shows the largest reduction, at approximately 30%. The same comparison for kernel execution time can be derived from Table 5. For the individual primitive operations, Vulkan SC kernels are faster than their CUDA counterparts, with relative runtimes of about 0.30 for SAXPY, 0.25 for ReLU and Sigmoid, 0.61 for Softmax, and 0.26 for Tanh. However, this advantage decreases as workload complexity increases. Chain_{10} is already slightly slower at 1.06, while Chain_{100} and Chain_{1000} reach approximately 1.21 and 1.23, respectively.

4.2 Timing Variability and Jitter

In Fig. 6a, the contribution of various operations to the MAJ is illustrated. In the CUDA results, the largest shares of jitter originate from memory management, followed by a measurable contribution from kernel execution. Conversely, Vulkan SC exhibits a minimal jitter contribution related to memory management. Furthermore, the kernel contribution to the total MAJ in Vulkan SC remains low even under high workloads and represents a smaller proportion compared to the CUDA implementation and is dominated by data transfer operations. The NN (Fig. 6b) picks up on the operations' trend with a small CUDA kernel jitter.

In Fig. 8, the skewness values of the total execution time distributions for CUDA and Vulkan SC are compared. CUDA consistently exhibits a positive skew across all operations, with values ranging from 0.08 to 6.06. In contrast, Vulkan SC shows a negative skew in all operations, with values ranging from −0.54 to −1.69. The NN had a skew of 1.270597 for CUDA and -0.001317 for Vulkan SC.

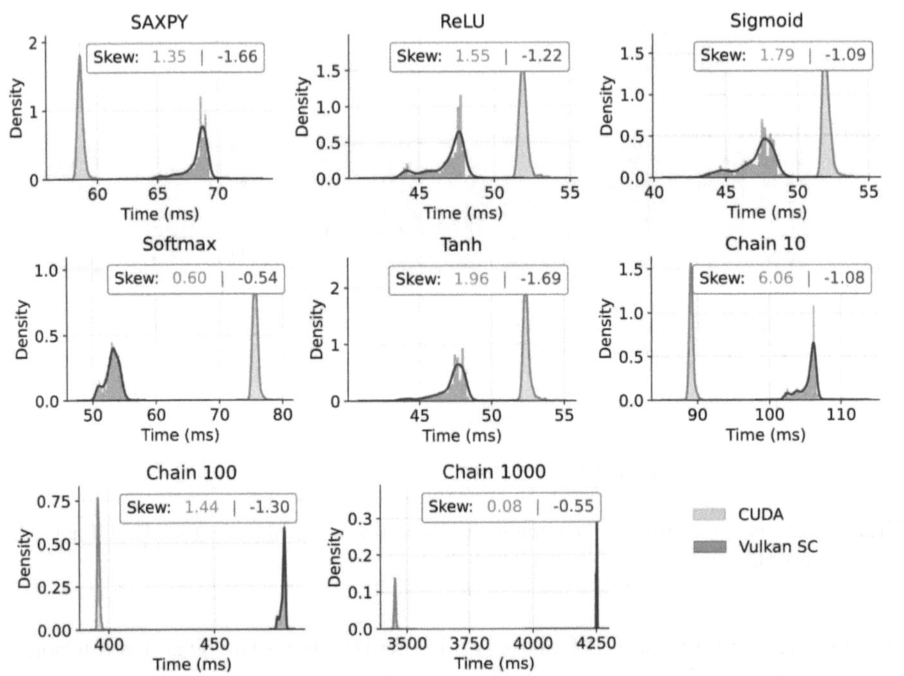

Fig. 8. total execution time skewness for all synthetic operations comparing CUDA and Vulkan SC

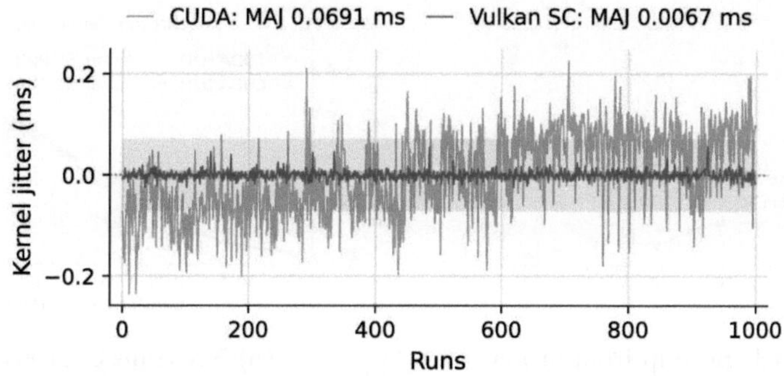

Fig. 9. Kernel MAJ and kernel jitter over 1000 post-warmup runs for CUDA and Vulkan SC

Figure 9 showcases the kernel jitter for both CUDA and Vulkan SC for the fused NN with the MAJ of CUDA being approximately x10 that of Vulkan SC.

4.3 Fused and Unfused Kernel and Transfer Timing Behavior

As detailed in Sect. 2.4 the comparison is based on scaling the timing of the synthetic and primitive operations by their operation count in the NN. Figure 10 showcases the mean kernel time vs. the input size for the unfused scaled baseline and the NN in a log-log scale for both CUDA in Fig. 10a and Vulkan SC in Fig. 10b. Figure 11a shows the mean unfused to fused kernel time (`K_unfused/K_fused`) vs. N and Fig. 11b shows the relationship between host and device transfer times vs. the data size corresponding to N in megabytes (MB).

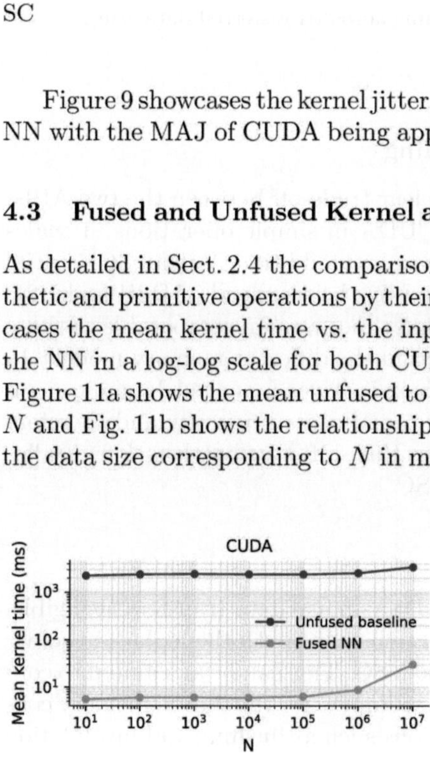

(a) CUDA

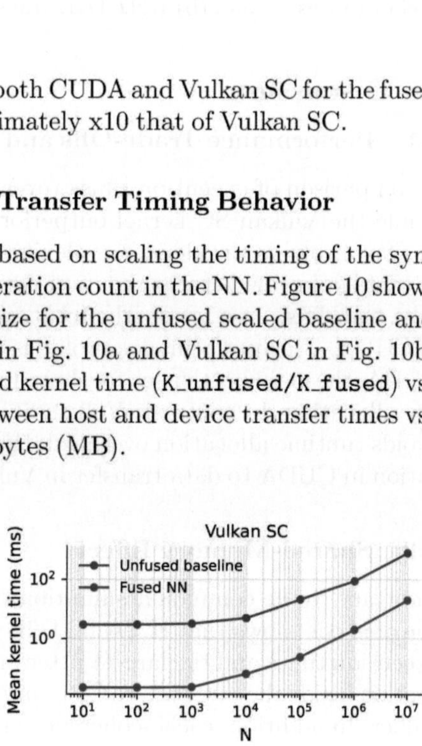

(b) Vulkan SC

Fig. 10. Mean kernel time of the operation-count-equivalent unfused primitive baseline and the measured fused NN implementation across input sizes N

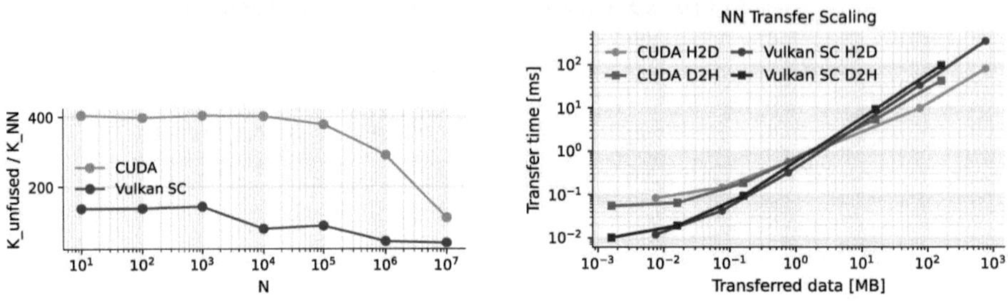

(a) Kernel speedup from fusion (b) NN transfer growth

Fig. 11. Fused NN growth behavior across input sizes ((a) `K_unfused/K_NN` across parallel instances N, and (b) H2D/D2H transfer time across transferred data size).

5 Discussion

5.1 Performance Trade-Offs and Scaling

A comparison of execution times reveals a clear trade-off between the two APIs. While the Vulkan SC kernel outperforms CUDA in simple operations, it scales less favorably as workload complexity increases. In addition, Vulkan SC's initial advantage in raw kernel execution is reduced by the overhead of D2H and H2D data transfers. As a result, the total execution times remain comparable to those of CUDA. This performance profile is rooted in the different memory models. In CUDA, the main bottleneck for short-lived tasks is driver-managed dynamic memory allocation. In contrast, Vulkan SC uses pre-allocated memory and therefore avoids runtime allocation overhead. However, this shifts the bottleneck from allocation in CUDA to data transfer in Vulkan SC.

5.2 Shared-Memory Effects

Contrary to the common assumption that UMAs make data transfers negligible, the analysis shows that H2D and D2H operations have both the highest absolute execution times and the largest jitter in Vulkan SC. This can be attributed to contention inside the SoC, where CPU and GPU compete for the shared memory controller. In addition, cache-coherence operations such as flushing and invalidation introduce further delays. Their duration depends on the transient state of the system and can therefore vary from run to run. While CUDA often hides these effects through driver-level heuristics and implicit pointer mapping, Vulkan SC's explicit synchronization model makes the underlying hardware latencies directly visible. This improves transparency, but it also makes transfer overhead much more apparent than in the CUDA case.

5.3 Execution Variability and Jitter

A primary source of jitter in CUDA is memory management. The analysis of the MAJ shows that a large share of timing fluctuations originates from malloc and free operations. In contrast, Vulkan SC shows almost no jitter related to memory management, since dynamic allocation is avoided during the critical execution path. The results also show that Vulkan SC maintains nearly jitter-free kernel execution, even as complexity increases in the compute chains. In the CUDA implementation, kernel execution contributes a visible part of the total MAJ, especially for shorter operations. The small kernel contribution in Vulkan SC suggests that its driver and execution model provide highly stable scheduling and execution on the GPU. The skewness results further support the observed difference in timing behavior. CUDA consistently shows a positive skew, which indicates a tendency toward higher-runtime outliers and latency spikes. Vulkan SC, in contrast, shows a negative skew across all operations, which means that deviations occur mainly toward shorter runtimes. This is the safer behaviour for real-time systems.

5.4 Relationship Between Synthetic and Fused NN

Figure 10 shows that, for small and medium input sizes, the CUDA kernel time is nearly independent of N for both the fused NN and the unfused baseline. In the log-log plot, this corresponds to $T_{\text{kernel}}(N) = CN^p$ with $p \approx 0$, and therefore $T_{\text{kernel}}(N) \approx C$. In this region, both curves are separated by an approximately constant multiplicative factor. For larger inputs, especially beyond about $N = 10^5$, this dependence changes. For Vulkan SC, the slopes vary more noticeably, indicating less uniform growth with N. The gap between the fused and unfused curves reflects the difference between isolated primitive kernels and a fused kernel/shader implementation, where intermediate layer values do not need to be fetched between separate primitive kernels. This is also visible in Fig. 11a, where the `K_unfused/K_NN` ratio decreases with increasing N. Finally, Fig. 11b shows that H2D/D2H transfer timing becomes data-size dominated for larger transfers. In the log-log plot, $T_{\text{transfer}}(B) = CB^p$, and the near-45° slope indicates that transfer time scales almost linearly with transferred data size. Small transfers remain more affected by fixed overheads and synchronization, while Vulkan SC shows higher transfer times than CUDA at larger data sizes.

5.5 Coding Complexity and Documentation

While CUDA provides a higher level of abstraction and more implicit resource management, Vulkan SC requires a much more explicit implementation style. This increases development effort and also increases the risk of configuration errors. Although the Khronos Group provides documentation, implementation on the target hardware remains challenging, especially when selecting and integrating the correct driver. In our case, the complete pipeline for SAXPY requires around 40 lines of code (LOC) in CUDA, while the corresponding Vulkan SC implementation requires around 800 LOC.

6 Conclusion and Future Work

This study provided a comparative analysis of CUDA and Vulkan SC on the Jetson UMA platform with a focus on timing determinism for safety-critical ML applications. The results show a clear trade-off. For simple operations, Vulkan SC achieves lower kernel execution times and avoids runtime allocation overhead, this is also true for the analysed NN. However, this advantage is reduced at the total runtime level by explicit H2D and D2H transfers. As workload complexity increases, the difference in total runtime becomes smaller and, for deeper chains, CUDA scales more favorably. From the perspective of timing stability, Vulkan SC provides the more robust execution profile. Dynamic memory management introduces a large share of jitter in CUDA, whereas Vulkan SC shows only minor variability related to memory management and a consistently small kernel contribution to total jitter. The contrast in skewness further supports this. While CUDA shows a positive skew and therefore a tendency toward latency spikes, Vulkan SC shows a negative skew, meaning that deviations occur mainly toward shorter runtimes. This behavior fits the needs of safety-critical real-time systems much better. The fused NN comparison shows that the operation-count-equivalent unfused baseline remains above the measured fused kernel time for both CUDA and Vulkan SC. The ratio `K_unfused`(N)/`K_NN`(N) decreases with increasing N, while H2D/D2H timing becomes increasingly data-size dominated for larger transfers. Vulkan SC's advantages come at the cost of substantially higher implementation effort and lower tooling convenience. Nevertheless, for domains in which predictable timing behavior matters more than development comfort, Vulkan SC is the more suitable choice. Our findings are in line with the reviewed literature. The high jitter and architectural limitations of CUDA, as also discussed in [6], remain a weak basis for safety-critical use. Vulkan SC, on the other hand, showed a robust timing profile and, together with analytic approaches such as [1], provides a promising basis for the deterministic integration of complex ML workloads into real-time systems.

Future work should include the deployment of more complex ML models, including more extensive arithmetic (e.g. convolutions) on safety-critical-grade hardware and tests under varying system load (i.e. UMA contention) or clearer internal separation to limit shared-memory interference.

Acknowledgments. Prof. Bjoern Annighoefer contributed input regarding the assessment of aircraft-system feasibility, and the writing of the paper. He has also provided technical expertise for addressing the reviewers' comments in the revised version.

References

1. Aleman, J.L.M., Agenjo, A., Carretero, S., Kosmidis, L.: On the mc/dc code coverage of vulkan sc gpu code. In: 2022 IEEE/AIAA 41st Digital Avionics Systems Conference (DASC), pp. 1–9 (2022). https://doi.org/10.1109/DASC55683.2022.9925766
2. Benito, M., Trompouki, M.M., Kosmidis, L., Garcia, J.D., Carretero, S., Wenger, K.: Comparison of gpu computing methodologies for safety-critical systems: an avionics case study. In: 2021 Design, Automation & Test in Europe Conference & Exhibition (DATE), pp. 717–718 (2021). https://doi.org/10.23919/DATE51398.2021.9474060
3. KhronosGroup: vulkan graphics and compute for the safety-critical systems. https://www.khronos.org/vulkansc/. Accessed 11 Feb 2026

4. KhronosGroup: Vulkan®sc 1.0.20 - a specification (with all registered vulkan sc extensions) : based on vulkan 1.2.330. Technical Report, KhronosGroup (2025). https://registry.khronos.org/VulkanSC/specs/1.0-extensions/html/vkspec.html. Accessed 02 Feb 2026
5. Luettig, B., Dallmann, J., Annighoefer, B.: ADIMA: automatic configuration by peripheral detection and adaptive distributed task execution for integrated modular avionics platforms. In: 2022 IEEE/AIAA 41st Digital Avionics Systems Conference (DASC) (2022)
6. Perez-Cerrolaza, J., Abella, J., Kosmidis, L., Calderon, A.J., Cazorla, F., Flores, J.L.: Gpu devices for safety-critical systems: a survey. ACM Comput. Surv. **55**(7) (2022). https://doi.org/10.1145/3549526
7. Trompouki, M.M., Kosmidis, L.: Brook auto: High-level certification-friendly programming for gpu-powered automotive systems. In: 2018 55th ACM/ESDA/IEEE Design Automation Conference (DAC), pp. 1–6 (2018). https://doi.org/10.1109/DAC.2018.8465869

Testing

Uncertainty-Aware Threat-Informed Test Case Prioritization with Human-in-the-Loop

K. Darshana Abeyrathna(✉), Juan Camilo Guevara Gomez, Meine van der Meulen, Per Myrseth, and Andreas Hafver

Group Research and Development, Det Norske Veritas (DNV), Høvik, Norway
{Darshana.Abeyrathna.Kurug,Juan.Camilo.Guevara.Gomez,
Meine.van.der.Meulen,Per.Myrseth,Andreas.Hafver}@dnv.com

Abstract. Security testing frameworks provide large collections of executable test cases, but running all of them is often impractical. Existing test case prioritization (TCP) methods typically rely on historical execution outcomes, coverage information, or project-specific internal signals, which are often unavailable or insufficient for security testing. We propose a threat-informed and uncertainty-aware TCP framework that instead prioritizes tests using external threat intelligence and features extracted directly from test descriptions.

Using Atomic Red Team as a case study, we enrich each test with global vulnerability severity evidence through ATT&CK–CVE mappings and combine this with local operational features derived from the test definitions. Because many tests lack linked vulnerability evidence, we introduce a learned global severity imputer trained on evidence-backed techniques and used to estimate severity for unmapped ones. Regression models, including Random Forest, ExtraTrees, and quantile gradient boosting, are then trained to predict priority scores and estimate uncertainty. To make uncertainty actionable, we learn symbolic rules with a Tsetlin Machine that separates high- and low-uncertainty cases and supports human review.

Results show that uncertainty profiles differ across models and that high uncertainty is frequently associated with sparse or inconsistent vulnerability evidence, including cases where CVEs exist but supporting fields are incomplete. These findings support a practical human-in-the-loop workflow in which uncertain high-priority tests are flagged for expert review and threat-intelligence enrichment. Overall, the framework provides a practical alternative to history-based TCP by enabling learned, threat-informed prioritization under incomplete evidence. While the current work focuses on learning a prioritization policy from threat intelligence and test-case characteristics, the resulting models provide a foundation for future system-aware prioritization using deployment-specific contextual information.

Keywords: Test Case Prioritization · Software Testing · Machine Learning · Uncertainty-Aware Machine Learning · Human-in-the-Loop AI

© The Author(s), under exclusive license to Springer Nature Switzerland AG 2027
E. Troubitsyna et al. (Eds.): SAFECOMP 2026, LNCS 16868, pp. 349–364, 2027.
https://doi.org/10.1007/978-3-032-34867-8_22

1 Introduction

Modern software systems evolve under strong time and resource constraints, driven by frequent releases, continuous integration/continuous deployment (CI/CD), and rapidly changing operational requirements. Each change triggers regression and security test campaigns that must provide actionable feedback quickly. In practice, however, executing all available tests is often infeasible due to limited time budgets, infrastructure cost, and the growing size of test suites. This makes *test case prioritization* (TCP) a central resource-optimization problem: given a finite test execution budget, which tests should be executed first to maximize early discovery of faults or vulnerabilities.

TCP seeks an ordering of a test suite that maximizes objectives such as early fault detection, typically measured with the Average Percentage of Fault Detection (APFD) metric [5,14]. Prior work includes coverage- and history-based heuristics, cost-aware variants, and more recently ML-based approaches for adaptive prioritization [15,16]. However, these methods largely target functional regression failures and rely on internal project signals, whereas security regression testing must also account for the external threat landscape and adversary behaviors relevant "right now".

Security testing motivates a shift toward *threat-informed* prioritization. Threat intelligence frameworks such as MITRE ATT&CK provide a standardized and widely adopted taxonomy of adversary tactics, techniques, and procedures (TTPs), supporting adversary emulation and defense gap analysis [10]. At the same time, executable security test repositories such as Atomic Red Team provide large collections of tests mapped to ATT&CK techniques, enabling realistic and interpretable security regression testing [13]. However, these suites are large and heterogeneous, and without prioritization they can lead to inefficient test campaigns in which scarce resources are spent on low-impact behaviors while high-risk attack paths are delayed.

A natural source of global risk signals is public vulnerability intelligence from NVD and CVSS, which is available at scale [6,12]. Because CVSS reflects technical severity rather than exploitation likelihood, the framework is designed to incorporate additional signals, such as EPSS and the CISA KEV catalog, when available [3,7]. An operational TCP mechanism should therefore handle incomplete vulnerability evidence and remain maintainable as external risk information evolves.

Finally, once prioritization is delegated to learned models, *uncertainty awareness* becomes essential: overconfident but wrong priorities can waste testing budget and delay the discovery of critical weaknesses. In security testing, uncertainty may stem from irreducible noise in the data or from limited training coverage and distribution shift [9]. We therefore treat uncertainty as a triage signal, flagging tests predicted as high priority but with high uncertainty for expert review.

In this paper, we propose a *threat-informed and uncertainty-aware* TCP framework based on probabilistic machine learning [11]. Rather than predicting observed incidents, the model learns to generalize a transparent, threat-informed priority score under incomplete evidence. The framework is test-suite agnostic and is instantiated here using Atomic Red Team as a realistic case study. Specifically, we (1) enrich security test cases with external threat intelligence and local operational features, (2) compute an interpretable composite score, including a learned severity imputer for tests lacking vulnerability evidence, and (3) train uncertainty-aware regression models with TM-based symbolic explanations to support human-in-the-loop prioritization.

To the best of our knowledge, this is the first work to study TCP in this way: learning a threat-informed security prioritization policy from external intelligence and test-description-derived features, while explicitly handling missing evidence through learned imputation and using uncertainty explanations to guide selective expert intervention. The result is a practical workflow in which automated ranking is the default, and human attention is directed only to uncertain high-priority cases.

2 Methodology

The proposed methodology consists of five main steps: (1) data collection and enrichment (Sect. 2.1), (2) local feature extraction from test cases (Sect. 2.2), (3) baseline score construction (Sect. 2.3), (4) learned global severity imputation to address missing evidence (Sect. 2.3), and (5) learning with uncertainty estimation and uncertainty explanation (Sect. 2.4 and 2.5). The goal is to output a prioritized list of tests and, for tests with unreliable ranks, provide actionable explanations that guide review and re-ranking.

2.1 Linking Atomic Test Cases with Threat Intelligence

To build a solid threat-aware foundation, the first step was to semantically enrich test cases by linking them with well-established threat intelligence databases. In this study, this process is demonstrated using Atomic Red Team (ART) test cases. The linkage process was automated through a Python-based pipeline.

This pipeline retrieves, merges, and processes metadata from several key sources:

MITRE ATT&CK: ART test cases are directly mapped to corresponding TTPs (e.g., T1059.001), providing semantic labels and context for adversarial behavior.

NIST National Vulnerability Database (NVD): By identifying Common Vulnerabilities and Exposures (CVEs) that mention the ATT&CK techniques to which the ART test cases were mapped, the pipeline extracts associated severity scores (CVSS), exploitability metrics, and descriptions.

This multi-source mapping enriches each test case with adversary-context information from ATT&CK, linked vulnerabilities, associated CVSS severity, and platform-related metadata.

To ensure up-to-date and reproducible results, the metadata extraction pipeline interfaces directly with public APIs, including MITRE's TAXII service and the NVD REST API. These data are kept in a database and are referred to throughout this paper as technique-based features, or global-level data.

However, we found that many tests have no linked vulnerability evidence. This can happen because a technique is not tied to specific CVEs, because the mapping dataset is limited, or because public sources have gaps. Instead of removing these tests, we keep them and treat missingness explicitly in the scoring step using a single parameter α. This allows the dataset to stay complete and allows the prioritization policy to be stated clearly.

2.2 Local Feature Extraction from Test Cases

To enrich the prioritization framework with operational context, the project incorporated additional attributes derived directly from the test case descriptions. In this work, these descriptions originate from the ART repository, but the same feature extraction process applies to test cases from other testing frameworks and domains. This phase involved downloading the full set of ART tests to a local environment and parsing them to extract descriptive characteristics using simple yet effective natural language processing (NLP) techniques.

From these, four key attributes were derived:

- Privilege Requirement: This describes the execution context needed for the test, such as User-level or Admin-level access. Privilege level is a key factor in security testing since high-privilege attacks typically imply greater severity.
- Execution Complexity: Using the length and structure of each test's command description, complexity levels were assigned as Low, Medium, or High. Tests that consist of multiple steps or require complex scripting were labeled higher in complexity.
- Detection Difficulty: Descriptions were analyzed for keywords like "stealth," "bypass," "obfuscate," and "evade." The presence of these terms indicated the use of stealth tactics, and tests were accordingly labeled with higher detection difficulty.
- Target Platform: Each test case was tagged according to the operating systems it was designed for–such as Windows, Linux, or macOS. This information is critical for future steps involving context-aware testing based on the system under evaluation.

These local attributes provide valuable insight into the practical and operational implications of executing each test, which is not typically captured by external threat intelligence data alone. By combining this local context with the global threat data (as described in Sect. 2.1), we develop a richer prioritization strategy.

2.3 Priority Score Computation

We compute a transparent composite score for each test case. This score serves as a baseline ranking, and a supervised target for machine learning based ranking models. The score has a local component and a global component.

Local Component: For a selected test case j, the local importance score, S_j^{local} integrates the following factors: the privilege score, S^{p} is 1 when user-level privileges are required and 2 if administrative privileges are needed. The complexity score, S^{c} is 1 for low complexity, 2 for medium, and 3 for high complexity. Detection score, S^{d} is the final factor, assigned as 1, 2, or 3 for low, medium, or high detection difficulty, respectively.

$$S_j^{\text{local}} = w^{\text{p}} \cdot S_j^{\text{p}} + w^{\text{c}} \cdot S_j^{\text{c}} + w^{\text{d}} \cdot S_j^{\text{d}}. \tag{1}$$

Setting the weights, w^{p}, w^{c}, and w^{d} should be done by considering that privilege and detection difficulty have a higher impact than complexity.

Global Component with Learned Severity Imputation: The global component is derived from linked vulnerability severity evidence. When a test is linked to one or more vulnerabilities, we convert CVSS-based severity indicators (e.g., severity is low, medium, high) into numeric values and sum them.

Let k_j be the number of linked vulnerabilities for test j and let s_i be the numeric severity contribution of linked evidence. When CVSS evidence exists, we define:

$$S_j^{\text{global}} = \sum_{i=1}^{k_j} s_i. \tag{2}$$

When CVSS evidence is missing, instead of using a single constant fallback or dropping the case, we impute a global severity contribution using a learned regression model trained on techniques with CVSS evidence. Let $\hat{g}(z)$ denote the imputer prediction using technique-level metadata z (e.g., tactics, data sources, defenses bypassed, and permissions required). For missing-evidence cases we set:

$$S_j^{\text{global}} = \hat{g}(z_j). \tag{3}$$

This yields a fully populated global component while preserving dataset coverage.

Baseline Policy for Sensitivity Checks: For transparency, we also consider a simple baseline policy using a constant severity α for missing evidence in a limited sensitivity analysis:

$$S_j^{\text{global}} = \sum_{i=1}^{k_j} s_i + \alpha \cdot \mathbb{1}_{k_j=0}. \tag{4}$$

where $\mathbf{1}_{\{k_j=0\}}$ is an indicator function that equals 1 when test j has no linked vulnerabilities ($k_j = 0$), and 0 otherwise. Thus, when no CVSS evidence exists, a constant fallback severity α is assigned.

In the main pipeline and results, we use learned imputation rather than manual α. The complete process of data collection is summarized in Fig. 1.

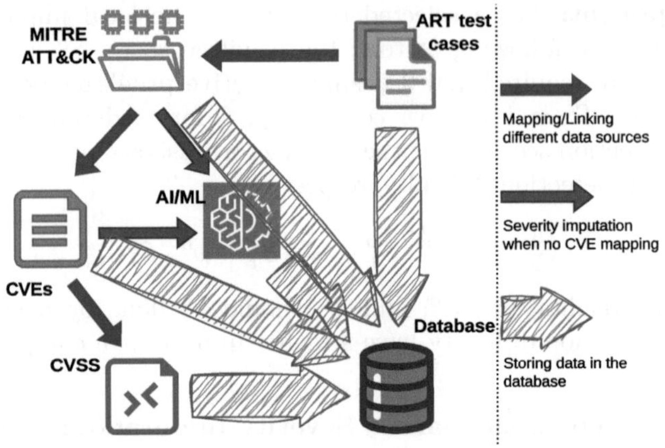

Fig. 1. Overview of global enrichment and local feature extraction. Atomic tests are mapped to ATT&CK techniques, enriched with vulnerability evidence (CVE/CVSS when available), and combined with local operational features to build a learning dataset.

Final Priority Score: The final priority score combines the local and global components through a weighted aggregation:

$$S_j^{\text{final}} = \lambda S_j^{\text{local}} + \gamma S_j^{\text{global}}, \tag{5}$$

where λ and γ are weighting parameters controlling the relative contribution of the local operational factors and the global vulnerability severity information, respectively.

For this initial study, we set $\lambda = \gamma = 1$, resulting in equal treatment of both components and yielding the additive formulation used throughout the evaluation. This choice was made to provide a simple and interpretable baseline without introducing additional calibration parameters. While the additive formulation offers a practical starting point, the optimal weighting may depend on organizational priorities, threat landscapes, and the availability of historical validation data.

In practice, the global component often contributes more strongly to the final ranking because it aggregates vulnerability severity evidence, whereas the local component acts as an operational adjustment reflecting factors such as attack complexity, required privileges, and detection difficulty. Future work will investigate calibration strategies for λ and γ, including optimization against historical attack outcomes, expert-driven risk assessments, and data-driven learning approaches.

This formulation supports both direct ranking and the construction of training targets for downstream learning. As threats evolve and new test cases are added, the model can be dynamically updated by refreshing the global and local scores and recalibrating the weighting parameters when sufficient validation data becomes available.

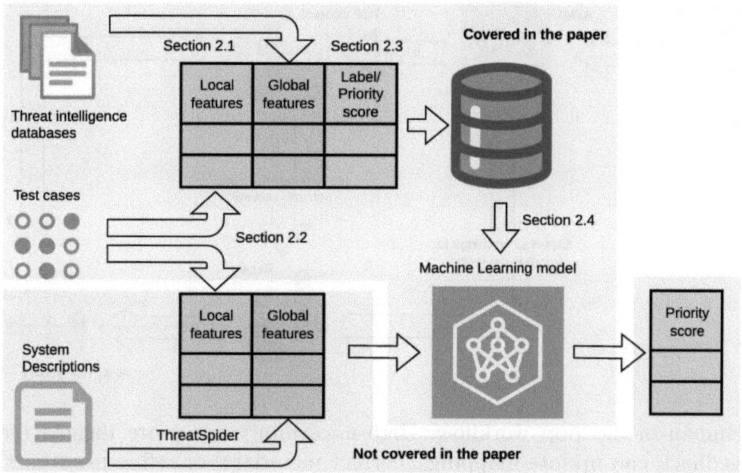

Fig. 2. Overview of the proposed framework and its intended deployment workflow. The yellow-highlighted components are implemented and evaluated in this work, while the blue-highlighted components represent future integration of system-specific context for operational test prioritization. (Color figure online)

Operational Deployment Perspective: The priority score serves both as a transparent baseline ranking mechanism and as a supervisory signal for training machine learning models. While the current study focuses on developing and evaluating this prioritization policy, the long-term objective is not to repeatedly compute the handcrafted score during deployment. Instead, once trained, the model can directly infer priority scores from contextual information. As illustrated in Fig. 2, future extensions will incorporate system-specific context extracted from system descriptions, using tools such as ThreatSpider [2]. These features, together with the local test-case attributes used in this work, will be provided as inputs to the trained model, enabling system-aware and threat-informed prioritization without requiring manual CVSS aggregation or explicit score construction.

2.4 Learning, Uncertainty Estimation, and Human-in-the-Loop Use

We train regression models to predict S_j^{final} from the engineered feature set. The regression output provides a predicted score $\hat{s}(x)$ for each test. Sorting tests by $\hat{s}(x)$ produces an automated prioritized list.

We also compute an uncertainty value $u(x)$ for each prediction. We use three regression methods: Random Forest (RF), Extra Trees (ET), and Quantile Gradient Boosting (GBR-Q). For RF and ET, we use variance across trees as an uncertainty proxy. For GBR-Q, we train quantile models corresponding to the 0.10 and 0.90 quantiles and compute uncertainty as the width of the resulting prediction interval.

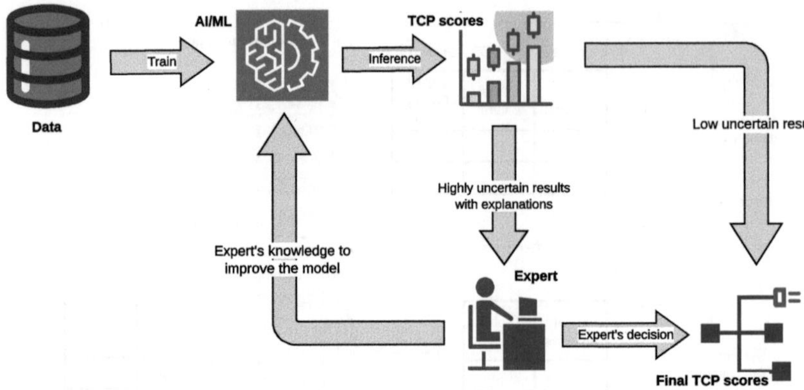

Fig. 3. Human-in-the-loop workflow. High-uncertainty cases are flagged for review. Expert feedback can update mappings, correct metadata, or refine prioritization policies.

In the proposed workflow, uncertainty is used as a *decision aid* rather than a replacement for scoring. High-uncertainty tests with high predicted scores are flagged for review. As illustrated in Fig. 3, an expert can confirm or adjust the ordering and, when appropriate, enrich mappings or metadata. This creates a human-in-the-loop TCP loop.

2.5 Uncertainty Explanation with Tsetlin Machines

Uncertainty values can be hard to interpret. We therefore explain uncertainty by learning symbolic rules that separate high-uncertainty from low-uncertainty cases.

We first convert uncertainty into a binary label. We compute low and high quantile thresholds and assign each test in a held-out set to one of three groups: low, mid, or high uncertainty. Specifically, samples below the lower quantile threshold are labeled as low uncertainty (class 0), while samples above the upper quantile threshold are labeled as high uncertainty (class 1). We then train a Tsetlin Machine (TM) classifier [8] to learn this separation.

Importantly, in our TM setup, **Clauses with positive polarity vote for class 1** (high uncertainty) and **Clauses with negative polarity vote for class 0** (low uncertainty). Let v^+ be the number of satisfied positive clauses and v^- the number of satisfied negative clauses. We define the vote difference:

$$\Delta = v^+ - v^-. \tag{6}$$

The predicted class is determined by the following rule:

$$\hat{y}_{\text{UQ}} = \begin{cases} 1 & \text{if } \Delta \geq 0, \\ 0 & \text{if } \Delta < 0. \end{cases} \tag{7}$$

Table 1. Baseline sensitivity analysis over default imputed severity α.

α	Train MAE	Test MAE	Pred. Mean	Pred. Std
0	0.040	0.287	9.191	11.060
2	0.040	0.292	10.738	10.785
4	0.041	0.306	12.277	10.556
6	0.042	0.313	13.821	10.415
8	0.038	0.289	15.368	10.329
10	0.039	0.291	16.915	10.323

This decision rule is useful because it can be explained on a single test case: we can show which clauses fired, compute Δ, and then interpret the outcome as "high uncertainty" or "low uncertainty", with reasons extracted from activated clauses. In the discussion section, we use this rule to show how a TM explanation can support reprioritization.

3 Experimental Evaluation

We evaluate three aspects: a brief sensitivity check of manual α, learning and uncertainty under the imputed dataset, and stability of uncertainty explanations across folds and across regression methods.

We extracted 1770 atomic tests from 328 ART technique folders without parsing or fetch failures. The imputer was trained on techniques with CVSS evidence and applied to techniques without such evidence. Downstream regression is evaluated using both a fixed train/test split and 5-fold cross-validation, with mean absolute error (MAE) as the primary error metric.

3.1 Brief Sensitivity Check for Manual α

We retain α only as a baseline sanity check to show that prioritization is not arbitrarily fragile under simple missing-evidence assumptions. Table 1 summarizes MAE trends for different α values on a fixed split. In the remainder of the paper, we use learned global severity imputation rather than manual α.

Limitation of the α Baseline: While Table 1 provides a useful sanity check, the α policy treats missing vulnerability evidence as a single homogeneous event. In practice, missingness arises from multiple causes: the technique genuinely has no linked CVEs or the mapping coverage is incomplete (e.g., technique-to-CVE gaps). Because α collapses these cases into the same default severity, it cannot distinguish structurally different evidence patterns, which later emerges as a key driver of uncertainty explanations.

Table 2. Imputer-based regression performance and uncertainty behavior.

Method	MAE (single split)	CV MAE mean
ExtraTrees	0.061	0.116
RF	0.093	0.149
GBR-Q	0.121	0.182

3.2 Regression Methods and Uncertainty Signals

Using the imputed dataset, we compare RF, ExtraTrees, and GBR-Q under both single split evaluation and cross-validation (CV). ExtraTrees achieves the lowest MAE (e.g., single-split MAE $\approx$ 0.061; CV MAE mean $\approx$ 0.116), while RF is slightly higher (single-split MAE $\approx$ 0.093; CV MAE mean $\approx$ 0.149). GBR-Q yields a higher MAE (single-split MAE $\approx$ 0.121; CV MAE mean $\approx$ 0.182).

We observe that uncertainty profiles differ markedly by method: tree-disagreement uncertainty (RF/ET) can be sparse with spikes, while quantile-based uncertainty (GBR-Q) is smoother, making it easier to threshold robustly and to use as a decision aid for reprioritization. In our latest experiments, GBR-Q provided the most usable uncertainty signal for downstream uncertainty classification and rule learning (Table 2).

3.3 Uncertainty Explanation and Rule Stability

We train a TM to separate high- and low-uncertainty cases and extract symbolic clauses [1]. We evaluate explanation stability in two ways:

- **Feature overlap:** overlap of features mentioned in the clause sets.
- **Clause overlap:** overlap of exact clause strings.

Interpretability Stability Across Folds: Across 5 folds, we observe moderate to high feature-level stability but low clause-string stability. Many pairwise overlaps remain moderate to high (e.g., between approximately 0.44 and 0.71), indicating that the *drivers* of uncertainty are consistent across folds. This pattern is illustrated in Fig. 4 (left), which shows the Jaccard overlap of features appearing in TM clauses across cross-validation folds.

Interpretability Stability Across Methods: We also compare uncertainty explanations across RF, ExtraTrees, and GBR-Q on a fixed split. Here we again observe that feature overlap is more stable than clause overlap. Feature overlap across methods is moderate but not identical. In this dataset, RF and GBR-Q show slightly higher overlap than pairs involving ExtraTrees, suggesting that their uncertainty partitions highlight more similar drivers (Fig. 4, right). Clause overlap remains low across all method pairs, reflecting differences in uncertainty definitions and partitions.

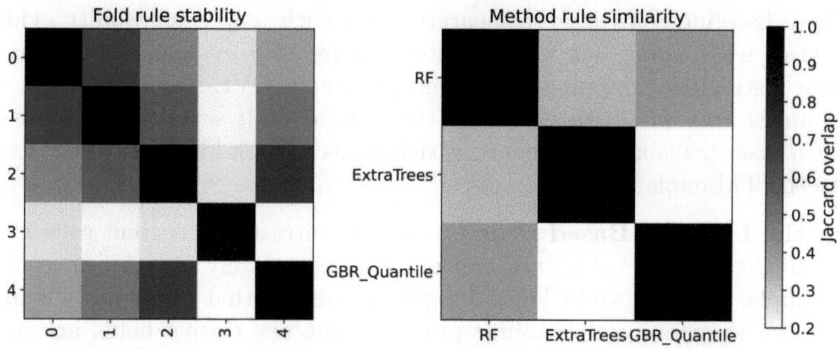

Fig. 4. Stability of TM uncertainty explanations based on feature overlap (Jaccard similarity). Left: overlap of features appearing in TM clauses across cross-validation folds. Right: overlap of features across regression methods on a fixed split. Darker shades indicate higher overlap.

3.4 Interpreting Representative GBR-Q Rules and the HITL Rationale

A representative set of clauses (Fold 3, GBR-Q) is:

Clause 1: `No. of CVEs is more than 1 AND There is no CVE description`

Clause 2: `No. of CVEs is 0 AND Complexity is High AND CVE list is empty AND There is CVE description`

Clause 3: `CVE list is not empty AND There is no CVE description`

Clause 4: `No. of CVEs is 0 AND Complexity is Medium AND CVE list is empty`

Comparison to the Earlier α-Baseline Uncertainty Rules: In the earlier pipeline (manual α handling), the TM explanations were dominated by coarse missingness flags and occasionally by specific CVE tokens, e.g.:

Clause 1: `Vulnerability is CVE-2014-0295`

Clause 2: `There are no vulnerabilities`

Clause 3: `Clause has no selected literals (always-1)`[1].

Clause 4: `There are no vulnerabilities`[2]

[1] In a Tsetlin Machine, a clause may end up with no selected literals during learning. Such a clause evaluates to true for all inputs (always-1), effectively acting as a constant vote in the decision function [4].

[2] Tsetlin Machines learn clauses independently, and multiple clauses may converge to the same logical condition. Such duplicate clauses simply reinforce the same pattern by contributing additional votes.

These baseline rules largely separate cases with *any* vulnerability evidence from cases with *none*, but they do not capture *why* evidence is missing nor do they distinguish mapping gaps from genuinely CVE-free techniques. This limitation is inherent to α: it assigns the same default severity to all missing-evidence cases, causing heterogeneous situations to look identical to the learner and to the TM explainer.

What the Imputer-Based Rules Add: In contrast, the current rules explicitly highlight *evidence structure and sparsity*, not merely absence of evidence. For example, Clause 1 and Clause 3 together indicate that uncertainty is higher when more than one vulnerability presents, but descriptive fields are incomplete or inconsistently populated (captured by `cve list` is empty and no `cve description`). In other words, uncertainty is not simply "no CVEs"; it also arises when CVEs exist but the supporting evidence is too sparse to ground a stable prioritization.

This is precisely where a human-in-the-loop step is valuable: an expert can review the flagged tests, validate whether the missing evidence is due to mapping gaps or benign causes, and adjust ordering or trigger enrichment (e.g., improved mappings or additional intelligence sources).

4 Discussion and Future Work

Our results support a practical view of threat-informed TCP: prioritization is feasible at scale, but the limiting factor is not the ranking model itself, it is the *quality and completeness of the evidence layer* used to construct targets and features. We therefore focus the discussion on three connected questions: (1) how to handle missing global evidence, (2) how to use uncertainty as an operational control signal, and (3) how to close the loop when uncertainty indicates evidence gaps.

4.1 From Manual α to Learned Severity Imputation

Missing vulnerability evidence is unavoidable in threat-informed TCP. A constant fallback α is transparent but collapses heterogeneous cases into a single assumption. This collapse is visible in the baseline TM uncertainty explanations, which were dominated by coarse missingness indicators and occasionally a specific CVE token, providing limited actionable insight for reprioritization.

Learned severity imputation improves practicality by preserving a simple policy while reducing one-size-fits-all bias. The imputer is trained only on techniques with CVSS evidence and is applied systematically to techniques without evidence, producing a filled global severity contribution that is still interpretable as a *consistent evidence-driven estimate*. In effect, this yields a more informative and less brittle global feature layer, enabling downstream learning and uncertainty estimation to reflect evidence structure rather than only evidence absence.

4.2 Why We Prefer Actionable Uncertainty (GBR-Q) Over Lowest MAE

Across regressors, ExtraTrees achieved the lowest MAE, but its uncertainty signal was often sparse (near-zero for many samples with occasional spikes), making quantile thresholding less stable and leading to less consistent uncertainty labeling. In contrast, quantile gradient boosting (GBR-Q) produced smoother uncertainty (interval width) that was more robust for downstream "low vs. high uncertainty" separation and symbolic rule learning.

This motivates a pragmatic design choice: when uncertainty is used as a decision aid, the most useful model is not necessarily the one with the smallest point-error, but the one with uncertainty that is *usable* for operational decisions. In our experiments, GBR-Q best satisfied this requirement and therefore served as our primary uncertainty estimator for explanation.

4.3 What the Tsetlin Machine Rules Reveal and Why Human-in-the-Loop is the Right Intervention

TM explanations translate uncertainty into symbolic conditions that can be acted upon. A consistent finding is that uncertainty is strongly associated with *evidence sparsity and inconsistency.* The representative GBR-Q clauses highlight patterns such as: CVEs exist, but key evidence fields are missing or unevenly populated (e.g., `cve list` is empty, no `cve description`). Operationally, these are exactly the cases where automated ranking should be treated as less reliable: the model is signaling that the evidence does not sufficiently constrain the prioritization.

This leads naturally to a human-in-the-loop (HITL) policy: if a test is predicted high priority but classified as high uncertainty with evidence-sparsity rules, it should be flagged for expert review. The expert can (1) validate whether the missingness reflects mapping gaps or benign causes, (2) correct or enrich mappings, and/or (3) adjust the test's position if organizational context indicates higher or lower urgency. Figure 3 illustrates this loop: uncertainty becomes a trigger for targeted review, rather than a reason to distrust the entire ranking.

4.4 Stability of Explanations: Trust the Drivers, Not the Exact Strings

Our overlap analysis shows an important interpretability principle. Across folds, feature-mention overlap is high while exact clause-string overlap is low. Across methods, overlap is moderate at the feature level and remains low at the clause-string level. This is expected: symbolic models often express similar concepts with different logical forms depending on sampling variation and partition boundaries. The practical implication is that the most reliable explanation is at the level of *drivers* (e.g., "evidence sparsity in CVE fields"), not the exact clause text. In deployment, we therefore recommend summarizing frequently occurring features across runs and using clause strings as local, run-specific explanations for individual decisions.

4.5 Future Work: Evidence Quality, Feedback, and Calibration

The results suggest three high-impact directions for improving threat-informed TCP:

Improve Linkage Coverage and Evidence Richness: Since uncertainty is repeatedly tied to evidence completeness, improving ATT&CK to CVE coverage and retaining richer evidence fields should directly reduce uncertainty and increase prioritization reliability. Combining multiple intelligence sources (e.g., updated mappings, Exploit Prediction Scoring System (EPSS), Known Exploited Vulnerabilities (KEV)) and keeping provenance metadata are promising steps.

Feedback-Driven Refinement through HITL: Expert feedback on uncertain high-priority cases can update mapping quality, imputation behavior, and local feature extraction heuristics. Over time, this should reduce both uncertainty rates and explanation volatility, and produce organization-specific prioritization that remains grounded in public evidence.

Uncertainty Calibration and Robust Thresholding: Future work will explore calibrated uncertainty and alternative thresholding strategies that are stable under sparse signals, as well as reporting explanation stability via feature-level summaries alongside run-specific clause examples.

Overall, the combined findings support a practical workflow where automated ranking is the default, uncertainty is a safety and triage signal, and symbolic explanations identify when the limiting factor is evidence quality, enabling targeted enrichment and human review rather than ad hoc reordering.

5 Conclusion

We presented a threat-informed and uncertainty-aware framework for test case prioritization that integrates external threat intelligence with local operational features. To address incomplete vulnerability evidence, we introduced a learned global severity imputer that estimates severity contributions for techniques without CVSS mappings, preserving dataset coverage without relying on a fixed fallback assumption.

We trained regression models to predict priority scores and estimate uncertainty, and we used a Tsetlin Machine to generate symbolic explanations separating high- and low-uncertainty cases. Our analysis shows that uncertainty is often driven by sparse or inconsistent vulnerability evidence, motivating a practical human-in-the-loop workflow in which tests with high-uncertain priority scores are flagged for expert review and enrichment. We found that feature-level drivers of uncertainty are stable across folds and more consistent across methods than exact clause strings, supporting the use of explanations as actionable guidance rather than fixed logical rules.

Overall, the framework provides a practical pathway for integrating threat intelligence, machine learning, and human expertise in security test prioritization. While the current study focuses on learning and evaluating a threat-informed prioritization policy, future work will incorporate system-specific contextual information and threat-modelling outputs to enable system-aware prioritization in operational environments.

References

1. Abeyrathna, K.D., El Mekkaoui, S., Edward, L.Y., Hafver, A., Granmo, O.C.: Extension of regression tsetlin machine for interpretable uncertainty assessment. In: International Joint Conference on Rules and Reasoning, pp. 19–33. Springer (2023)
2. Amro, A., Kavallieratos, G.: Threatspider: Cti-driven semi-automated threat modelling for cybersecurity certification. In: 2025 IEEE International Conference on Cyber Security and Resilience (CSR), pp. 619–625. IEEE (2025)
3. Cybersecurity and infrastructure security agency: known exploited vulnerabilities (KEV) catalog. https://www.cisa.gov/known-exploited-vulnerabilities-catalog (2025). Accessed 28 Jan 2026
4. Darshana Abeyrathna, K., Granmo, O.C., Zhang, X., Jiao, L., Goodwin, M.: The regression tsetlin machine: a novel approach to interpretable nonlinear regression. Philos. Trans. Royal Soc. Math. Phys. Eng. Sci. **378**(2164) (2020)
5. Elbaum, S., Malishevsky, A.G., Rothermel, G.: Test case prioritization: a family of empirical studies. IEEE Trans. Software Eng. **28**(2), 159–182 (2002). https://doi.org/10.1109/32.988497
6. FIRST.org: common vulnerability scoring system (CVSS) v3.1: specification document. https://www.first.org/cvss/specification-document (2019). Accessed 28 Jan 2026
7. FIRST.org: exploit prediction scoring system (EPSS). https://www.first.org/epss/ (2025). Accessed 28 Jan 2026
8. Granmo, O.C.: The tsetlin machine–a game theoretic bandit driven approach to optimal pattern recognition with propositional logic. arXiv preprint arXiv:1804.01508 (2018)
9. Kendall, A., Gal, Y.: What uncertainties do we need in Bayesian deep learning for computer vision? In: Advances in Neural Information Processing Systems (NeurIPS), vol. 30 (2017). https://arxiv.org/abs/1703.04977
10. MITRE corporation: getting started with ATT&CK. Technical Report, The MITRE Corporation (2019). https://www.mitre.org/sites/default/files/2021-11/getting-started-with-attack-october-2019.pdf
11. Murphy, K.P.: Probabilistic Machine Learning: An Introduction. MIT Press (2022). https://probml.github.io/pml-book/
12. National institute of standards and technology: national vulnerability database (nvd). https://nvd.nist.gov/ (2025). Accessed 28 Jan 2026
13. Red canary: atomic red team. https://github.com/redcanaryco/atomic-red-team (2025). Accessed 28 Jan 2026
14. Rothermel, G., Untch, R.H., Chu, C., Harrold, M.J.: Test case prioritization: an empirical study. In: 1999 International Conference on Software Maintenance (ICSM 1999), pp. 179–188 (1999). https://doi.org/10.1109/ICSM.1999.792604

15. Spieker, H., Gotlieb, A., Marijan, D., Mossige, M.: Reinforcement learning for test case prioritization. arXiv preprint arXiv:2011.01834 (2020), https://arxiv.org/abs/2011.01834
16. Yoo, S., Harman, M.: Regression testing minimization, selection and prioritization: a survey. Softw. Testing Verification Reliabil. **22**(2), 67–120 (2012). https://doi.org/10.1002/stvr.430

ML-Powered Test Case Prioritisation: Leveraging Code Changes for Optimal Testing

Loui Al Sardy(✉), Niklas Markert, and Reinhard German

Friedrich-Alexander-Universität Erlangen-Nürnberg, Computer Networks and Communication Systems (Informatik 7), Martensstr. 3, 91058 Erlangen, Germany
{loui.alsardy,niklas.markert,reinhard.german}@fau.de

Abstract. The increasing number of test cases and limited time frames often prevent the execution of complete regression test suites following software changes. This paper addresses this challenge in an industrial software testing context, focusing on daily system tests at Siemens Healthineers. We propose a machine learning-based approach for test case prioritisation that ranks test cases according to their relevance to recent source code changes. Two models, a neural network and a random forest, are developed and evaluated using a dataset comprising over 10,000 test case executions. The experimental results show that both models outperform traditional prioritisation techniques, achieving up to a 50% reduction in the number of test cases required to detect all defects. The random forest model consistently delivers superior prioritisation performance compared to conventional methods. The findings demonstrate that machine learning can significantly improve regression test prioritisation in industrial environments, with feature engineering and model selection emerging as key performance factors. Although evaluated in a single industrial setting, the proposed approach is architecture-independent and applicable across diverse software systems and domains.

Keywords: Machine Learning · Test Case Prioritisation · Regression Testing · Secure Software Development · Source Code Changes

1 Introduction

Maintaining software quality becomes increasingly challenging as systems evolve. Frequent updates demand comprehensive regression testing to prevent unintended side effects, yet executing full test suites is often time and resource-intensive, especially under rapid release cycles in agile and DevOps environments.

Test Case Prioritisation (TCP) addresses this by reordering test execution to maximise early fault detection [23]. However, traditional TCP methods face critical limitations:

© The Author(s), under exclusive license to Springer Nature Switzerland AG 2027
E. Troubitsyna et al. (Eds.): SAFECOMP 2026, LNCS 16868, pp. 365–380, 2027.
https://doi.org/10.1007/978-3-032-34867-8_23

- **Manual Selection:** Relies on subjective judgment, making it inconsistent and impractical for large-scale projects.
- **Code Coverage Analysis:** Tools like SonarQube [26] and dotCover [10] ignore code change relevance or risk, limiting effectiveness.
- **Historical Results:** Focusing solely on past failures overlooks new defects from recent code modifications.

Machine Learning (ML) offers a data-driven alternative by adapting to evolving codebases and leveraging past execution patterns [17,20]. ML models can prioritise test cases dynamically based on change impact, improving testing efficiency, scalability, and accuracy.

Beyond software quality, timely regression testing is also a security concern. In security-critical and safety-regulated domains such as medical systems, delayed detection of regression faults can directly introduce exploitable vulnerabilities or undermine existing security guarantees. Regression faults may unintentionally weaken access controls, error handling, or safety mechanisms, thereby expanding the system's attack surface. Efficient test case prioritisation, therefore, constitutes not only a quality assurance problem, but a preventive security mechanism within the secure software development lifecycle, supporting early identification of security-relevant regressions before deployment. In this context, it is important to distinguish between conventional functional regression testing and vulnerability-oriented testing. While functional testing primarily relies on correctness-based test oracles, security and vulnerability testing often require specialised oracles capable of identifying unintended security behaviours or policy violations. Extending ML-based prioritisation toward vulnerability-focused testing therefore introduces additional challenges that remain outside the scope of the present work.

This paper provides one of the first industrial-scale evaluations of ML-based test case prioritisation driven explicitly by source code changes, demonstrating that lightweight ensemble models can outperform more complex neural architectures under realistic constraints.

1.1 Problem Statement

Regression testing ensures that recent software changes do not compromise existing functionality. However, in large-scale industrial software systems, executing complete regression test suites after every software change is often infeasible due to limited time and computational resources. In practice, regression testing is frequently subject to strict time constraints, meaning that only a subset of the full test suite can be executed. In such scenarios, the order of test execution becomes critical, since faults remain undetected if the corresponding test cases are not executed within the available time. Therefore, prioritisation directly impacts the effectiveness of testing by increasing the likelihood that critical defects are identified [23]. Existing prioritisation approaches, such as manual selection, code coverage analysis, or historical execution results, frequently fail to account for the impact of recent source code changes. This creates the need for adaptive and data-driven prioritisation techniques capable of identifying high-risk test cases efficiently within realistic CI/CD environments.

1.2 Objectives

This research aims to advance the application of machine learning in software quality assurance by addressing the scalability and adaptability challenges of regression testing in large industrial systems. The specific objectives are:

- To design and implement ML-based techniques for test case prioritisation informed by recent source code changes.
- To empirically compare the performance of neural networks and random forest models in predicting test case failure likelihoods.
- To improve fault detection efficiency in continuous testing environments by prioritising high-risk test cases, thereby supporting early identification of critical issues and enhancing system reliability.

Existing test case prioritisation approaches either rely on static heuristics, such as code coverage or historical failures, or do not explicitly incorporate the impact of recent source code changes in a systematic, data-driven manner. In contrast, this work introduces and evaluates a machine learning-based prioritisation approach that directly leverages fine-grained code change information in an industrial setting, demonstrating improved fault detection efficiency and practical applicability.

1.3 Structure of the Paper

This paper is structured as follows: Sect. 2 reviews the existing literature on test case prioritisation, source code analysis, and machine learning applications in software testing. Section 3 details the methodology, including data collection, feature selection, and model training. Section 4 presents the experimental results and discussions. Finally, Sect. 5 concludes the paper with a summary of findings, contributions, and future work.

2 Literature Review

2.1 Traditional Test Case Prioritisation Techniques

Traditional techniques for TCP have focused on several methods, including manual selection, code coverage analysis, and historical test results [18,25]. Manual selection relies on expert judgment to prioritise test cases, but this method is often subjective and not scalable for large projects. Code coverage analysis attempts to measure the extent to which the source code is exercised by the test suite, providing insights into which parts of the code are tested. However, code coverage does not account for the importance or recency of changes in the code, making it less effective in dynamic environments [4]. Historical test results are another common approach, where test cases are prioritised based on their past execution outcomes [13]. While this method can be useful, it typically overlooks the specific impact of recent code changes and may not provide the most relevant prioritisation for the current build. These traditional methods, although widely used, face significant limitations in handling the complexities of modern software development, where rapid changes and large codebases are common [8,12].

2.2 Source Code Analysis

Analysing source code changes is crucial for understanding their impact on the system and effectively prioritising test cases. Techniques such as static and dynamic analysis, dependency graphs, and change impact analysis help identify the areas of the codebase most affected by recent changes. Static analysis examines the source code without executing it, identifying potential issues and dependencies [15]. Dynamic analysis, on the other hand, involves executing the code to observe its behaviour and interactions in real-time [3]. Dependency graphs map out the relationships between different parts of the code, allowing for a clearer understanding of how changes in one area might impact others. Change impact analysis uses these techniques to determine which parts of the code are most likely to be affected by recent modifications. However, integrating these techniques into an automated prioritisation process remains challenging, particularly in large-scale software projects where the complexity and volume of changes can be overwhelming [6,20].

2.3 Machine Learning in Software Testing

Machine learning is showing significant promise in various aspects of software engineering and testing, including defect prediction, automated test generation, and test case prioritisation [14,21]. ML models can learn from historical data to identify patterns and make predictions about future test outcomes, optimising the testing process. In the context of TCP, ML can analyse past test executions and source code changes to predict the likelihood of test case failures, enabling more effective prioritisation [1,16]. Several studies have explored the application of ML in TCP. For instance, neural networks have been used to predict test case priorities based on historical data and source code changes [9,24]. These models can capture complex relationships between features, making them suitable for dynamic environments where the impact of code changes can vary widely. Random forests, an ensemble learning method, have also been used to predict the outcome of a test suite, offering robustness and interpretability by combining multiple decision trees to make predictions [19]. Research by Abdelkarim and ElAdawi (2022) introduced the TCP-Net, a deep neural network specifically designed for test case prioritisation based on source code changes. The TCP-Net demonstrated excellent results in prioritising test cases by predicting failure probabilities, significantly outperforming traditional methods [1,2]. However, the TCP-Net approach relies on detailed unit test coverage information, which may not be available or easily obtainable in many industrial contexts. In contrast, the proposed method does not require such coverage data, instead leveraging source code changes and historical execution data, thereby increasing its applicability in environments with limited test instrumentation.

Other studies have used various ML techniques, including support vector machines, gradient-boosted decision trees, and reinforcement learning, to improve TCP [11,20].

3 Methodology

3.1 Concept

Our approach to enhancing TCP involves leveraging ML to predict test case outcomes based on underlying code modifications. This is formulated as a classification task, where historical execution data serves as the training foundation. During inference, the trained model generates a probability score ranging from 0 to 1, representing the likelihood of a test case failure. The test cases are then prioritised based on their predicted failure probability, aligning with the core principle of software testing by identifying failures as early as possible. Figure 1 illustrates the concept followed in this paper.

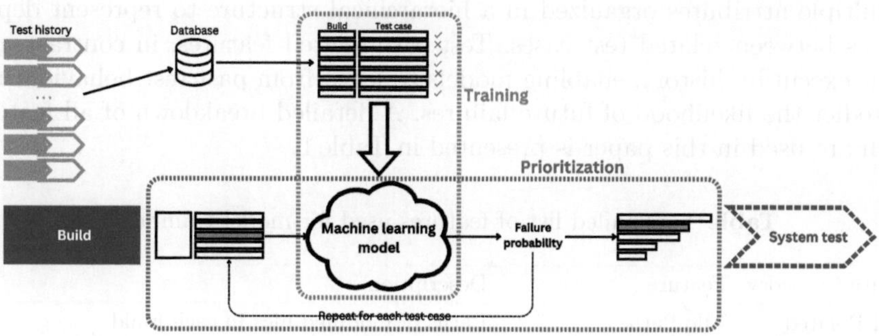

Fig. 1. Overview of the ML-based test case prioritisation pipeline.

3.2 Data Collection

The data required for this study were collected from Siemens Healthineers software system tests. This dataset includes historical test execution records, source code repositories, and metadata from automated build systems. Specifically, the data encompassed:

- **Test Case Results:** Historical execution results for more than 10,000 test cases, including pass/fail statuses and execution times.
- **Source Code Changes:** List of files that have been modified, including file paths, file types and the associated commit messages.

– **Build Information:** Metadata about each build, such as build IDs, timestamps, and branch information.

To ensure the integrity, relevance, and consistency of the data, classic preprocessing steps were taken. In accordance with the methodology presented in [1], the file paths of the altered files were transformed using a bag-of-words encoding to improve the interpretability of the machine learning models. This encoding scheme additionally preserves information about the spatial relationships between files that are located in close proximity to one another.

3.3 Feature Selection

Feature selection is a critical step in ML, ensuring that the most relevant data points are used for training models. In this study, the goal was to identify features most strongly correlated with test case execution results. The selected features were categorised into three main groups: build-related features, test case-defining features, and test case-related features.

Build-related features remain constant for a given software build and provide contextual information about the modifications introduced in that build, with a particular focus on code changes. Test case-defining features focus on uniquely identifying test cases and establishing relationships between them. They consist of multiple attributes organized in a hierarchical structure to represent dependencies between related test cases. Test case-related features, in contrast, represent execution history, enabling models to learn from past test behaviour and to predict the likelihood of future failures. A detailed breakdown of all features which are used in this paper is presented in Table 1.

Table 1. Detailed list of features used for model training.

Feature Category	Feature	Description
Build Related	File Paths	Paths of modified files in each build
	File Extensions	Types of files changed
	Branches	Branches from which changes originated
	Commit Types	Nature of commits (e.g., bug fixes, enhancements)
Test Case Defining	Test Family	Categorization of tests based on functionality
	Test Name	Unique identifier for a group of test cases
	Protocol/Program	Specific protocol or program being tested
Test Case Related	Failure Rate	Historical failure rate of a test case
	Last Execution Result	Outcome of the last execution (pass/fail)

3.4 Machine Learning Models

This study evaluates two ML models: a neural network and a random forest. These models were selected for their ability to capture complex relationships and generate reliable predictions.

Neural Network. The neural network architecture, illustrated in Fig. 2, is inspired by the "TCP-Net" [1] and consists of a primary fully connected network, referred to as the "main network". Additionally, two smaller "compression networks" are used to preprocess and reduce the dimensionality of the high dimensional input features before they are passed to the main network.

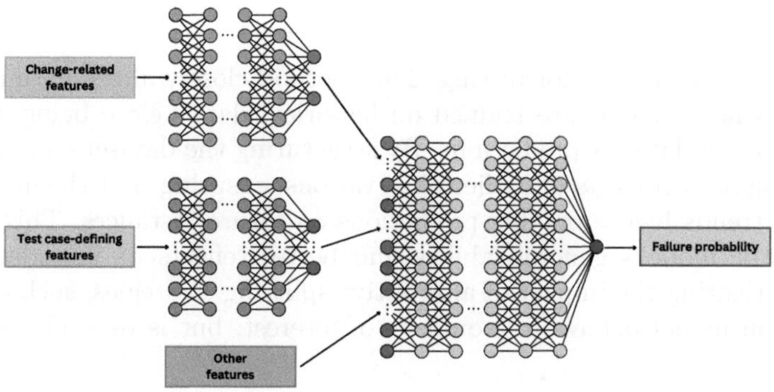

Fig. 2. Architecture of the neural network developed for this study.

Random Forest. The random forest model serves as a simpler, more lightweight alternative for comparison with the neural network. It has been widely recognised for its effectiveness in predicting test outcomes [20]. The random forest utilises the same input features as the neural network but without additional splitting or preprocessing.

3.5 Model Training and Evaluation

Dataset Description. The dataset utilised in this study was collected from Siemens Healthineers' software system tests and comprises over 10,000 test case executions, as presented in Table 2. It includes comprehensive logs detailing source code modifications along with metadata from automated build systems. A key characteristic of the dataset is the significant imbalance between passed test cases (negative samples) and failed test cases (positive samples). To address this issue, a weighted loss function was employed. This approach adjusts the loss function so that errors from the underrepresented class (failed test cases) carry a higher weight, ensuring they have a greater influence on the learning process.

The models were trained using the collected historical data. The training process involved:

Splitting the Data. To ensure a realistic training process, the dataset was divided into three subsets: Training, validation, and test set. Unlike conventional random splitting, a chronological approach was adopted, where the earliest 70% of the data was assigned for training, followed by the next 15% for validation,

Table 2. Summary of the dataset used in this study.

Data Type	Description	Quantity
Test Case Results	Pass/fail statuses, execution times	10,000+
Source Code Change	Different file paths	10,000+
Test Cases	Specific test cases	1,000+

and the most recent 15% for testing. This method closely aligns with real-world scenarios, where models are trained on historical data before being applied to newly incoming data for predictions. By structuring the dataset chronologically, the training data consists of earlier observations, ensuring that the model learns from past trends before making predictions on future instances. This approach enhances the model's generalizability and better reflects its intended application. Investigating the impact of alternative splitting strategies, such as random splitting, on model behaviour would be of interest, but is beyond the scope of this work.

Hyperparameter Tuning. To determine the optimal configuration for each model, various hyperparameters were systematically tested. For the neural network, this included the number of hidden layers and neurons per layer, while for the random forest, the primary parameters were the number of trees and the maximum depth of each tree. Beyond these parameters, feature selection was incorporated into the hyperparameter tuning procedure for each model. The outcomes of this process offered further insight into the relative importance of individual features in determining model performance. The hyperparameter tuning was conducted using grid search, a method that systematically evaluates different hyperparameter combinations to identify the most effective configuration. A schematic representation of the hyperparameter tuning process as applied in this study is presented in Fig. 3.

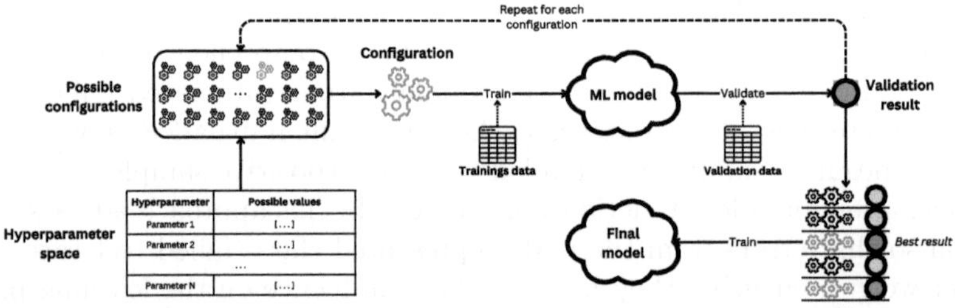

Fig. 3. The process of hyperparameter tuning used in the experiment.

This process was applied independently to both models, as each has distinct hyperparameters influencing its learning and predictive performance. The most significant hyperparameters from the final selected configurations of the neural network and random forest models are listed in Table 3.

Table 3. Optimal hyperparameters for the neural network and random forest models found in the hyperparameter tuning.

Model	Hyperparameter	Optimal Value
Neural Network	Number of Layers (Main Network)	5
	Number of Layers (Compression Network)	5
	Learning Rate (Compression Network)	0.0001
Random Forest	Number of Trees	400
	Maximum Depth	no max. depth

Evaluation Metrics: To assess model performance, several evaluation metrics were employed, emphasizing the effectiveness of TCP. Conventional classification measures, including accuracy, precision, and recall, were intentionally omitted, since the aim of this work is not to evaluate classification correctness but to determine how effectively the models produce an appropriate prioritisation of test cases.

1. **APFD (Average Percentage of Faults Detected)** is one of the most widely used metrics for evaluating TCP techniques [7,11,20]. It measures how quickly faults are detected during test execution, producing a score between 0 and 1, where higher values indicate a more effective prioritisation. APFD is specifically designed for comparing different prioritisation strategies applied to the same set of test cases.

   The APFD metric is defined by the following equation [5]:

$$APFD = 1 - \frac{\sum_{i=1}^{m} TF_i}{nm} + \frac{1}{2n} \tag{1}$$

   Where:
   - n = Amount executed test cases
   - m = Amount detected faults
   - TF_i = Position of first test case that detects fault i

   The fundamental principle behind APFD, which emphasises detecting failures as early as possible, was first introduced in [22,23].
2. **Failure Coverage** quantifies the proportion of test cases that must be executed, following a given prioritisation order, before all failing test cases are identified. A well-optimised prioritisation strategy aims to minimise this value, ensuring that failures are detected with fewer test executions.

By using these evaluation metrics, the models' ability to efficiently prioritise test cases was systematically analysed. The employed evaluation metrics assume homogeneous test execution costs and therefore do not account for varying execution times, representing a deliberate level of abstraction that enables a focused comparison of prioritisation effectiveness independent of system-specific runtime characteristics.

4 Experimental Results

4.1 Model Performance

The neural network and random forest models were evaluated based on their ability to effectively prioritise test cases across multiple software builds. For each build, predicted test case orderings were compared against known execution outcomes to assess prioritisation effectiveness.

To contextualise performance, the trained models were benchmarked against established prioritisation strategies: random prioritisation as a baseline, perfect prioritisation as an upper bound, a history-based approach relying on past execution results [13], and the current industrial prioritisation method based on recent test outcomes within a four-week window.

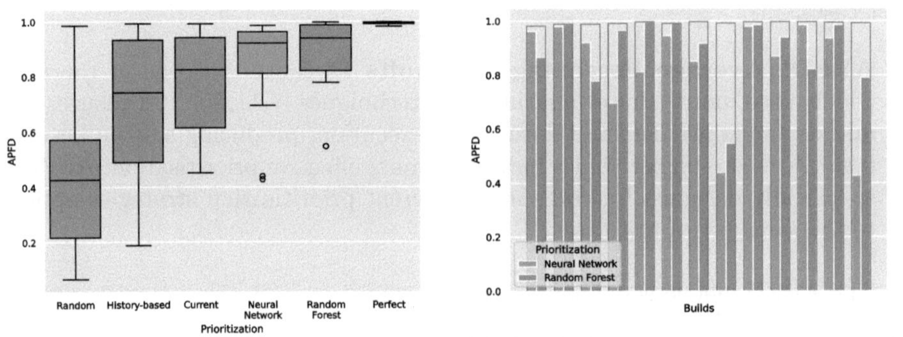

(a) Overview of APFD scores achieved by different methods.

(b) Detailed comparison between neural network and random forest for specific software builds.

Fig. 4. Graphical representation of experimental results.

Both machine learning models outperformed traditional techniques. The neural network demonstrated robust performance by capturing complex relationships between source code changes and historical test data. The random forest consistently achieved higher APFD scores, indicating superior early fault detection and more effective prioritisation of failing test cases.

A detailed comparison of APFD scores across all prioritisation methods is provided in Table 4, while the distribution of results is illustrated using boxplots in Fig. 4a. While the proposed models achieve higher average performance,

Table 4 shows that alternative methods, such as history-based prioritisation, can outperform them on individual builds. This highlights that no single method consistently dominates across all scenarios, making average performance across multiple test sets the more relevant indicator of overall effectiveness.

Table 4. APFD scores of the different methods for different test case sets.

Build	Random	History-based	Current	Neural Network	Random Forest	*Perfect*
2024/03/13	0.5720	0.6831	0.4506	**0.9619**	0.8663	*0.9815*
2024/03/14	0.2602	0.4898	**0.9884***	0.9797	**0.9884***	*0.9884*
2024/03/15	0.4263	**0.9321**	0.8251	0.9205	0.7775	*0.9884*
2024/03/17	0.5104	0.7406	0.6151	0.6948	**0.9651**	*0.9906*
2024/03/18	0.3408	0.1879	0.9904	0.8121	**0.9968***	*0.9968*
2024/03/19	0.4673	0.9454	0.9871	0.9454	**0.9911***	*0.9911*
2024/03/20	0.2167	**0.9300**	0.8500	0.8500	0.9167	*0.9967*
2024/03/28	0.1154	0.3923	**0.5615**	0.4385	0.5462	*0.9923*
2024/04/01	0.5994	**0.9915**	0.9403	0.9801	0.9858	*0.9972*
2024/04/02	0.6831	**0.9913**	0.6134	0.8692	0.9390	*0.9971*
2024/04/10	**0.9854**	0.1901	0.9035	**0.9854**	0.8216	*0.9971*
2024/04/11	0.1453	0.5447	0.5978	0.9358	**0.9860**	*0.9944*
2024/04/15	0.0652	**0.8913**	0.8188	0.4275	0.7899	*0.9928*
Average	0.4144	0.6854	0.7802	0.8308	**0.8900**	*0.9926*

* Corresponds to the best possible prioritization

4.2 Comparison Neural Network and Random Forest

Across 13 builds, the random forest model outperformed the neural network in 10 cases, although both models achieved comparable results overall. A closer analysis reveals distinct differences in their prioritisation strategies:

- **Random Forest:** This model consistently deprioritised test cases with a failure rate of 0 and a last result of "Passed", effectively focusing on higher-risk cases. While generally effective, this strategy underperformed in scenarios where previously stable tests failed unexpectedly, as seen in Build 2024/03/15.
- **Neural Network:** The neural network exhibited more complex and less interpretable behaviour, likely influenced by multi-feature interactions rather than a dominant heuristic.

Figure 4b presents a comparison of APFD scores across builds.

4.3 Expert Evaluation

To assess the practical effectiveness of the ML models, an evaluation was conducted with an experienced software testing professional. The expert manually prioritised test cases for three software builds, using the same input data provided to the neural network and random forest models. The resulting rankings were compared in terms of ordering similarity and APFD scores.

Notably, the expert often prioritised critical test cases based on domain knowledge, even when those tests had consistently passed in prior builds. In contrast, the ML models focused on failure likelihood and did not account for strategic test importance. Despite these differences, both models consistently achieved higher APFD scores than the expert, indicating superior fault detection efficiency.

The manual prioritisation process was also time-intensive, requiring roughly two hours per build, compared to near-instantaneous ML-based rankings. This underscores the scalability and practical value of the proposed approach in real-world testing pipelines.

4.4 Discussion

This section discusses key findings, model behaviour, limitations, and implications for real-world applications.

Limitations. The study was conducted on a single industrial MedTech dataset, which limits the generalisability of the findings. Although the proposed approach is not tied to a specific software architecture or domain, further validation across additional application domains and software systems is required. A known limitation is the reliance on historical test execution data, which may restrict applicability in new projects or modules with limited history. In such cases, bootstrapping with synthetic data or applying transfer learning could mitigate the cold-start problem. Additionally, the dataset contained noisy entries, such as test failures caused by hardware faults or unstable test environments rather than code changes, which may have reduced model accuracy. A direct comparison with approaches such as TCP-Net was not feasible due to the lack of publicly available implementations and limited reproducibility, which constrained the scope of the evaluation.

Model Insights. The random forest consistently outperformed the neural network due to its effective use of historical features, particularly the last test result and failure rate. These features were confirmed through hyperparameter tuning to be essential for accurate prioritisation. However, this reliance also limited the model's capacity to capture more complex relationships, especially in high-dimensional input spaces like source code paths. Neural networks, while better suited in theory for learning such interactions, underperformed in this context, likely due to limited data volume and the sparsity of signal in early test case executions.

Implications. These results reflect the inherent trade-offs between model expressiveness and interpretability. Decision trees favour high-information-gain features, which leads to effective and transparent decision-making but reduces their capacity for nuanced reasoning across feature combinations. Addressing these challenges will require broader feature sets, improved data quality, and possibly hybrid or ensemble models that combine interpretability with flexibility.

In safety-critical contexts, relying solely on data-driven prioritisation may introduce risks if safety-relevant tests are lower prioritised due to low predicted failure likelihood. Therefore, the proposed approach should be applied as a complementary technique rather than a standalone solution. In practice, it can be combined with predefined safety-criticality levels or constraints that ensure the regular execution of safety-critical tests, for example by always executing them or enforcing their execution after a defined time interval. Although the proposed ML-based test case prioritisation approach may support earlier identification of security-relevant regressions, its effectiveness for dedicated vulnerability-oriented testing has not been evaluated in this study. The applicability of the approach to security testing, therefore, remains a promising direction for future work requiring further empirical investigation.

Comparison with Traditional Methods. Compared to traditional TCP approaches, such as manual selection, code coverage, or historical failure analysis, ML-based methods offer significant advantages. Manual selection lacks scalability and consistency. Coverage tools (e.g. SonarQube, dotCover) identify untested code but overlook change criticality. Historical test outcomes alone fail to address newly introduced bugs. These limitations highlight the need for dynamic, data-driven solutions capable of adapting to evolving software systems.

Advantages of Machine Learning Models. ML models offer adaptive, scalable, and accurate test case prioritisation. By learning from historical execution data and recent code changes, they ensure that high-risk test cases are executed earlier, improving fault detection rates. In this study, both the neural network and random forest models significantly outperformed traditional methods, achieving higher APFD scores and reducing the number of tests required to detect all faults. However, maintaining their effectiveness requires periodic retraining to align with evolving test suites and code structures, an engineering consideration important for real-world CI/CD integration.

5 Conclusion

This study presented a machine learning-based approach to test case prioritisation, leveraging historical execution data and source code changes to enhance regression testing efficiency. Two models, a neural network and a random forest, were evaluated with both outperforming traditional methods. The random forest achieved the highest average APFD score (0.8900), demonstrating strong performance and interpretability.

The approach supports dynamic adaptation to recent code changes, improving early fault detection and aligning with agile and DevOps practices. Its architecture-agnostic design suggests broad applicability across diverse software systems, though further validation beyond the current industrial context is needed. Practically, the findings highlight the importance of data quality, feature engineering, and model selection, balancing interpretability with predictive flexibility. These insights offer actionable guidance for integrating ML-based prioritisation into real-world testing pipelines.

From a security perspective, the proposed prioritisation approach may support reducing the temporal exposure of newly introduced vulnerabilities by accelerating the execution of high-risk tests after code changes. This aligns with secure-by-design and shift-left security principles, where rapid feedback is essential to prevent security regressions from reaching production systems.

Future work includes incorporating real-time code coverage and telemetry, extending the method to distributed systems, and evaluating cross-domain generalisability. While the original dataset cannot be shared due to confidentiality, a synthetic or illustrative implementation will be made available in future work to support reproducibility.

Acknowledgments. We sincerely thank Siemens Healthineers for providing the dataset and valuable domain expertise that made this research possible. Their support in granting access to real-world software system test data significantly contributed to the development and validation of our machine learning models for test case prioritisation.

Disclosure of Interests. The authors have no competing interests to declare that are relevant to the content of this article.

References

1. Abdelkarim, M., ElAdawi, R.: TCP-net: test case prioritization using end-to-end deep neural networks. In: 2022 IEEE International Conference on Software Testing, Verification and Validation Workshops (ICSTW), pp. 122–129 (2022). https://doi.org/10.1109/ICSTW55395.2022.00034
2. Abdelkarim, M., ElAdawi, R.: TCP-Net++: test case prioritization using end-to-end deep neural networks - deployment analysis and enhancements. In: 2023 IEEE International Conference On Artificial Intelligence Testing (AITest), pp. 99–106 (2023). https://doi.org/10.1109/AITest58265.2023.00024
3. Ball, T.: The concept of dynamic analysis. In: Nierstrasz, O., Lemoine, M. (eds.) ESEC/SIGSOFT FSE -1999. LNCS, vol. 1687, pp. 216–234. Springer, Heidelberg (1999). https://doi.org/10.1007/3-540-48166-4_14
4. Cai, X., Lyu, M.R.: The effect of code coverage on fault detection under different testing profiles. ACM SIGSOFT Softw. Eng. Notes **30**(4), 1–7 (2005). https://doi.org/10.1145/1082983.1083288
5. Elbaum, S., Malishevsky, A.G., Rothermel, G.: Test case prioritization: a family of empirical studies. IEEE Trans. Softw. Eng. **28**(2), 159–182 (2002). https://doi.org/10.1109/32.988497

6. German, D.M., Hassan, A.E., Robles, G.: Change impact graphs: determining the impact of prior codechanges. Inf. Softw. Technol. **51**(10), 1394–1408 (2009). https://doi.org/10.1016/j.infsof.2009.04.018
7. Greca, R., Miranda, B., Bertolino, A.: State of practical applicability of regression testing research: a live systematic literature review. ACM Comput. Surv. **55**(13s), 274:1–274:36 (2023). https://doi.org/10.1145/3579851
8. Hao, D., Zhang, L., Mei, H.: Test-case prioritization: achievements and challenges. Front. Comput. Sci. **10**(5), 769–777 (2016). https://doi.org/10.1007/s11704-016-6112-3
9. Jahan, H., Feng, Z., Mahmud, S.H., Dong, P.: Version specific test case prioritization approach based on artificial neural network. J. Intell. Fuzzy Syst. **36**(6), 6181–6194 (2019). https://doi.org/10.3233/JIFS-181998
10. JetBrains: dotcover –.net unit test runner and code coverage tool (2026). https://www.jetbrains.com/dotcover/. Accessed 21 May 2026
11. Khatibsyarbini, M., et al.: Trend application of machine learning in test case prioritization: a review on techniques. IEEE Access **9**, 166262–166282 (2021). https://doi.org/10.1109/ACCESS.2021.3135508
12. Khatibsyarbini, M., Isa, M.A., Jawawi, D.N., Tumeng, R.: Test case prioritization approaches in regression testing: a systematic literature review. Inf. Softw. Technol. **93**, 74–93 (2018). https://doi.org/10.1016/j.infsof.2017.08.014
13. Kim, J.M., Porter, A.: A history-based test prioritization technique for regression testing in resource constrained environments. In: Proceedings of the 24th International Conference on Software Engineering. ICSE 2002, pp. 119–129. IEEE, Orlando, FL, USA (2002). https://doi.org/10.1109/ICSE.2002.1007961
14. Kolawole, I., Fakokunde, A.: Machine learning algorithms in devops: optimizing software development and deployment workflows with precision. Int. J. Res. Publ. Rev. **6**(1), 247–264 (2025). https://doi.org/10.55248/gengpi.6.0125.0209
15. Lenarduzzi, V., Pecorelli, F., Saarimaki, N., Lujan, S., Palomba, F.: A critical comparison on six static analysis tools: detection, agreement, and precision. J. Syst. Softw. **198**, 111575 (2023). https://doi.org/10.1016/j.jss.2022.111575
16. Machalica, M., Samylkin, A., Porth, M., Chandra, S.: Predictive test selection. In: 2019 IEEE/ACM 41st International Conference on Software Engineering: Software Engineering in Practice (ICSE-SEIP), pp. 91–100 (2019). https://doi.org/10.1109/ICSE-SEIP.2019.00018
17. Mece, E.K., Paci, H., Binjaku, K.: The application of machine learning in test case prioritization - a review. Eur. J. Electr. Eng. Comput. Sci. **4**(1) (2020). https://doi.org/10.24018/ejece.2020.4.1.128
18. Mukherjee, R., Patnaik, K.S.: A survey on different approaches for software test case prioritization. J. King Saud Univ. Comput. Inf. Sci. **33**(9), 1041–1054 (2021). https://doi.org/10.1016/j.jksuci.2018.09.005
19. Pan, C., Pradel, M.: Continuous test suite failure prediction. In: Proceedings of the 30th ACM SIGSOFT International Symposium on Software Testing and Analysis, pp. 553–565. ISSTA 2021, ACM, New York (2021). https://doi.org/10.1145/3460319.3464840
20. Pan, R., Bagherzadeh, M., Ghaleb, T.A., Briand, L.: Test case selection and prioritization using machine learning: a systematic literature review. Empir. Softw. Eng. **27**(2), 1–43 (2021). https://doi.org/10.1007/s10664-021-10066-6
21. Pandhare, H.V.: Future of software test automation using AI/ML. Int. J. Eng. Comput. Sci. **13**(5), 27159–27182 (2025). https://doi.org/10.18535/ijecs/v14i05.5139

22. Rothermel, G., Untch, R.H., Chu, C., Harrold, M.J.: Test case prioritization: an empirical study. In: Proceedings IEEE International Conference on Software Maintenance (ICSM 1999), pp. 179–188. IEEE, Oxford (1999). https://doi.org/10.1109/ICSM.1999.792604
23. Rothermel, G., Untch, R.H., Chu, C., Harrold, M.J.: Prioritizing test cases for regression testing. IEEE Trans. Softw. Eng. **27**(10), 929–948 (2001). https://doi.org/10.1109/32.962562
24. Sharif, A., Marijan, D., Liaaen, M.: DeepOrder: deep learning for test case prioritization in continuous integration testing. In: 2021 IEEE International Conference on Software Maintenance and Evolution (ICSME), pp. 525–534. IEEE (2021). https://doi.org/10.1109/ICSME52107.2021.00053
25. Singh, A., Singhrova, A., Bhatia, R., Rattan, D.: A systematic literature review on test case prioritization techniques. In: Agile Software Development: Trends, Challenges and Applications, pp. 101–159. Wiley, 1 edn. (2023). https://doi.org/10.1002/9781119896838.ch7
26. SonarSource: Sonarqube – continuous code quality (2026). https://www.sonarsource.com/products/sonarqube/. Accessed 21 May 2026

Author Index

© The Editor(s) (if applicable) and The Author(s), under exclusive license
to Springer Nature Switzerland AG 2027

E. Troubitsyna et al. (Eds.): SAFECOMP 2026, LNCS 16868, pp. 381–382, 2027.
https://doi.org/10.1007/978-3-032-34867-8

MIX
Papier aus verantwortungsvollen Quellen
Paper from responsible sources
FSC® C105338

If you have any concerns about our products,
you can contact us on
ProductSafety@springernature.com

In case Publisher is established outside the EU,
the EU authorized representative is:
Springer Nature Customer Service Center GmbH
Europaplatz 3, 69115 Heidelberg, Germany

Printed by Libri Plureos GmbH
in Hamburg, Germany